# MARSH-ESTUARINE SYSTEMS SIMULATION

The Belle W. Baruch Library in Marine Science

THE BELLE W. BARUCH LIBRARY IN MARINE SCIENCE NUMBER 8

# Marsh-Estuarine Systems Simulation

Edited by Richard F. Dame

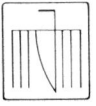

Published for the Belle W. Baruch Institute for Marine Biology and

Coastal Research by the

UNIVERSITY OF SOUTH CAROLINA PRESS

COLUMBIA, SOUTH CAROLINA

Published in Columbia, South Carolina by the
University of South Carolina Press, 1979

Manufactured in the United States of America

**Library of Congress Cataloging in Publication Data**

Main entry under title:
Marsh-estuarine systems simulation.
    (The Belle W. Baruch library in marine science;
no. 8)
    "Papers in this book were presented in January,
1977, at a symposium organized by the Belle W.
Baruch Institute for Marine Biology and Coastal
Research."
    Includes index.
1.  Tidemarsh ecology—Data processing—Congresses.
2.  Estuarine ecology—Data processing—Congresses.
3.  Ecology—Data processing—Congresses.
4.  Computer simulation—Congresses.   I.   Dame,
Richard F.   II.   Belle W. Baruch Institute for
Marine Biology and Coastal Research.   III.   Series.
QH541.5.S24M37        574.5′263        78-31554
ISBN  0-87249-375-X

# PARTICIPANTS

S. MARSHALL ADAMS
Oak Ridge National Laboratories

ROBERT AMBROSE, JR.
Environmental Protection Agency

LEONARD M. BAHR, JR.
Louisiana State University

RONALD D. BONNELL
University of South Carolina

JOHN CONOMOS
U.S. Geological Survey

RICHARD F. DAME
Coastal Carolina College-USC

TUTOR DAVIES
Environmental Protection Agency

WILL DAVIS
Johns Island, SC

JOHN W. DAY, JR.
Louisiana State University

CAROL DROWOTA
Baptist College

RICHARD M. ECKENROD
Louisiana State University

RICHARD T. EDWARDS
Baruch Marine Field Station-USC

RANDOLPH L. FERGUSON
National Marine Fisheries Service

STUART FINDLAY
University of South Carolina

RICHARD GARNAS
Environmental Protection Agency

JONATHAN GRANT
University of South Carolina

C. T. HACKNEY
Mississippi State University

CHARLES HALL
Cornell University

KATHRYN HATCHER
Maryland Water Resources Adm.

DONALD R. HEINLE
University of Maryland

ROGERS T. HUFF
University of Maryland

WILLIAM V. JOHNSON
Baruch Marine Field Station-USC

VIC KENNEDY
University of Maryland

WILEY M. KITCHENS
Baruch Marine Field Station-USC

BJÖRN KJERFVE
University of South Carolina

JAMES N. KREMER
University of Southern California

KENNETH H. MANN
Dalhousie University

HENRY MCKELLAR
University of South Carolina

CLAY MONTAGUE
University of Georgia

RICHARD MOORE
Coastal Carolina College-USC

DOUGLAS NELSON
Coastal Carolina College-USC

SCOTT W. NIXON
University of Rhode Island

HOWARD T. ODUM
University of Florida

JOSEPH ZIEMAN
University of Virginia

CANDACE A. OVIATT
University of Rhode Island

MIKE PACE
University of Georgia

BERNARD C. PATTEN
University of Georgia

DAVID PETERSON
U.S. Geological Survey

MICHAEL PILSON
University of Rhode Island

GAIL PORTER
Coastal Carolina College-USC

FRANCOIS C. RONDAY
University of Liege, Belgium

ARDIS SAVORY
University of South Carolina

NANCY STEVENS
University of Georgia

STUART STEVENS
University of Georgia

KEVIN SUMMERS
University of South Carolina

ROBERT E. ULANOLWICZ
University of Maryland

F. JOHN VERNBERG
University of South Carolina

RICHARD L. WETZEL
Virginia Institute of
Marine Science

RICHARD G. WIEGERT
University of Georgia

# PREFACE

The computer simulation of environmental processes at an eco-system level has gained a great deal of attention in recent years. Many of these simulation efforts have centered on salt marshes and estuaries because of man's impact on these highly productive and valuable systems. *Marsh-Estuarine Systems Simulation* was conceived as a means of documenting and summarizing progress in this field.

The papers in this book were presented in January, 1977, at a symposium organized by the Belle W. Baruch Institute for Marine Biology and Coastal Research. The contributors to the symposium were invited by a steering committee composed of R. Dame, T. Davies, K. Mann, S. Nixon, and B. Patten. Travel funds for contributors and printing costs were provided by a National Science Foundation Grant DEB76-81699 and an Environmental Protection Agency Grant R804407010.

The symposium and the subsequent publication of *Marsh-Estuarine Systems Simulation* would not have taken place were it not for the guidance of F. John Vernberg to whom I am eternally grateful. I would like to thank Carol J. Drowota for the professional editing and indexing and Dot Knight for management of the production aspects of the book. Finally, I would like to thank the numerous students, graduate and undergraduate, who aided in this successful symposium.

RICHARD F. DAME
Coastal Carolina College
University of South Carolina
Conway, SC
November 1978

# CONTENTS

# Contents (cont.)

# MARSH-ESTUARINE SYSTEMS SIMULATION

# A Historical Perspective of Marsh-Estuarine Systems Simulation

**Richard F. Dame**
Coastal Carolina College
University of South Carolina
Conway, South Carolina 29526

## Introduction

Estuaries and their bordering marshes are one of the most productive and stressed systems on our planet. A simple, but meaningful model of these energy-subsidized systems is shown in Figure 1. Because they are valuable, considerable effort has been expended by ecologists to understand the individual as well as holistic processes at work in these complex marsh-estuarine systems. In order to sort out and synthesize the complexities of biological, chemical, and physical interactions in marshes and estuaries, ecologists have resorted to simplified models or simulations of marsh-estuarine systems. These simulation models provide important feedback indicating what kind of new data are needed for understanding and managing these systems.

Scientists utilize models as a fundamental component of the scientific method. In the first phase of the scientific method, insight, experience, and logic are used to build a model of some part of nature. Next a hypothesis is developed which asserts that there is no difference in the response of the model and the observed response of nature. Experiments are devised to test or disprove the hypothesis. When a sufficient number of observations have been made to disprove the hypothesis, a new general statement consistent with the observed facts is developed. This statement is called a theory. The new or revised theory is used to modify our model of nature. Thus, in any scientific research effort, there should be a realistic integration of field observations, laboratory experimenta-

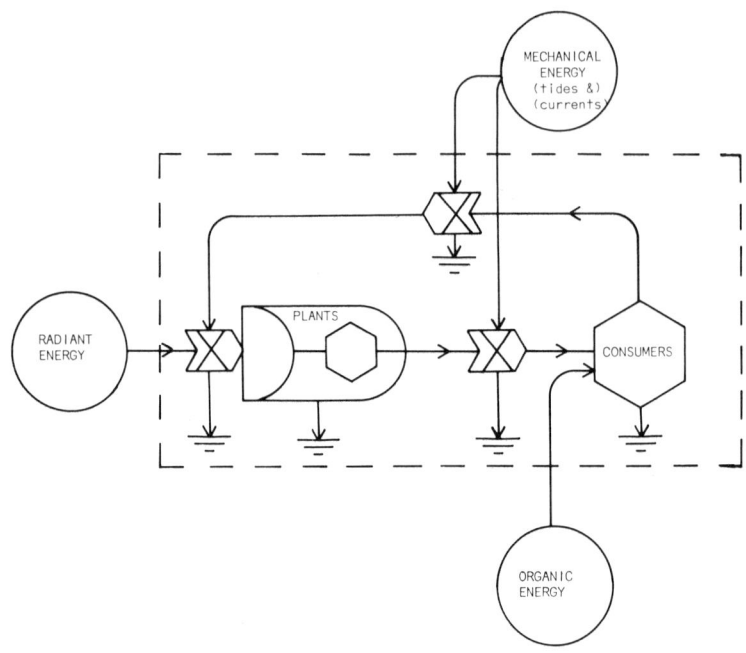

Fig. 1. A simple conceptual model of an estuarine ecosystem (after H. T. Odum).

tion, and theoretical model building to better aid us in our understanding of nature. I believe we find these attributes built into
the majority of marsh-estuarine system simulations. In ecosystem
level models, models, hypotheses, and theories are developed and
tested on individual processes within the systems model and for the
entire system as a whole.

Marsh-estuarine ecosystem level modeling usually involves teams
of researchers working toward a common goal. Team members are not
always ecologists or biologists but can include chemists, engineers,
geologists, physicists, mathematicians, economists, lawyers and many
other professionals. It is in the interaction of team members that
many new concepts of nature are developed.

The simulation of marsh-estuarine systems has developed along
two general and often concurrent paths: computer models of ecosystems which can handle large amounts of data, and microcosms which
are small scale models of marsh-estuarine systems. Each of these
model types will be discussed in turn.

COMPUTER SIMULATIONS

The first suggestion that ecosystems might be simulated by com-
puters was made by H.T. Odum (1960). This idea was soon followed
by Teal's (1962) conceptual model of energy flow in a Georgia salt
marsh ecosystem. These two works were milestones in the development of marsh-estuarine models in the late 1960's and early 1970's.

Simulation models of marsh-estuarine processes usually emphasize
one of three major classes of variables: biological, chemical, and
physical. By definition all three classes tend to be found in eco-

system level models, although, as is the case in this symposia, ecosystem models are usually biologically oriented. Invariably, estuarine ecosystem models have biotic state variables of differing degrees of complexity, which include phytoplankton and/or macroflora, zooplankton, benthos, nekton and microbial decomposers, and abiotic components of temperature, salinity, nutrients, suspended sediments, and detritus.

Wiegert (1975) has reviewed computer simulation models in ecosystems. In general, simulation models fall into two categories: theoretical models which attempt to further our understanding of nature, and empirical models which are typically applied to practical problems and faithfully reproduce the behavior of a given system.

### Theoretical models

One of the earliest estuarine ecosystem level models was that of Pomeroy et al. (1972), which simulated phosphorus flux in several Georgia estuaries. This report was soon followed by a simulation of a New England salt marsh embayment, Bissel Cove, by Nixon and Oviatt (1973). Bahr (1974) directed his attention to the development of both linear and non-linear simulation models of energy flux in Georgia oyster beds. A similar exercise on South Carolina oyster beds was pursued by Dame et al. (1977) and emphasized sensitivity analysis. The first attempt at simulating a Georgia salt marsh was by Wiegert et al. (1975) and focused on the major food chains of this system. Kremer and Nixon (1975) developed a numerical simulation model for Narragansett Bay, Rhode Island, which emphasized the planktonic community. A simulation model of the Crystal River on the west coast of Florida emphasized biological metabolism and the influences of a power plant on the metabolism of the entire ecosystem (Odum, 1974; McKellar, 1977). Finally, Vernberg (1977) with others has developed a simple water column model of the North Inlet marsh-estuarine ecosystem in South Carolina to direct future research emphasis.

### Empirical models

The utilization of empirical simulation models of estuarine processes has tended towards the prediction of water quality. Many of these models use definitions of physical processes and statistical curve fitting to produce predictive power within a given data set. O'Conner et al. (1972) developed a model of phytoplankton dynamics in the Sacramento-San Joaquin Bay-Delta, while Chen and Orlob (1975) have developed a simulation model of the San Francisco Bay-Delta system. Recently Penumalli et al. (1976) have reported on an estuarine water quality simulation model applicable to the Texas coast estuaries. There are probably other applied (empirical) estuarine models in existence, but these are not generally available in the literature.

## MICROCOSM SIMULATIONS

To trace the origins of microcosm research in marsh-estuarine systems, we must again look to H.T. Odum, who with his students began to use "multiple-seeded" or "derived" systems in the mid 1950's (Odum and Hoskins, 1957). The microcosm approach has developed to the point where numerous research efforts are utilizing large outdoor tanks to experiment upon systems intermediate between laboratory cultures and real world estuaries. The microcosm approach has been strongly recommended as part of any evaluation of foreign stress on total ecosystems.

The carbon metabolism of estuarine ponds receiving treated sewage wastes was investigated by Day (1971). Oviatt *et al.* (1977) are studying scaled benthic water column microcosms of Narragansett Bay. In the Netherlands, de Wilde and Kuipers (1977) have developed a large indoor microcosm of a mudflat. Finally, Ulanowicz *et al.* (1975) have reported on naturally eutrophic, mesohaline planktonic microcosms. This latter study is significant because it also reports an empirical computer simulation model of the same system.

## Conclusions

The preceding review has in a short space indicated what has been done in the field of simulation models. The implication is clear that computer and microcosm simulation models are pertinent if not crucial components of marsh-estuarine ecosystem analysis. The microcosm is one way we can simulate experiments on natural marsh-estuarine systems by using actual biotic and abiotic components. The computer simulations require data and thoughtful model building before simulations can offer insight into possible processes or accurate predictions, as the case may be. Through the remainder of this volume we will see the present status of marsh-estuarine systems simulation. We hope that readers will criticize the models as useful or useless, not as true or false. In any case, these works can only be considered a part of MESS.

## References

Bahr, L. 1974. Aspects of the structure and function of the intertidal oyster reef community in Georgia. Ph. D. dissertation, University of Georgia.

Chen, C. W. and G. T. Orlob. 1975. Ecological simulation for aquatic environments. In: *Systems Analysis and Simulation in Ecology* (B. C. Patten, ed.). Vol. 3, pp. 475-583. Academic Press, New York.

Dame, R. F. and S. A. Stevens. 1977. A linear systems model of an intertidal oyster community (unpub. MS.).

Dame, R. F., F. J. Vernberg, and R. J. Bonnell. 1977. Analysis of the North Inlet marsh-estuarine ecosystem. Helgolander wiss. meeresunters. 30:343-356.

Day, J. 1971. Carbon metabolism of estuarine ponds receiving treated sewage wastes. Ph. D. dissertation, University of North Carolina.

Kremer, J. N. and S. W. Nixon. 1975. An ecological simulation model of Narragansett Bay. The plankton community. In: *Estuarine Research* (L. E. Cronin, ed.). Vol. 1, pp. 672-690. Academic Press, New York.

McKellar, H. 1977. Metabolism and model of an estuarine bay ecosystem affected by a power plant. Ecol. Model. 3 (in press).

Nixon, S. W. and C. A. Oviatt. 1973. Ecology of a New England salt marsh. Ecol. Monogr. 43: 463-498.

O'Connor, D. J., D. M. DiToro, and J. L. Mancini. 1972. Mathematical model of phytoplankton dynamics in the Sacramento - San Joaquin Bay Delta. Hydroscience, Inc., Westwood, New Jersey.

Odum, H. T. 1960. Ecological potential and alalogue circuits for the ecosystem. Am. Sci. 48: 1-8.

Odum, H. T. 1974. Energy cost benefit models for evaluating thermal plans. In: *Thermal Ecology* (J. W. Gibbons and R. R. Sharitz, eds.). AEC Symposium.

Odum, H. T. and C. M. Hoskin. 1957. Metabolism of a laboratory stream microcosm. Cont. Mar. Sci. 4: 115-133.

Oviatt, C. A., W. K. Pereg and S. W. Nixon. 1977. Multivariate analysis of experimental marine ecosystems. Helgolander wiss. meeresunters. 30: 30-46.

Penumalli, B. R., R. H. Flake and E. Gus Fruh. 1976. Large scale systems approach to estuarine water quality modeling with multiple constituents. Ecol. Model. 2: 101-115.

Pomeroy, L. R., L. R. Shenton, R. D. H. Jones, and R. J. Reimold. 1972. Nutrient flux in estuaries. Linol. Oceanogr. Special Sym. 1: 274-291.

Teal, T. M. 1962. Energy flow in the salt marsh ecosystem of Georgia. Ecology 43: 614-624.

Ulanowicz, R. E., D. A. Flemer, D. R. Heinle and C. D. Mobley. 1975. The *a posteriori* aspects of estuarine modeling. In: *Estuarine Research* (L. E. Cronin, ed.). Vol. 1, pp. 602-616. Academic Press, New York.

Vernberg, F. J. 1977. The dynamics of an estuary as a natural ecosystem. EPA-600/3-77-016. National Technical Information Service, Springfield, Virginia.

Wiegert, R. G. 1975. Simulation models of ecosystems. Ann. Rev. Ecol. Syst. 6: 311-338.

Wiegert, R. G., R. R. Christian, J. L. Gallagher, J. R. Hall, R. D. Jones and R. L. Wetzel. 1975. A preliminary ecosystem model of coastal Georgia *Spartina* marsh. In: *Estuarine Research* (L. E. Cronin, ed.). Vol. 1, pp. 583-601. Academic Press, New York.

de Wilde, P. A. W. and B. R. Kuipers. 1977. A large indoor tidal mud-flat ecosystem. Helgolander wiss. meeresunters. 30:384-342.

# Simulation Experiments With a Fourteen-Compartment Model of a Spartina Salt Marsh

**Richard G. Wiegert**

Department of Zoology
University of Georgia
Athens, Georgia 30602

**Richard L. Wetzel**

Wetlands Research Section
Virginia Institute of Marine Science
Gloucester Point, Virginia 23064

A simple 14-compartment simulation model of a Georgia *Spartina* salt marsh was constructed initially to summarize the existing data and literature on the ecology of the salt marsh ecosystem. Carbon was chosen as the unit of compartmental exchange. Evolution of the model has resulted in three versions, two of which are compared and contrasted in detail. New data, revision of parameter values and incorporation of new exchange pathways and mathematical representation have resulted in the third and final version of the simple 14 compartment model. Where possible, field measures corroborate simulation results and suggest increased general model fidelity. The results of model simulation experiments suggest complementary field experiments and/or sampling regimes that might be used to simultaneously test the model predictions and suggest further changes in an experimental manipulation with the model. Comparison of field experiments with similar perturbations to the model suggests the utility of the approach but only by incorporating in the mathematical structure of the model biologically realistic interaction equations that do not violate general ecological principals and can be explained in ecological terms.

## Introduction

Salt marshes dominated by cordgrass (*Spartina alterniflora*) comprise a large portion of the marine wetlands of the eastern coast of the U.S.A. Because of its extent and great productivity, this community has the potential to influence profoundly the foodchains in the coastal and possibly offshore marine waters. Thus knowledge of the dynamics of this community is requisite for any rational plan of conservation and/or management in coastal wetlands.

For more than two decades the marsh near the Marine Institute of

the University of Georgia, located on Sapelo Island, has been the subject of a variety of ecological investigations. (For a brief summary of
some of the work done during the first half of this period, see Teal,
1962.) In October 1972 a group of researchers associated with the
University of Georgia began a long-term cooperative effort to describe
and explain the pathways and dynamics of carbon flow in the marsh, a
model whose first version would simply incorporate the then-existing
information (Wiegert *et al*., 1975).

The general view of carbon flow in the marsh adopted by these
investigators is summarized in Figure 1. The ecosystem is seen as
three distinct zones of ecological interaction and flux: 1) terrestrial interactions in the above-water part of the marsh, 2) aquatic
interactions within the water in the tidal creeks and on the marsh
during high tide, and 3) the interactions within the largely anaerobic
sediments. Carbon is seen as either particulate and dissolved organic
matter, i.e. POC and DOC, or gaseous carbon, i.e., $CO_2$ and $CH_4$. The
major processes responsible for the flux of carbon between these zones
and forms of carbon are summarized in the diagram.

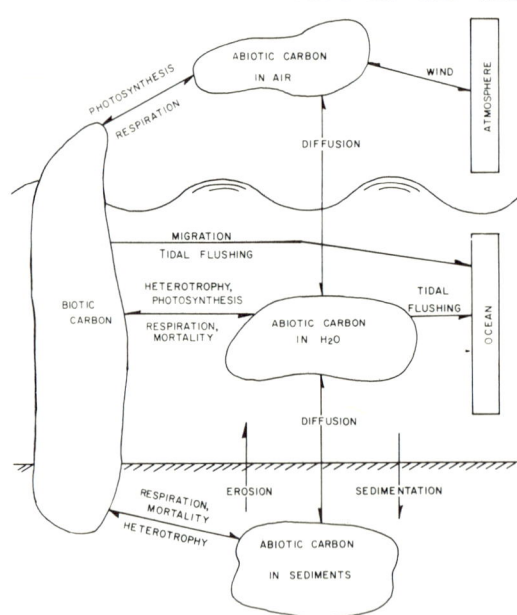

Fig. 1. Diagram of carbon flow in a salt marsh. Carbon is
viewed as either living or non-living and the marsh is divided into
the air, water and sediment zones.

From this diagram, a more detailed food web was outlined containing the 14 compartments that were to form the basis of the first
model (Fig. 2). There were 14 of these, 7 biotic and 7 abiotic.

The purpose of this first version of the carbon flow model was
simply to guide the proposed research program. Although the parameters of the model, as well as all of the feedback controls, were
defined or structured in as realistic a manner as possible following
the procedures first proposed in 1973 by Wiegert and elaborated in
subsequent publications (see Wiegert, 1975), the initial model was in

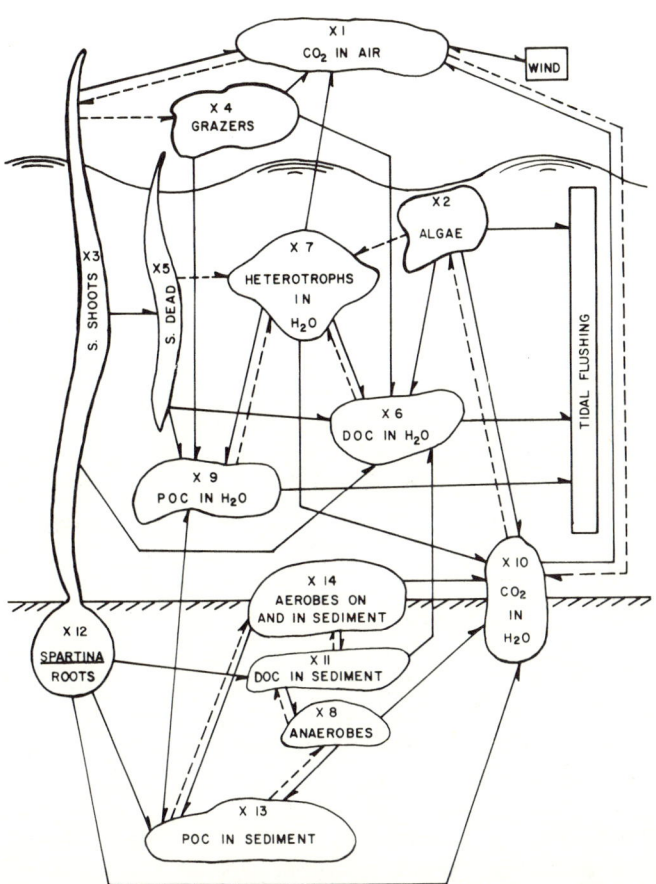

Fig. 2. Pathways of carbon flux between the components of a 14-compartment salt marsh model continuing the structural and functional view outlined in Figure 1. Solid lines designate donor-controlled pathways. Dotted lines designate fluxes that respond to changes in either or both donor and recipient.

no sense regarded as a necessarily valid predictor of the effects of perturbation on the marsh as a whole. Instead the model was used to indicate important parameters, important in the sense that small changes in a given parameter corresponded to relatively large changes in the simulated nominal dynamics of carbon flow within the marsh. Based on the information obtained from model simulations, research studies would then be designed to improve the accuracy of these important, i.e., "sensitive," parameters or to revise the thinking about the functional form of particular feedback relationships. This new information would be incorporated into the model and a new set of simulations and revisions begun. In this manner, a continuous feedback between modeling and field research would increase the efficiency with which the research effort and money were allocated and at the same time result in improved capabilities of the model for prediction.

To date, this process has resulted in three versions of the first salt marsh-carbon model (MRSH1V1, MRSH1V2, MRSH1V3). The initial development and sensitivity analysis of version 1 were described by Wiegert *et al.* (1975). Version 2 differed from version 1 only in the provision of a mechanism for simulating tidal export from certain compartments. These mechanisms were retained and other changes in parameter values and feedback structure were added to produce version 3. Our information about the ecology of the marsh has now reached the point where further development of the model will require a major restructuring of the compartments and at least a threefold expansion in size and complexity.

Thus a final report on the third and last version of model 1 is now appropriate. This is our objective in this paper where we describe the operation of marsh model 1, version 3 (MRSH1V3) by way of several simulated experimental perturbations.

## MODEL STRUCTURE

All three versions of model 1 contain the same 14 compartments shown in Figure 2 (Definitions of state variables and parameter symbols are given in Appendix 1). Model 1, version 1 (MRSH1V1) contains 81 parameters and 39 flows, 5 less than the number shown in Figure 2, which is a diagram of version 3 (MRSH1V3). Descriptions of the compartments and the rationale for constructing each of the feedback control relationships of MRSH1V1 as well as for choosing the parameter values were given by Wiegert *et al.* (1975). However, between MRSH1V1 and MRSH1V3 there were several important structural changes as well as changes in parameter values. The number of parameters increased to 91 and the fluxes to 44. Because MRSH1V2 was a transient intermediate model, this paper will focus on the differences between MRSH1V1 and MRSH1V3.

A most important change was to give MRSH1V3 the capability to simulate the tidally-mediated export of organic carbon from three compartments, algae ($X2$), DOC ($X6$), and POC ($X9$). This is achieved by multiplying each of these compartments by a tidal export coefficient (TIDE, Table 1 and Appendix 1). All algae are suspended in water; about half are benthic forms, of which only a certain proportion are resuspended each high tide and thus subjected to the influence of possible tidal washout or export. A coefficient BENTH (Table 1, Appendix 1) was used to simulate this resuspension. Thus in MRSH1V3 the export pathways of Figures 1 and 2 are at least simulated, if only crudely, by F0200, F0600 and F0900 (Appendix 1).

Table 1. Changes and deletions in parameter values between models MRSH1V1 and MRSH1V3. Parameters with 4 values listed are those changed seasonally and the values are given in the order Winter (Jan.1–Mar.31), Spring (April 1–June 30), Summer (July 1–Sept.30) and Fall (Oct.1–Dec.31).

| Parameter | MRSH1V1 Value | Units | MRSH1V3 Value | Units | Sources[***] |
|---|---|---|---|---|---|
| T0103 | .22 | $day^{-1}$ | .062 | $day^{-1}$ | 2 |
| | .31 | " | .106 | " | |
| | .17 | " | .058 | " | |
| | .12 | " | .034 | " | |
| T0312 | .094 | " | .03 | " | 2 |
| | .136 | " | .071 | " | |
| | .075 | " | .029 | " | |
| | .054 | " | .01 | " | |

Table 1 (cont.)

| Parameter | MRSH1V1 Value | Units | MRSH1V3 Value | Units | Sources *** |
|---|---|---|---|---|---|
| T0507 | .8 | " | .054 | " | 3 |
| T0607 | .8 | " | .64 | " | 3 |
| T0907 | .35 | " | .054 | " | 3 |
| T1002 | .082 | " | .071 | " | 5,7 |
|  | .257 | " | .3 | " |  |
|  | 1.5 | " | 1.3 | " |  |
|  | .171 | " | .148 | " |  |
| T1108 | .8 | " | .02 | " | 1 |
| T1114 | .8 | " | .06 | " | 1 |
| T1203 | .002 | " | .0001 | " | 2 |
|  | .006 | " | .003 | " |  |
|  | 0. | " | .0001 | " |  |
|  | 0. | " | .001 | " |  |
| T1308 | .179 | " | .62 | " | 1 |
| T1314 | .35 | " | .32 | " | 1 |
| R0210 | .007 | $day^{-1}$ | .013 | $day^{-1}$ | 5,7 |
|  | .021 | " | .041 | " |  |
|  | .125 | " | .24 | " |  |
|  | .014 | " | .027 | " |  |
| R0301 | .094 | " | .0035 | " | 2 |
|  | .121 | " | .0048 | " |  |
|  | .071 | " | .0026 | " |  |
|  | .053 | " | .0019 | " |  |
| R0701 | --- | --- | .004 | " | 6 |
|  | --- | --- | .01 | " |  |
|  | --- | --- | .01 | " |  |
|  | --- | --- | .004 | " |  |
| R0710 | .1 | $day^{-1}$ | .01 | " | 6 |
| R0810 | .1 | " | .25 | NONE* | 1 |
| R1210 | .058 | " | .00072 | $day^{-1}$ | 2 |
|  | .088 | " | .0094 | " |  |
|  | .053 | " | .0056 | " |  |
|  | .041 | " | .0043 | " |  |
| R1410 | .1 | " | .40 | NONE* | 1 |
| AE0206 | .01 | $day^{-1}$ | .007 | $day^{-1}$ | 5,7 |
|  | .01 | " | .021 | " |  |
|  | .01 | " | .125 | " |  |
|  | .01 | " | .014 | " |  |
| AE0811 | .8 | " | --- | --- | 1 |
| AE1211 | .002 | " | .000053 | " | 2 |
|  | .003 | " | .00008 | " |  |
|  | .002 | " | .000053 | " |  |
|  | .001 | " | .000027 | " |  |
| AE1411 | .01 | " | .005 | " | 1 |
| EMU0305 | .015 | $day^{-1}$ | .023 | $day^{-1}$ | 2 |
|  | .017 | " | .015 | " |  |
|  | .016 | " | .012 | " |  |
|  | .024 | " | .038 | " |  |
| EMU0813 | .01 | " | .005 | " | 1 |
| EMU1213 | .018 | " | .0031 | " | 2 |
|  | .025 | " | .0043 | " |  |
|  | .020 | " | .0035 | " |  |
|  | .013 | " | .0023 | " |  |
| EMU1413 | .01 | " | .005 | " | 1 |
| A0202 | 14. | $g \cdot c \times m^{-2}$ | 1. | $g \cdot c \times m^{-2}$ | 5,7 |
| A0207 | 11. | " | 1.4 | " | 6 |
| A0303 | 150. | " | 145. | " | 2 |
| A0304 | 1. | " | 10. | " | 9 |

Table 1 (cont.)

| Parameter | MRSH1V1 Value | Units | MRSH1V3 Value | Units | Sources[***] |
|-----------|---------------|-------|---------------|-------|---------|
| A0808 | 180. | " | 30. | " | 1 |
| A1308 | .001 | NONE** | .0025 | NONE** | 1 |
| G0202 | 15. | $g \cdot c \times m^{-2}$ | 2. | $g \cdot c \times m^{-2}$ | 1 |
| G0207 | 1. | " | .8 | " | 5,7 |
| G0707 | 10. | " | 21.3 | " | 6 |
| G0808 | 200. | " | 60. | " | 4,6 |
| G1308 | 2200. | " | 3019. | " | 1 |
| G1414 | 10. | " | 9. | " | 1 |
|  |  |  |  |  |  |
| EPS1108 | --- | NONE* | .31 | NONE* | 1 |
| EPS1308 | .8 | " | .31 | " | 1 |
| EPS1314 | .25 | " | --- | " | 1 |
|  |  |  |  |  |  |
| TIDE | --- |  | .25 | $day^{-1}$ | 8 |
|  |  |  |  |  |  |
| BENTH | --- |  | .5 | NONE | 5,7 |

* Represents a proportion of ingested carbon that is lost (respired or egested).

** Represents the maximum ratio of anaerobic bacteria in sediments to available particulate carbon permissible without a limiting effect.

*** For details see Appendix 2.

Two additional minor additions were made to the fluxes. A pathway F0001 (the only flux permitted to assume a negative value) was provided whereby the daily balance of carbon in the air over the marsh could be reinstated by exchange with the surroundings (the gaseous C pool of the general atmosphere). This redress was made each iteration in order to maintain the gaseous C pool (X1) constant at 875 $gCxm^2$, the calculated weight of C in a 1 $m^2$ column of air reaching to the limits of the atmosphere, assuming $CO_2$ content to be 0.03% of the air. The parameter CLOSED (Appendix 1) was used to specify whether this constant exchange was in operation or not, thus giving the model the capability, if desired, of simulating the $CO_2$ content of the air over the marsh if all lateral exchange were prevented, the hypothetical equivalent to putting a giant cylinder or dome over the marsh.

A flux from the heterotrophs in the water to the air was added to better define respiratory carbon loss because many of these organisms are exposed at low tide and some respire directly to the air. However, this change had no effect on the overall dynamics predicted by the model.

A fundamental change made in the feedback control equations was to link the effects of resource scarcity multiplicatively with the effects of scarcity of space rather than linearly as had been done previously. The rationale behind these controls and the various functions used to represent common types of ecological interactions have been described in detail previously (see summary and citations in Wiegert, 1975) and need not be repeated here. The actual feedback control equations used in MRSH1V3 are in Appendix 1.

An example will make clear the change that was made and its fundamental importance. Assume a population faces a scarcity of an essential resource to such a degree that by spending the usual amount of time and effort searching for nourishment, it acquires only one-half the normal amount, i.e., ingestion is reduced by 50%. Assume the same population to be so crowded, i.e. limited by space, that direct intraspecific interference with its ordinary activities also tends to double the amount of time or effort needed to acquire the normal amount of sustenance. Under the previous method of summing the negative effects of the two types of controls, the net ingestion would have been 100% - (50% + 50%) = 0. However, the logical (and most parsimonious) assumption, in the absence of other evidence, is that two doublings of the effort or time required to ingest a given amount under optimum conditions would produce an ingestion rate of 100% x [(1-0.5) x (1-0.5)] = 25% of the optimum, not zero. Thus the equations were revised to represent this effect of combining two or more simultaneously limiting factors. Although this is a very important and consequential change in the way the model structure is viewed, it had, fortunately, no real effect on the dynamics of the salt marsh model predictions, for all of the biotic compartments were limited in a major way by only one factor, resource or space, and thus the method of combining the negative effects of two or more factors mattered little.

Changes were made in the number, definition, and particularly the value of the model parameters of MRSH1V1 during its evolution to MRSH1V3. These changes represent the incorporation of new information from the field and laboratory research conducted at Sapelo Island or from the recent literature. The values of all 81 parameters used in MRSH1V1 were given by Wiegert *et al.* (1975) who also discussed the basis of the 21 most important, i.e., sensitive parameters. The parameter values of MRSH1V3 are given in subroutines DATA1-DATA5 of Appendix 1. A comparison of the two models (MRSH1V1 vs. MRSH1V3) is given in Table 1 which shows all parameters of MRSH1V3 that were deleted or changed plus all new parameters added to MRSH1V3. The source for each change or addition is given in the far right column of the table.

Simulation of carbon dynamics with the model depends on the numerical solution of the model equations because the model is non-linear with discontinuities in the first derivatives. A simple Euler numerical solution was found satisfactory (Wiegert and Wetzel, 1974). A 0.1 day iteration interval was used for all simulations reported in this paper.

Assuming a flux of carbon ($F_{ij}$) from the donor ($X_i$) to the recipient ($X_j$), the various parameters used in the model (see Appendix 1) are:

1) (T) the maximum specific rate of carbon ingestion that can either be achieved under optimum conditions by a biotic component or by specific rate of transfer from an abiotic component to another abiotic component.

2) (R) The specific rate of carbon loss by biotic components via $CO_2$ or methane production.

3) (ETA)  The specific rate of carbon loss by biotic components
          via the excretion of dissolved organic compounds, usually
          nitrogenous waste products.
4) (EMU)  the minimal specific rate of loss of carbon by nonpreda-
          tory (physiological) mortality.
5) (A)    either the threshold response density of the standing
          stock of the recipient component above which direct inter-
          ference competition begins ($A_{jj}$), or the threshold response
          density of the donor below which the donor component
          exerts control on ingestion by being perceived as scarce
          by the recipient ($A_{ij}$)
6) (G)    either the maximum concentration (standing stock) of car-
          bon in the recipient component that can be maintained
          under otherwise optimal conditions ($G_{jj}$, an equilibrium
          density), or the minimum density of the donor standing
          stock ($G_{ij}$, a refuge) at or below which the donor is
          unavailable to the recipient.
7) (EPS)  the proportion of carbon ingested that is unassimilated.

The existence of more than one resource necessitates two more
parameters:

$PP_{i,j}$ represents the feeding preference of $X_j$ for resource
        $X_i$ when the level of $X_i$ is optimum, i.e., exceeds $A_{i,j}$
$P_{i,j}$ represents the average feeding preference of $X_j$ for
        resource $X_i$, taking into account the relative availability
        of all other potential resources of $X_j$.

In general, the functional relationship between a unit change in
$X_i$ or $X_j$ and the effect on ingestion by $X_j$ is assumed linear, but
sometimes the nature of the ecological interaction is such--as in
the case of the "ingestion" of carbon by anaerobic microorganisms
(X8)--that a much different functional relationship is needed and
the definitions of one or more of the above parameters may need to be
changed.

## SIMULATION EXPERIMENTS

With the model structure and parameter values changed, a simulation run of 5 years was made with the revised model MRSH1V3. This
provided a set of nominal predictions of seasonal change in standing
stock and flux of carbon against which the results of several experimental changes in the parameters and functions could be evaluated.
By the fifth year, all compartments and fluxes predicted by the model
showed a constant annual pattern of variation. Comparison of these
patterns with measurements of standing stocks provided one check on
the fidelity with which the model could reproduce the normal behavior
of the salt marsh ecosystem. The validity of this comparison is, of
course, only possible because field-measured standing stocks and
fluxes of carbon were not used in constructing the model.

Five different experimental manipulations of the model were
undertaken: 1) increased winter grazing on *Spartina* shoots, 2) the
effect of increased spring and summer grazing on algae, 3) harvesting *Spartina* shoots, 4) clipping and root-pruning *Spartina* in small

plots, and 5) varying the tidal export of algae, DOC and POC. The reasons for selecting each of these manipulations and the methods used in the simulation were as follows:

1) Grazing of *Spartina* shoots by biophagous arthropods, although large in absolute amount ingested per unit area (Wiegert and Evans, 1967), represents a small proportion of the extremely high total net productivity of *Spartina alterniflora* in the marshes (Smalley, 1960; Teal, 1962). Model MRSH1V1 showed, however, a great sensitivity of *Spartina* to factors tending to decrease gross production or increase losses. Furthermore, the major grazing pressure in the earlier model was exerted in summer when the standing crop and net productivity were very high. The field data on standing stocks of grazing insects, however (summarized by Teal, 1962), indicated a large population of small plant hoppers (*Prokelesia*) present in the winter when *Spartina* shoot standing stocks were very low. Thus we felt an increase in the winter grazing pressure on *Spartina* shoots would be a useful and potentially realistic manipulation of the model, showing whether, although still low in absolute amounts of carbon consumed, grazing had a significant role as a controller of primary productivity.

To carry out this manipulation, only one parameter of the model was changed: the winter mortality of grazers (EMU0409) was reduced to 0.01.

2) The manipulation of algal grazing had a rationale similar to the experiment on *Spartina*. Although the compartment X2 represents the combined standing stocks of both benthic algae and phytoplankton, the net productivity of the latter is very low. Ragotzkie (1959) claimed it was zero! The net production by the benthic algae, when prorated over the entire marsh is 190 gC x $m^{-2}$ (Pomeroy, 1959), more than 10% of the net primary production of *Spartina*. Most of the parameter values governing the dynamics of X2 are drawn from experiments on the benthic algae so that the compartment can be regarded as tracking the benthic algal dynamics rather than phytoplankton dynamics. (In the next major revision of the model benthic algae and phytoplankton are separated.)

A recent study was made of the effects of mud snails grazing on algae (Pace, 1977) and thus a model experiment along these lines was indicated. To simulate increased grazing on algae, two experiments were run. In the first, the maximum specific rate of ingestion by the heterotrophic grazers (the parameter T0207) was increased by 100%, from 0.8 to 1.6. Note, this does not necessarily increase the realized rate of ingestion by the same amount. In the second run the maximum specific ingestion rate remained the same, but the threshold response density (the parameter A0207) and the refuge level (the parameter G0207) for the algal prey were reduced to 0.8 and 0.4, respectively.

3) *Spartina alterniflora* marshes have in the past been grazed and otherwise disturbed by the removal of the annual production of shoots. An experiment of this type had been performed some years before by a group under the direction of L. R. Pomeroy. We felt the model should be asked to predict the effects of such a perturbation

to see if it had the capability to simulate this kind of drastic perturbation with accuracy.

To simulate the harvesting of all live, shoot standing stock of *Spartina* on June 30 in two successive years (the terms of the field experiment), appropriate IF statements were introduced into the model program to remove the accumulated shoot biomass at these times. The simulation otherwise proceeded normally through 6, not 5 years.

4) A more complex field experimental manipulation on *Spartina* was begun two years ago by a group directed by R. Christian. They set up plots from which all *Spartina* material was removed down to the sediment surface. The plots were clipped periodically to remove new growth. In addition, the roots and rhizomes were cut at the edge of the plot and a metal shield was inserted to a depth of 15 cm to ensure that no new roots or rhizomes grew into the plots from outside. They monitored standing stocks of POC, anaerobic bacteria and root biomass within the plots. This experiment obviously provided another interesting possibility for testing the predictive ability of the model, so a similar simulation experiment was set up.

Appropriate IF statements were inserted in the program to stop the shoot-root transport from reaching the roots, but still remove it from the shoot compartment. Thus a "normal" shoot component is maintained, as in the field experiment where the small plot is surrounded by normal *Spartina* shoot growth. Thus in both the field and model experiments, the production of *Spartina* POC and DOC that goes into the water and could settle out onto the plot is maintained.

5) A major objective of the entire salt marsh ecosystem study is the answer to the question of whether the marsh is a net importer or exporter of carbon. Currently, the model has a tidal exchange which moves material out of the marsh at a rate of 25% of the standing stock of POC and DOC in water per day and 12.5% per day of the algal standing stock. These were merely "best guesses." As a prelude to a major effort to study both the hydrology of the tidal creeks and the standing stocks of POC, DOC, and phytoplankton in the summer of 1977, a series of simulation experiments was performed with the model in which this tidal export (the parameter TIDE) was varied from 0 to 0.5 per day and compared with the nominal run of TIDE = 0.25 per day. Of particular interest was the effect of these perturbations on the standing stocks of POC, DOC in the water, and phytoplankton, as well as the effect on the POC in the sediments and the total annual export of carbon from each of the first three components.

## Results and Discussion

MODEL VALIDATION AND CORROBORATION

Annual variation in the standing crop of carbon in 13 of the 14 model compartments is shown in Figure 3 for the 5th year of the nominal simulation. The $CO_2$ content of air (X1) is kept constant by simulated instantaneous exchange with the surrounding air.

By the end of the 5th year, for any reasonable initial densities, the system reached constancy with respect to the year-to-year change

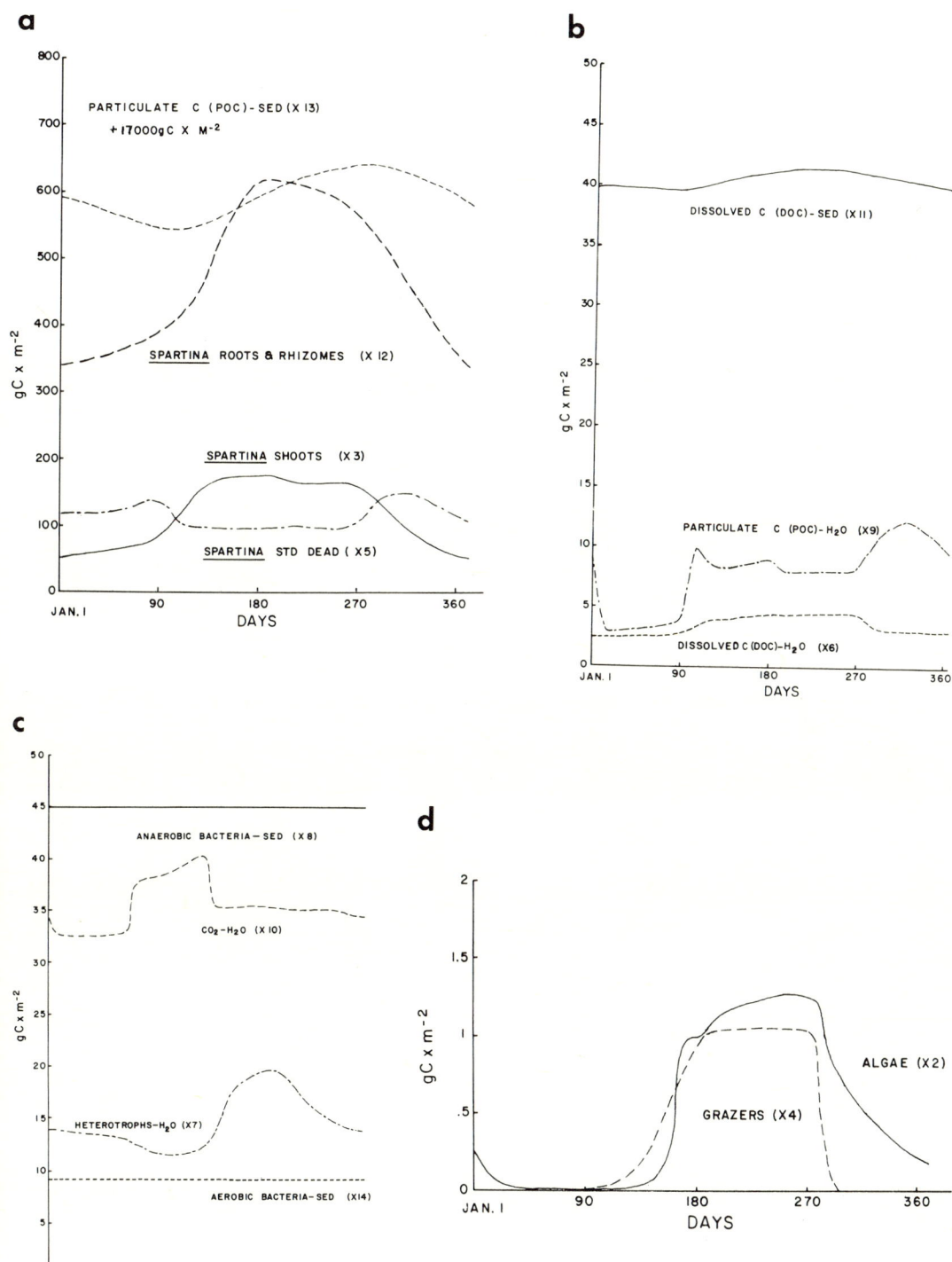

Fig. 3a–d.  Seasonal changes in the level of carbon in selected components of the 14-compartment salt marsh model (MRSH1V3).

in all compartments, except POC in the sediments (Fig. 3a-d and Table 2). The predicted standing stock of POC in the sediment, compartment X13, increased at an annual rate of 26 g C x m$^{-2}$ (704 minus 678 from the column and row sums of Table 2). The sediment contains, to a depth of 0.35 meter, 17 to 18 thousand g C x m$^{-2}$. Thus an annual increase of 26 g C x m$^{-2}$ represents an increase in sediment thickness of approximately $\approx$ 0.5 mm. This is in reasonable agreement with current geological evidence for the coastal Georgia marshes (Frey and Basan, in press) but less than the 2.5 mm per year deposited in the Flax Pond *Spartina* marsh on Long Island Sound (Flersa, Constantine, and Cushman, 1977).

The simulation data summarized in Figure 3a-d and Table 2, when compared with the available data from field measurements, permit an initial assessment of the ability of the model to predict the nominal behavior of the marsh and provide at least a first positive corroboration and test of validity.

Because most models of ecosystems unavoidably encompass both the predictive and theoretical objectives (Caswell, 1976), tests applicable to both kinds of models must be applied. The initial verbal statements of the compartments and modes of interaction, i.e. the state variables and the equations, form the qualitative structure of a model and constitute its theoretical basis. Thus, according to Caswell (1976), questions of number and character of stability points, the seasonal form of the standing crop changes, and the directional

Table 2. Chart of the simulated cumulative annual flow of carbon between components of a 14-compartment salt marsh model. Values represent g C x m$^{-2}$ going from components in column at left to components listed horizontally at top of table. All values are for the 5th year of a 5-year simulation and are rounded to the nearest gram. See text for discussion.

| From \ To | X0 | X1 | X2 | X3 | X4 | X5 | X6 | X7 | X8 | X9 | X10 | X11 | X12 | X13 | X14 | Total |
|---|---|---|---|---|---|---|---|---|---|---|---|---|---|---|---|---|
| X0 | | 1100 | | | | | | | | | | | | | | 1100 |
| X1 | | | | 2652 | | | | | | | 0 | | | | | 2652 |
| X2 | 24 | | | | | | 15 | 69 | | | 29 | | | | | 137 |
| X3 | | 145 | | | 25 | 867 | 93 | | | | | | 1708 | | | 2838 |
| X4 | | 14 | | | | | 1 | | | 9 | | | | | | 24 |
| X5 | | | | | | | 73 | 46 | | 748 | | | | | | 867 |
| X6 | 332 | | | | | | | 1 | | | | | | | | 333 |
| X7 | | 32 | | | | | 47 | | | 127 | 47 | | | | | 253 |
| X8 | | | | | | | | | | | 466 | 245 | | 80 | | 791 |
| X9 | 718 | | | | | 137 | | | | | | | | 29 | | 884 |
| X10 | | 1360 | 136 | | | | | | | | | | | | | 1496 |
| X11 | | | | | | | 104 | | 155 | | | | | 12 | | 271 |
| X12 | | | | 186 | | | | | | | 933 | 10 | | 579 | | 1708 |
| X13 | | | | | | | | | | 636 | | | | | 42 | 678 |
| X14 | | | | | | | | | | | 21 | 16 | | 16 | | 53 |
| Total | 1074 | 2651 | 136 | 2838 | 25 | 867 | 333 | 253 | 791 | 884 | 1496 | 271 | 1708 | 704 | 54 | 14085 |

(qualitative) responses to perturbations, all permit corroboration of the model. Quantitative considerations of magnitude of standing crop and/or fluxes are primarily tools of validation.

The data of Figure 3a-d and Table 2 showing predicted standing crop changes and annual carbon fluxes are compared with measured values from the field to assess the predictive capabilities of a model whose structure was initially established in part by deductive reasoning from a few accepted general statements about the growth and control of populations (Wiegert *et al.*, 1975). This initial corroboration can be extended by observing if the model successfully passes further testing by perturbation, the major objective of this paper.

Unfortunately, data on the seasonal changes in standing crop of many of the compartments are still scarce. Furthermore, the scope of the present report is not sufficient to justify lengthy comparisons. Instead, Table 3 has been prepared which lists the mean annual standing crop in each compartment predicted by models MRSH1V1 and MRSH1V3, together with the annual average of the latest field data

Table 3. Annual mean standing crop (g C x m$^{-2}$) of the 14 state variables. Predictions of MRSH1V1 and MRSH1V3 are compared with measurements from the field. For sources, see Appendix 2.

| Compartment | MRSH1V1 Annual Average | Annual Change | MRSH1V3 Annual Average | Annual Change | Field Data Annual Average | Annual Change | Sources |
|---|---|---|---|---|---|---|---|
| 1 | 782 | | 875 | | 875 | | 8 |
| 2 | 15 | | .5 | | 1 | | 5,7 |
| 3 | 125 | | 118 | | 135 | | 2 |
| 4 | .5 | | .4 | | 1 | | 9 |
| 5 | 98 | | 115 | | 130 | | 2 |
| 6 | 995 | +225 | 4 | | 5.6 | | 1 |
| 7 | 10 | | 12 | | | | 10 |
| 8 | 25 | + 1 | 44 | | 30-45 | | 1 |
| 9 | 204 | | 8 | | 9 | | 1 |
| 10 | 46 | | 35 | | ? | | 11 |
| 11 | .8 | | 36 | | 26 | | 1 |
| 12 | 137 | | 475 | | 450 | | 2 |
| 13 | 20291 | +589 | 17733 | +26 | 17500-19000 | +50* | 1 |
| 14 | 10 | | 9 | | 3-6 | | 1 |

* Based on an average annual increase in sediment of 1 mm.

and its source. Thus by comparing the output of two different models, the task of subjectively assessing the relative ability of the models is facilitated. Indeed, with more extensive field data, objective statistical means could easily be employed in this comparison.

Comparing the output of the two models to the means obtained from the field in Table 3 shows the generally more satisfactory fit of version 3. Clearly the greatest deficiency of MRSH1V1 was the lack of a mechanism to simulate tidal export; this caused the large accumulations of carbon as DOC, dissolved organic carbon, in the water (X6), POC, particulate organic carbon, in the water (X9) and POC, particulate organic carbon, in the sediments (X13). Carbon dioxide in the air was somewhat lower in MRSH1V1 because the corrections for loss or gain were made every 90 days instead of each iteration (0.1 day) as in MRSH1V3. In every case in Table 3 when the annual averages of MRSH1V1 differed greatly from those predicted by MRSH1V3, the values from the latter versions of the model were closest to the independent data obtained from the field. In addition, the large annual accretion of carbon in the sediments and of dissolved carbon in the water predicted by MRSH1V1 does not match reality. The annual sediment carbon buildup is very small, probably much closer to the value predicted by MRSH1V3. Thus the addition of a tidal export factor affecting carbon materials carried in the water and the parameter changes made in MRSH1V3 (Table 1) improved both the theoretical value and predictive effectiveness of the model.

### Grazing effects on Spartina

The latest information on the ingestion of *Spartina* by grazers shows both a high spring-summer and a high winter population of grazing insects (Pfeiffer, pers. comm.). The winter population comprises mostly small plant hoppers of the genus *Prokelesia*. Estimates of the winter standing stock in 1976–1977 were 0.2 g C x m$^{-2}$, substantiated by estimates from Smalley, 1960 (see also Teal, 1962 and Wiegert and Evans, 1967). The added ingestion due to this winter population is somewhat less than 0.1 g C x m$^{-2}$ x day$^{-1}$ (Pfeiffer, pers. comm.).

Winter grazing of *Spartina* was not included in the simulations of model MRSH1V1. To represent this additional drain, the winter mortality was relaxed and the standing crop of grazers thus increased as shown in Figure 4. This increase in the duration of grazing and the consequent increase of ingestion of *Spartina* by grazers had very little effect on the annual net primary production of *Spartina*, decreasing it by only 16 g C x m$^{-2}$ x yr$^{-1}$.

The increases in both the standing crop of grazers and in their ingestion produced by the simulation experiment were much greater than the field data on winter plant hoppers cited above would suggest. Thus the quantitative impact of arthropod grazing on *Spartina* productivity is small. Only if the grazers were utilizing some nutrient in short supply could they be assumed to play a controlling role in the carbon dynamics of the marsh. Research on the feeding of *Prokelesia* is currently underway.

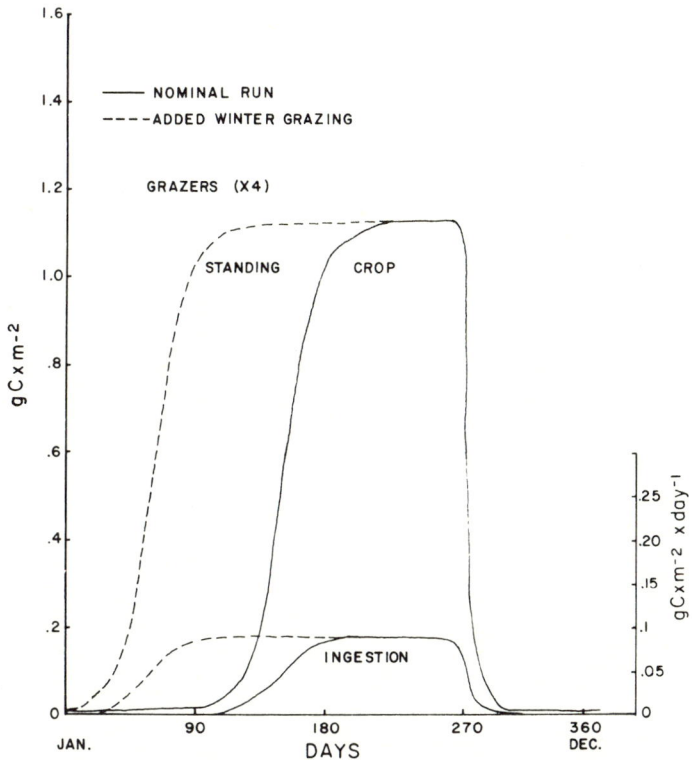

Fig. 4. Seasonal changes in standing crop of grazers (X4) and in their ingestion of *Spartina* shoots. Dotted lines represent the added effect obtained by allowing winter grazing.


## *Effects of grazing on algae*

In contrast to *Spartina* grazing, an increase in the accessibility of algae (X2) to grazers (heterotrophs in water, X7) in the model produced a marked decrease in the predicted standing crop of algae during the active summer–fall growth period (Fig. 5). There was also a decrease in the net primary production of the algae from 107 to 59 g C x m$^{-2}$ x yr$^{-1}$. Indeed, the simulated effect of increased grazing was so drastic that ingestion of algae by the grazers was also decreased, from 68 to 41 g C x m$^{-2}$ x yr$^{-1}$. Thus the relaxed strictures against grazing employed in the simulation increased the grazing pressure on the algal component and resulted in a decreased standing crop, decreased productivity, and a decrease, rather than an increase, in ingestion by grazers.

The above result, if confined to the benthic diatom-snail grazing relationship, does predict what is being observed in the field. Namely, a moderate to severe grazing pressure applied to the benthic diatoms results in decreased populations of diatoms, decreased primary production, and decreased ingestion by the snails (Pace, 1977). Although the diatom-snail interaction is found only on exposed mud-flats, the benthic algae within the marsh are grazed upon by the dense populations of fiddler crabs, mainly *Uca pugnax* (Montague, pers. comm.).

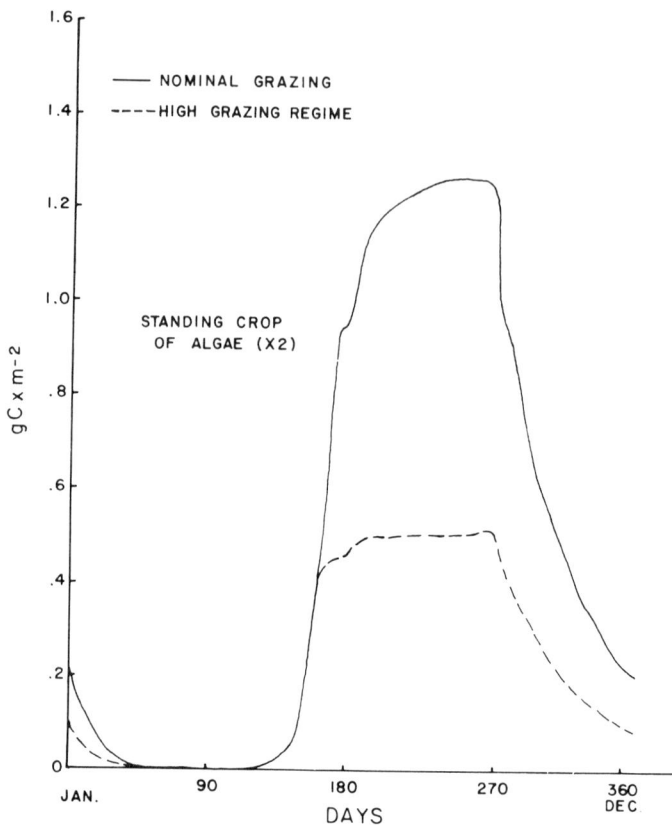

Fig. 5. Seasonal change in the standing crop of algae under both a normal grazing (solid line) and a high-grazing regime (dotted line).

Normally, the algal component (X2) is either uncontrolled in the sense that the maximum gross production rate (T1002) is so low that growth is prohibited (winter and spring), or else it is limited by intraspecific crowding effects via the factors A0202 and G0202 (see model equations in Appendix 1). In the fall the photosynthesis rate is large enough for growth to occur, but grazing causes a gradual reduction in the population size, so again no limiting factors are operational. Simulated grazing of the algae is, however, always strongly limited by scarcity of algae, relieved somewhat in summer when algal standing stocks are high. In addition, the grazers themselves, being included in the heterotrophs in water (X7) are controlled by direct interference type competition (FF0707 > .4, except during midsummer when high respiratory losses cause a slight decrease in the simulated standing stock in X7). Thus these biophages are not free in the model to respond to such perturbations as, for example, raising the maximum rate at which algae may be ingested (increasing PP0207). The model output is relatively insensitive to this kind of parameter change. Only by making algae more available through change in the thresholds governing grazing can increased biophagic ingestion of algae be observed, and this, of course, also introduces the possibility that the algal standing crop and productivity will also be lowered.

## Simulated effects of harvesting Spartina

Because of the dominant role played by *Spartina alterniflora* in the marsh system, the simulated effect of removing the shoots of this plant from plots in the marsh was of considerable interest. This experiment was performed several years ago by a group under the direction of L. R. Pomeroy. The final sampling remains to be completed this coming summer, so the results are not yet published. In general, harvesting twice in successive years led to a drastic decline in the standing crop of roots and rhizomes, with a general replacement of *Spartina* with *Salicornia*, a succulent halophyte. Although model MRSH1V3 has no capability to predict the replacement of species, we wanted to repeat this experiment with the model (1) to see if harvesting had such a drastic effect on *Spartina* as to make probable its partial replacement by a potential competitor and (2) to ascertain how completely the model would predict recovery of *Spartina* if the situation were not complicated by the presence of competing species.

Table 4 presents the results of this simulation experiment. Clearly, clipping the shoots had a drastic effect on *Spartina*. By the end of the second year the standing crop of shoots had been reduced to 10% of the nominal value on January 1 and roots-rhizomes were approximately 10% of normal. Interestingly, roots and rhizomes had begun to make a partial recovery during the second year, despite the second clipping, for they had increased to almost 20% of the usual value at the end of the first year. This is not only an unexpected conclusion from the model, but it is one that will repay investigation to see why the model predicts it and whether it is a real effect observable in the field. Annual net primary production is reduced by the end of the second year to approximately 20% of normal.

Given the drastic reduction in the *Spartina* components, not much imagination is needed to predict invasion by competitive species. Failing this, the model prediction is that 3 1/2 years after the second clipping the system will have returned to normal as did all other state variables and fluxes in the model (Table 4). The perturbation caused by harvesting *Spartina* can thus be considered a relatively transient phenomenon once clipping is stopped and if complications caused by the invasion of competitive species do not arise.

Table 4. Simulated effects of harvesting *Spartina* shoots June 30, for 2 consecutive years.*

| | Stc. Shoots | Stc. Roots | Annual Net Prim. Prod. (g C x m$^{-2}$) |
|---|---|---|---|
| Jan. 1, Yr. 1 | 54*clip | 345 | 1573 |
| Jan. 1, Yr. 2 | 11*clip | 34 | 880 |
| Jan. 1, Yr. 3 | 5 | 62 | 353 |
| Jan. 1, Yr. 4 | 31 | 130 | 603 |
| Jan. 1, Yr. 5 | 52 | 310 | 1447 |
| Jan. 1, Yr. 6 | 54 | 342 | 1573 |

*"Clipped" on June 30 in each of the first two simulated years.

An interesting question now is: Under what conditions may the latter take place?

Clearly the model prediction is that the *Spartina* marsh ecosystem cannot withstand continual annual harvesting at this time of the year. If one ignores the possibility of invasions and replacement of *Spartina*, the total yield of the system would be small, seeming to preclude any systematic commercial exploitation of the marsh productivity. In the present instance the harvest yield in the second year was $112$ g C x m$^{-2}$ versus $179$ g C x m$^{-2}$ the first year. The yield would decline further with continued harvesting. Other perturbation experiments are planned to ascertain when and if the yield would stabilize under annual harvesting and to investigate the predicted effects of harvesting at other times of the year.

### EFFECTS OF STOPPING CARBON FLOW TO ROOT-SHOOT COMPONENT

In 1975 a group headed by Robert Christian began an experiment in the high marsh at Sapelo designed to measure the effect of constantly clipping the *Spartina* shoots from selected plots and simultaneously preventing lateral input of carbon from growth of roots and rhizomes of plants outside the experimental plots. Again, this is a continuing experiment and the results are not published, but to date the overall results have been surprisingly negative. Eighteen months after the first clipping of shoots and the placement of the metal rings, the various sediment components remain much the same. There has been neither reduction of microbial standing crops (measured as ATP concentration) nor of level of metabolic activity (measured as energy charge). The standing crops of roots-rhizomes has declined, but not drastically. Many more detailed measurements of other variables were, of course, included in the field experiment, but the above three are the ones separately modeled in MRSH1V3.

The model perturbation produced similar results with one exception: the root-rhizome component disappeared rapidly (Table 5). All components and flows, in or affected by the sediment constituents, are given in Table 5, except fluxes from the root-rhizome component. Because the latter all but disappeared from the simulation in 18 months, the fluxes from roots-rhizomes also declined almost to zero. Note that all other compartments and fluxes remained the same as the nominal run or were only slightly affected by the cutoff of F0312 and the subsequent disappearance of the roots (compartments 8 and 11 are slightly decreased, along with their inputs to the water and DOC above the sediments). The energy for this maintenance of the normal microbial components and flows comes from a decrease in the sediment organic load. Thus the model, simplified as it is, presents the root-rhizome component as primarily a supplier of dissolved and particulate organic matter to the sediment system with little direct control of the processes occurring there. The degree to which this simplicity is mirrored by the field experiment is quite surprising and certainly pleasing. However, because the model predicts a much more rapid decline of the root-rhizome component than seems to be occurring in the field experiment, we need to find out (1) whether

our model root-rhizome mortality or respiration parameters in the
nominal run are in error or (2) whether the mortality and respiration
of roots and rhizomes decline once active connection with the shoots
is severed.  Furthermore, the field experiment must be monitored
further to determine if the microbial processes do indeed maintain
normal levels once the standing crop of living roots and rhizomes has
declined to the level seen in Table 6.

### EFFECTS OF VARYING TIDAL EXCHANGE IN THE SALT MARSH MODEL

The most important question about the model for the salt marsh
system concerns the fate of the organic carbon fixed by the autotrophs.
Is this organic carbon produced in such excess amounts as to necessi-
tate its export from the system to the offshore waters, or, alterna-
tively, is the secondary productivity of the marsh maintained at
least in part by the import of carbon from outside?  In short, is the
marsh system itself a source or a sink for organic carbon?

The first version of the Sapelo marsh model, MRSH1V1, answered
that the marsh was a source (Wiegert et al., 1975).  Simulations run
as an open system, i.e., with the $CO_2$ in air replenished each season,
predicted the steady buildup of organic matter in the system.  No
reasonable manipulation of the parameters of the model sufficed to
reverse this prediction; that is, the model could not easily be converted to one which predicted the marsh to be a carbon sink.

Therefore, the inclusion of a TIDE parameter to simulate the net
export of these organic carbon materials carried in the marsh waters

Table 5.  Comparison (with a normal run) of results after 18
months of simulating the effects of removing the carbon flow from
shoots to the root-rhizome compartment of a salt marsh model (MRSH1V3).
Only those components located in or affected by the sediment system
are given.  State variables are given in units of g C x m$^{-2}$, fluxes
in units of g C x m$^{-2}$ x day$^{-1}$.

| Component | Simulation Run | |
| --- | --- | --- |
| | Nominal | Experimental |
| *Spartina* roots-rhizomes (X12) | 627 | 1.8 |
| Anaerobes in sediments (X8) | 44 | 43 |
| Aerobes in sediments (X14) | 9 | 9 |
| DOC in sediments (X11) | 36 | 30 |
| POC in sediments (X13) | 17,654 | 16,946 |
| Flux sediments to water (F1106) | .3 | .2 |
| Flux anaerobes to water (F0810) | 1.3 | 1.2 |
| Flux anaerobes to DOC-sed (F0811) | .7 | .7 |
| Flux anaerobes to POC-sed (F0813) | .2 | .2 |
| Flux aerobes to water (F1410) | .1 | .1 |
| Flux aerobes to POC-sed (F1411) | .04 | .04 |
| Flux aerobes to POC-sed (F1413) | .04 | .04 |
| Flux DOC-sed to anaerobes (F1108) | .4 | .4 |
| Flux DOC-sed to aerobes (F1114) | .03 | .03 |
| Flux POC-sed to anaerobes (F1308) | 1.7 | 1.7 |
| Flux DOC-sed to aerobes (F1314) | .0 | .1 |

was an inevitable outgrowth of the first model, MRSH1V1. This tidal washout dramatically improved the predictive ability of the model, indeed, so dramatically that a more accurate estimate of the TIDE parameter became an important goal of the modeling effort. Unfortunately, attempts to simply measure the net import or export of materials from the marsh and thus directly determine the rate of exchange have met with little success. In fact, the entire concept of a simple, twice-a-day turbulent mixing and exchange of all water in the tidal creeks is being questioned (J. Imberger, pers. comm.). Thus, a more accurate estimate of TIDE and/or the mode of tidal action that must be incorporated into the model awaits further study. One avenue of exploration that proved useful, however, was simply to ask what effect the model would predict for different values of TIDE. Accordingly, the simulations were performed.

Table 6 summarizes the overall export of carbon materials and changes in carbon balance predicted for TIDE values of 0, 0.125, 0.25, and 0.5. Clearly, zero tidal exchange and transport of materials (the situation in MRSH1V1) produce a totally unrealistic situation with respect to the carbon balance. In terms of annual export alone, the three TIDE values leave little to choose among them, but in terms of carbon balance, the nominal value of TIDE = 0.25 or the increases to 0.5 seem in agreement with the geological sedimentation rate. However, the large carbon load in the sediments and the relatively infinitesimal rate of annual accretion make impossible a determination of the true value for TIDE by this approach.

Table 6. Simulated effects of varying the tidal exchange in a coastal Georgia salt marsh. Values are those of the final year of a 5-year simulation.

| Exchange (%/day) | Amount of Carbon Exported (g C x m$^{-2}$ x yr$^{-1}$) | Change in the Carbon Balance of the Marsh (g C x m$^{-2}$ x yr$^{-1}$) |
|---|---|---|
| 0 | 0 | 1012 |
| 12.5 | 1025 | 53.3 |
| 25 | 1074 | 26.1 |
| 50 | 1033 | 17.4 |

Plotting the change in seasonal standing crop of the three water-borne carbon components of the marsh (Figs. 6-8) produced some interesting and potentially helpful contrasts. Both POC and DOC in the water are passive abiotic components. By this is meant they have no dynamics of their own and the action of increasingly larger tidal export terms has no effect on the seasonal variation in standing crop but simply decreases the average annual value, thus maintaining a relatively constant export (Figs. 6, 7 and Table 6). Although the absolute difference between the standing crops of POC and DOC in the water are quite large at times, the field measurements of these materials are variable and do not permit resolution of which value of

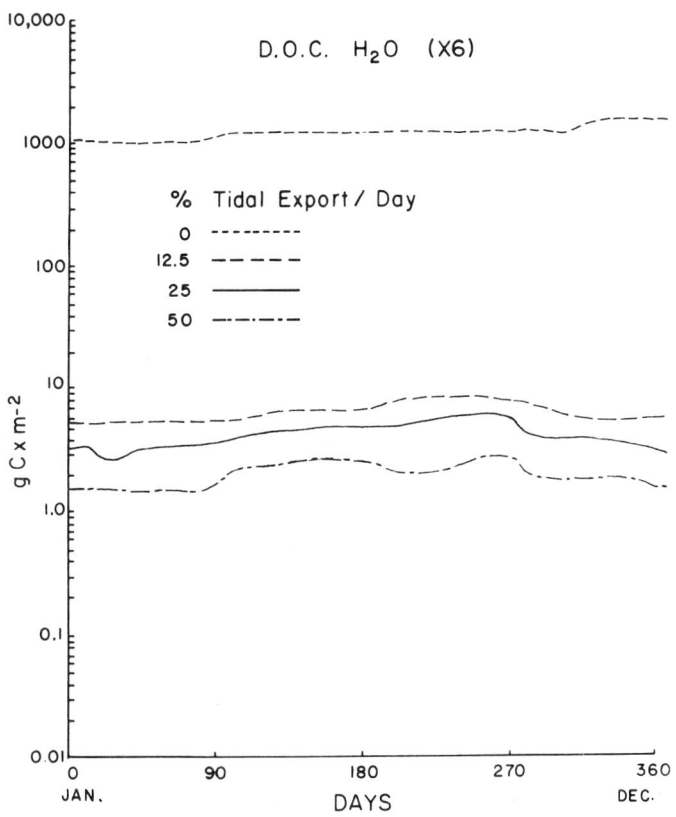

Fig. 6.  Effect of changing tidal export (TIDE) on the seasonal
distribution and level of dissolved organic carbon in the water.

TIDE 0.125, 0.25, or 0.5 is the most realistic value.  Reference to
Table 3, for example, shows that the best field estimates of DOC and
POC in water are annual averages of 6 and 9 g C x m$^{-2}$, respectively.
Comparison with Figures 5 and 6 shows that for X6 the closest estimate
is provided by a simulation using a TIDE value of 0.125, but in the
case of POC the closest prediction employs TIDE - 0.25, the nominal
value.  However, the variability of field estimates is wide and
statistically suitable sets of data are not available.

But in the case of the algae, X2, the effect of various tidal
export rates is combined with the dynamics of the algae to produce
not only differences in annual average standing crop, but also marked
seasonal shifts in distribution of biomass (Fig. 8).  In the case of
DOC and POC in the water, there was a marked difference between zero
tidal export and any one of the tidal export values (on the basis of
even the limited field data discussed).  For the algae, however, the
marked distinctions are between the TIDE values--0.125, 0.25, and
0.5--with the simulations for TIDE = 0 and TIDE = 1.25, 0.25, and
to each other.  Because a TIDE = 0 can be ruled out on the basis of
the evidence in Table 3 and Figures 6 and 7, the algae would appear
to offer a means of deciding between a tidal export of 12.5%, 25% or
50% per day.

Unfortunately, seasonal measurements of chlorophyll adequate to

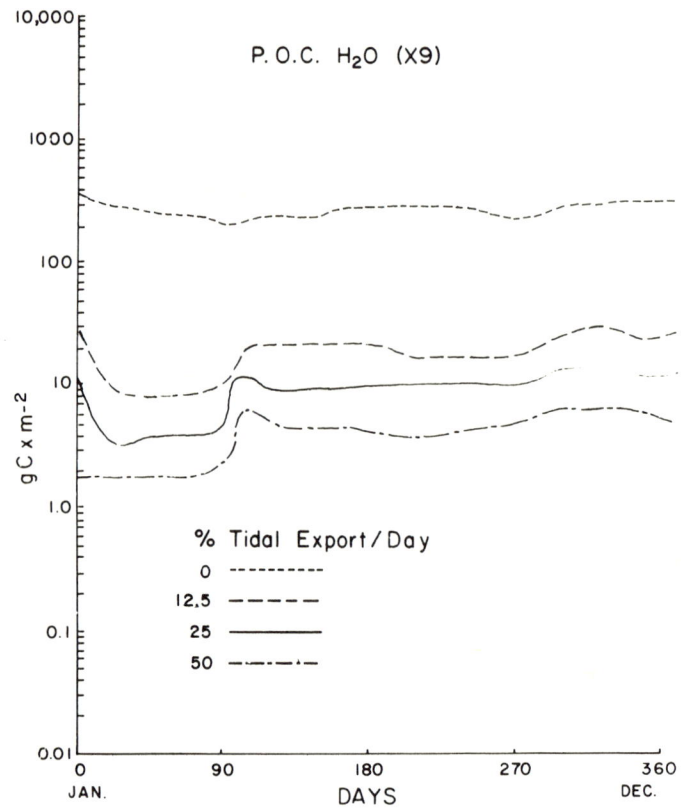

Fig. 7. Effect of changing tidal export (TIDE) on the seasonal
distribution and level of particulate organic carbon in the water.

make the comparison suggested by Figure 8 do not now exist for the
tidal creeks of the Duplin River drainage. However, the acquisition
of such data should present no great problems and such a test is
currently planned for the coming year. In the meantime, further simulations will be used to test the sensitivity of the patterns shown by
Figure 8 to changes in the parameters governing the dynamics of the
algae. If these patterns prove to be relatively insensitive to such
manipulations, the utility of this kind of simulation for testing
the adequacy of the TIDE parameter values will be enhanced.

## Conclusions

A simple 14-compartment model of a salt marsh ecosystem has been
used to perform a number of experiments, the results of which suggest
complementary field experiments and/or sampling regimes that might
be used to simultaneously test the model predictions and suggest
further changes in and experimental manipulations with the model.

The model, to be useful for the purposes discussed in this
paper, must be constructed in a realistic manner in the sense that all
parameters must have biological meaning and be measurable. The interaction equations must be explainable in ecological terms and must not
violate ecological principles. But beyond this the model need not go.
That is, preliminary versions of the model, even though omitting a

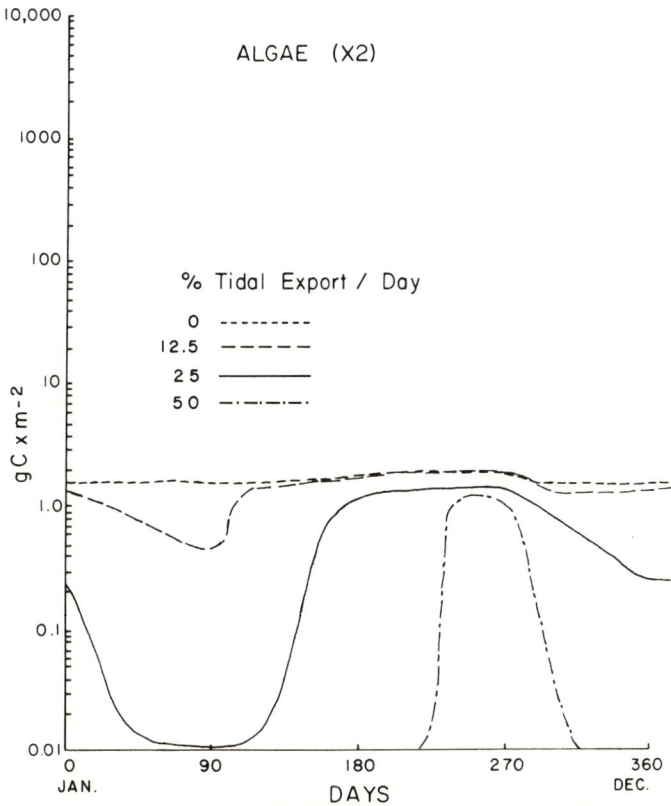

Fig. 8.  Effect of changing tidal export (TIDE) on the seasonal distribution and level of carbon in benthic diatoms and phytoplankton in the salt marsh.

vast store of the interactions and details of the real world, may perform useful service in pinpointing areas needing further work and suggesting or improving upon the design of field experimentation.

## Acknowledgments

A model such as this one necessarily summarizes the work of a large group of scientists.  Those who supplied published and unpublished data are acknowledged by name in Appendix 2.  We are very grateful to the many other colleagues who assisted us via comment and discussion.

For reading and commenting on the entire ms., we thank Alice Chalmers, Robert Christian, Evelyn Haines, Jorge Imberger, Lawrence Pomeroy and William Wiebe.  We are particularly indebted to Clay Montague, whose critical comments prompted a reevaluation of the general method of combining the feedback controls.

The work was supported by the NSF through a Coherent Area Grant, (DES 72-01605 A02).

## References

Caswell, H.  1976.  The validation problem.  In: B. Patten (ed.) *Systems Analysis and Simulation in Ecology*, vol. IV, pp. 313-325.  Academic Press, New York.
Christian, R. R.  1976.  Regulation of a salt marsh soil microbial community; a field experimental approach.  Ph.D. Dissertation, University of Georgia, Athens.  132 pp.

_____ and J. R. Hall. 1977. Experimental trends in sediment microbial heterotrophy: radioisotopic techniques and analysis. In: B. Coull (ed.) *Ecology of Marine Benthos*, pp. 66-88. University of South Carolina Press, Columbia.

_____, K. Bancroft, and W. J. Wiebe. 1975. Distribution of microbial adenosine triphosphate in salt marsh sediments at Sapelo Island, Georgia. Soil Science 119: 89-97.

Frey, R. W. and P. B. Basan. 1978. North American coastal salt marshes. In: R. A. Davis (ed.), *Coastal Sediments*. Springer-Verlag. (in press).

Flessa, K. W., K. J. Constantine, and M. Cushman. 1977. Sedimentation rates in a coastal marsh determined from historical records. Chesapeake Sci. 18: 172-176.

Gallagher, J. L. 1975. Effect of an ammonium nitrate pulse on the growth and elemental composition of natural stands of *Spartina alterniflora* and *Juncus roemerianus*. Am. J. Bot. 62(6): 644-648.

_____ and W. J. Pfeiffer. 1977. Aquatic metabolism of the communities associated with attached dead salt marsh plants. Limnol. Oceanogr. 22: 562-565.

_____, W. J. Pfeiffer, and L. R. Pomeroy. 1976. Leaching and microbial utilization of dissolved organic carbon from leaves of *Spartina alterniflora*. Estuar. and Coast. Mar. Sci. 4: 467-471.

_____, R. J. Reimold, and P. E. Thompson. 1972. Remote sensing and salt marsh productivity. Proc. 38th Annual Meeting American Society of Photogrammetry, Washington, D.C., pp. 338-348.

Guirgevich, J. R. and L. Dunn. 1977. Seasonal patterns of carbon metabolism and plant-water relations of *Juncus roemerianus* (Scheele) and *Spartina alterniflora* (Loisel) in a Georgia salt marsh. (in preparation).

Pace, M. L. 1977. The effect of macroconsumer grazing on the benthic microbial community of a salt marsh mudflat. M.S. Thesis, University of Georgia, Athens.

Payne, W. J. 1970. Energy yields and growth of heterotrophs. Ann. Rev. Microbiol. 24: 17-52.

Pomeroy, L. R. 1959. Algal productivity in the salt marshes of Georgia. Limnol. Oceanogr. 4: 367-386.

Ragotzkie, R. A. 1959. Plankton productivity in estuarine waters of Georgia. Inst. Mar. Sci. Univ. Texas 6: 146-158.

Smalley, A. E. 1960. Energy flow of a salt marsh grasshopper population. Ecology 41: 785-790.

Sottile, W. S. 1973. Studies of microbial production and utilization of dissolved organic carbon in a Georgia salt marsh estuarine ecosystem. Ph.D. Dissertation, University of Georgia, Athens.

Teal, J. M. 1962. Energy flow in the salt marsh ecosystem of Georgia. Ecology 43(4): 614-624.

Valiela, I., J. M. Teal, and N. Y. Persson. 1976. Production and dynamics of experimentally enriched salt marsh vegetation: Belowground biomass. Limnol. Oceanogr. 21: 245-252.

Wetzel, R. L. 1975. An experimental-radiotracer study of detrital carbon utilization in a Georgia salt marsh. Ph.D. Dissertation, University of Georgia, Athens.

_____. 1977. Carbon resources of a benthic salt marsh invertebrate *Nassarius obsoletus*. In: M. Wiley (ed.) *Processes*, vol. II, pp. 293-308. Academic Press, New York.

Wiegert, R. G. 1975. Simulation models of ecosystems. Ann. Rev. Ecol. Syst. 6: 311-338.

_____. 1978. Population models: experimental tools for analysis of ecosystems. In: (D. J. Horn, R. Mitchell and G. R. Stars (ed.) *Proceedings of Colloquium on Analysis of Ecosystems*. Ohio State University Press, Columbus. (in press).

_____ and F. C. Evans. 1967. Investigations of secondary productivity in grasslands. In: *Proceedings of Warsaw I.B.P. Meeting on Methods of Measuring Secondary Productivity in Grasslands*. pp. 499-518.

_____ and R. L. Wetzel. 1974. The effect of numerical integration technique on the simulation of carbon flow in a Georgia salt marsh. In: *Proc. Summer Computer Simulation Conference, Houston*. 2: 575-577.

_____, R. R. Christian, J. L. Gallagher, J. R. Hall, R. D. Jones, and R. L. Wetzel. 1975. A preliminary ecosystem model of coastal Georgia *Spartina* marsh. Estuarine Research 1: 583-601.

# Appendix 1

### Definitions of state variables and a listing of the FORTRAN IV program of MRSH1V3

The 14-compartment salt marsh model contains the following state variables (compartments) and parameters (expressed in units of carbon):

X1 -- Carbon dioxide in the air

X2 -- Algae, both benthic and phytoplankton

X3 -- Live shoots of *Spartina alterniflora*

X4 -- Grazers of *Spartina* shoots

X5 -- Standing dead *Spartina*

X6 -- Dissolved organic carbon (DOC) in water

X7 -- Heterotrophic organisms in water

X8 -- Anaerobic microorganisms in the sediment

X9 -- Particulate non-living organic matter (POC) in the water
X10-- Carbon dioxide and methane in the water
X11-- Dissolved organic carbon (DOC) in the interstitial water in the sediment
X12-- Roots and rhizomes of *Spartina alterniflora*
X13-- Particulate non-living organic carbon (POC) in the sediment (to depth of one meter)
X14-- Aerobic organisms on and in the sediments.

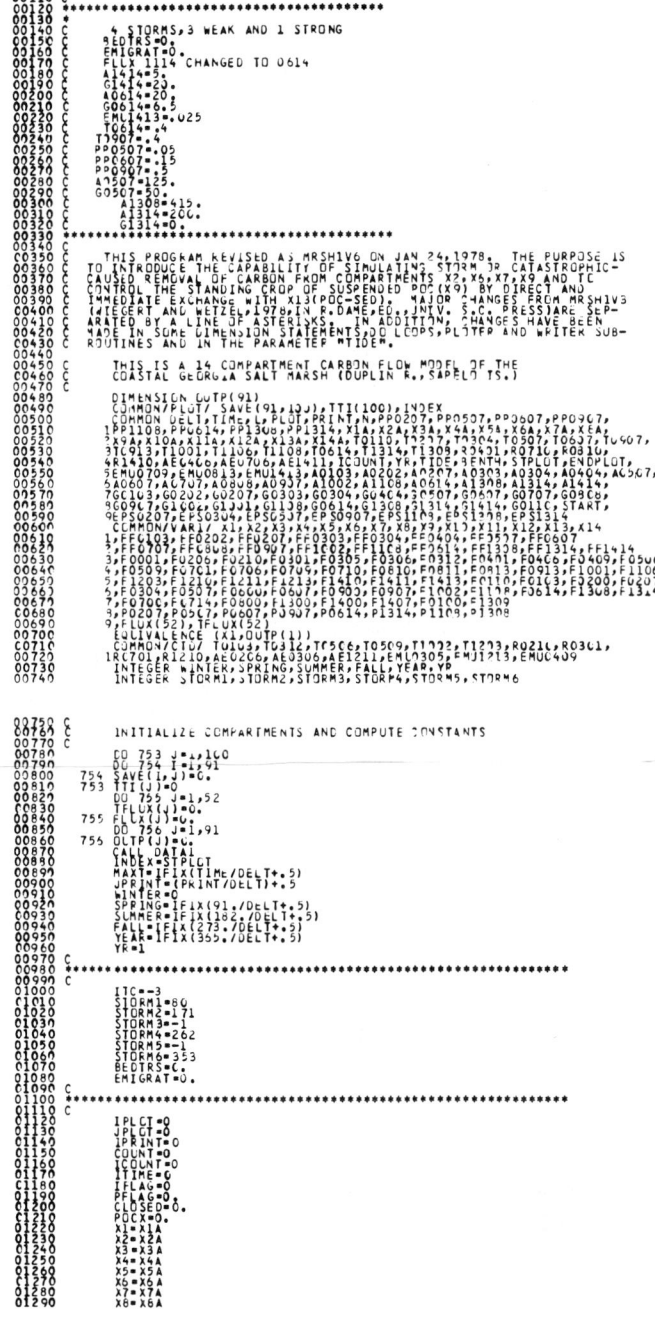

```
01300          X9=X9A
01310          X10=X10A
01320          X11=X11A
01330          X12=X12A
01340          X13=X13A
01350          X14=X14A
01360          START=0.
01370          CALL WRITER(1,1,0,0)
01380 C
01390 C        SAVE INITIAL VALUES TO BE PLOTTED
01400          IF(YR-SIPLOT) 27,26,27
01410       26 JPLOT=PLOT/DELT+.5
01420          DO 28 J=1,91
01430       28 SAVE(J,L)=OUTP(J)
01440          TTI(L)=COUNT
01450          L=L+1
01460          LPLOT=0
01470       27 CONTINUE
01480 C
01490 C        ENTER MAIN DO LOOP,SAVE INITIAL VALUES TO BE PLOTTED
01500 C        AND INITIALIZE SEASONALLY VARYING PARAMETERS
01510 C
01520          DO 17 K=1,MAXT
01530 C
01540 C        EVALUATE SEASONALLY-CHANGING PARAMETERS
01550 C
01560          IF(ITIME-WINTER) 39,31,32
01570       31 CALL DATA2
01580          GO TO 36
01590       32 IF(ITIME-SPRING) 39,33,34
01600       33 CALL DATA 3
01610          GO TO 38
01620       34 IF(ITIME-SUMMER) 39,35,36
01630       35 CALL DATA 4
01640          GO TO 38
01650       36 IF(ITIME-FALL) 39,37,39
01660       37 CALL DATA 5
01670       39 CONTINUE
01680 C
01690 C        COMPUTE CONSTANTS TO BE USED AN ENTIRE SEASON
01700 C
01710          EML35=EMU0305
01720          C0103=1.-(RG301+AEO306+TO312)/TO1C3
01730          CC207=1.-(AEO706+RO701+RO710)/(TC207*(1.-EPS0207))
01740          C0304=1.-(RO401+AEO406)/(TO304*(1.-EPS0304))
01750          C0507=1.-(AEO706+RO701+RO710)/(TO507*(1.-EPS0507))
01760          C0607=1.-(AEO706+RO701+RO710)/(TO6C7)
01770          C0907=1.-(AEO706+RO701+RO710)/(TO0907*(1.-EPS0907))
01780          C1002=1.-(AEO206+RO213)/T1002
01790          C1108=1.-EMU0813/T1108*(1.-EPS1108-R0810)
01800          C0614=1.-EMU1413/T0614*(1.-R1413)
01810          C1308=1.-EMU0813/T1308*(1.-EPS1308-R0810)
01820          C1314=1.-EMU1413/T1314*(1.-R1413)
01830       39 CONTINUE
01840 C
01850 C        COMPUTE FEEDBACK CONTROL TERMS.
01860 C
01870          FF0103=DIM(1.,DIM(A0103,X1)/(A0103-G0103))
01880          FF0202=DIM(1.,DIM(A0202,X2)/(G0202-A0202))
01890          FF0207=DIM(1.,DIM(A0207,X2)/(A0207-G0207))
01900          FF0303=DIM(X3,A0303)/(G0303-A0303)
01910          FF0304=DIM(1.,DIM(A0304,X3)/(A0304-G0304))
01920          FF0404=DIM(X4,A0404)/(G0404-A0404)
01930          FF0507=DIM(1.,DIM(A0507,X5)/(A0507-G0507))
01940          FF0607=DIM(1.,DIM(A0607,X6)/(A0607-G0607))
01950          FF0707=DIM(X7,A0707)/(G0707-A0707)
01960          FF0808=DIM(X8,A0808)/(G0808-A0808)
01970          FF0907=DIM(1.,DIM(A0907,X9)/(A0907-G0907))
01980          FF1002=DIM(1.,DIM(A1002,X10)/(A1002-S1002))
01990          FF1108=DIM(1.,DIM(A1108,X11)/(A1108-G1108))
02000          FF0614=DIM(1.,DIM(A0614,X6)/(A0614-G0614))
02010 C
02020 C        RESOURCE CONTROL OF MICROBES IN SEDIMENT EMPLOYS A
02030 C        RATIO OF AVAILABLE POC TO THE STANDING CROP OF ANAER-
02040 C        OBES.  THIS IS IN CONTRAST TO THE USUAL THRESHOLD
02050 C        LEVEL EXPRESSED AS GC/SQUARE METER
02060 C
02070 C
02080          FF1308=DIM(A1308,((X13-G1308)/X6))/(A1308-A1308*.70)
02090          FF1314=DIM(A1314,((1.+X13+250.)/X14))/(A1314-A1314*.67)
02100          FF1414=DIM(X14,A1414)/(G1414-A1414)
02110 C
02120 C        COMPUTE FEEDING PREFERENCE OF COMPARTMENTS HAVING
02130 C        MULTIPLE RESOURCES
02140 C
02150          SUMFF=PP0207*FF0207+PP0507*FF0507+PP0507*FF0607+
02160        1PP0907*FF0907+1.E-15
02170          P0207=PP0207*FF0207/SUMFF
02180          P0507=PP0507*FF0507/SUMFF
02190          P0607=PP0607*FF0607/SUMFF
02200          P0907=PP0907*FF0907/SUMFF
02210          SUMFF=PP0614*FF0614+PP1314*(1.-FF1314)+1.E-15
02220          P0614=PP0614*FF0614/SUMFF
02230          P1314=PP1314*(1.-FF1314)/SUMFF
02240          SUMFF=PP1108*FF1108+PP1308*(1.-FF1308)+1.E-15
02250          P1108=PP1108*FF1108/SUMFF
02260          P1308=PP1308*(1.-FF1308)/SUMFF
02270 C
02280 C        COMPUTE FLUXES THAT MAY BE CONTROLLED BY RECIPIENT AND/OR
02290 C        DONOR COMPARTMENTS
02300 C
02310          F0110=TO110*DIM(G0110,X10)
02320          F0103=TO103*X3*FF0103*DIM(1.,C0103*FF0303)
02330          F0207=P0207*TO207*X7*FF0207*DIM(1.,C0207*FF0707)
02340          F0304=TO3C4*X4*FF0304*DIM(1.,C0304*FF0404)
02350          IF(ITIME.EG.SPRING)F0304=.01/DELT
02360          IF(ITIME.EG.SUMMER)F0304=.01/DELT
02370          F0507=P0507*TO5C7*X7*FF0507*DIM(1.,C0507*FF0707)
02380          F0607=P0607*TO607*X7*FF0607*DIM(1.,C0607*FF0707)
02390          F0907=P0907*TO9C7*X7*FF0907*DIM(1.,C0907*FF0707)
02400          F1002=TO1002*X2*FF1002*DIM(1.,C1002*FF0202)
02410          F1108=P1108*TO1108*X8*FF1108*DIM(1.,C1108*FF0808)
02420          F0614=P0614*TO614*X14*FF0614*DIM(1.,C0614*FF1414)
02430          F1308=P1308*TO1308*X8*DIM(1.,C1308*FF1308)
02440          F1314=P1314*TO1314*X14*DIM(1.,C1314*FF1314)
02450 C
02460 C        COMPUTE FLUXES THAT ARE CONTROLLED SOLELY BY THE DONOR
02470 C        COMPARTMENT
02480 C
02490 C       *********************************************************
02500 C
02510 C        TIDAL TRANSFERS OUT OF THE SYSTEM ARE BASED ON AN OVERALL
02520 C        DIFFUSIVE EXCHANGE OF 4% OF THE WATER MASS PER DAY.  THIS IS
02530 C        MULTIPLIED BY THE DIFFERENCE BETWEEN THE CONCENTRATION OF THE
02540 C        PARTICULATE CARBON COMPOUND IN WATER MASS 4 VS. WATER MASS 3.
02550 C        THESE DIFFERENCES ARE 50% OF PHYTOPLANKTON,33% OF DOC-H23,7%
02560 C        OF HETEROTROPHS IN H23 AND 40% OF POC-120.
02570 C       *********************************************************
02580 C
02590 C
02600 C        FLUXES F0001 AND F0100 ARE COMPUTED AFTER XX1
02610 C
02620          F0200=TI00*BENTH*DIM(X2,.5)
02630          F0206=AEO206*X2
02640          F0210=RO210*X2
02650          F0301=RO301*X3
02660          IF(T1203*X12.GT.0.) GO TO 68
02670          EMU0305=.5
02680          GO TO 69
02690       68 F0305=EMU0305*X3
02700          EMU0305=EMU35
02710          F0306=AEO306*X3
02720          F0312=TO312*X3
02730          F0401=RO401*DIM(X4,.001)
02740          F0406=RO406*DIM(X4,.001)
02750          F0409=EMU0409*DIM(X4,.001)+EPS0304*F0304
```

```
02760      F(506=F0506*X5
02770      F(509=FC509*X5
02780      FG600=TILE*DIM((X6-6.),5.)
02790      FC701=R0701*X7
02800      FC706=AEU706*X7
02810      F0709=EMU0709*X7+EPSG507*F0507+EPS0907*F0907+EPS0207*F02)7
02820      FC710=R0710*X7
02830      F(80C=.01*BEUTR6*X13
02840      F0810=RG610*(F110d+F1J401+(R0610/15.)*X8
02850      F(810=EPS1106*F110d+EPS1306*F130E
02860      F(813=EMU0813*X8
02870 C
02880 C***********************************************************
02890 C
02900 C     COMPARTMENT X9(POC-H20) IS CONSTRAINED TO THE LIMITS 7.5-30.
02910 C     GC/M2 REPLENISHED FROM X13(POC-SED). EVERY 4TH ITERATION
02920 C     THE VALUE CF X9 IS BROUGHT UP TO 30 GC/M2.
02930 C
02940      PUCX=(X9-7.5)/DELT+F0409+F0509+F0709-F0900-F0907
02950      ITC=ITC+1
02960      IF(ITC.LT.116G IC 29
02970      PUCX=(X9-3C.)/DELT+FU+J9+F0509+FC709-F0900-F0907
02980      ITC=-3
02990   29 FC913=DIM(PUCX,0.)
03000      F13C9=DIM(G.,POCX)
03010      FC900=TILE*DIM(X9,7.5)
03020      F(700=.05*F0400+EMIGRAT*.C5*Y7
03030 C
03040 C***********************************************************
03050 C
03060      F0714=.01*F0913
03070      F1001=TILC01*DIM(X10,G10J1)
03080      F1106=I1106*X11
03090      F12C3=I1203*X12
03100      F1210=R1210*X12
03110      F1211=AEI211*X12
03120      F1213=X12*EMU1213
03130      F1360=BEUTR5*X13
03140      F1400=(BEUTRS*EMIGRAT)*X14
03150      F1407=.C1*F1309
03160      F1410=R1+10*(F0614+F1314)+(R1410/15.)*X14
03170      F1411=AEI411*X14
03180      F1413=EMU1413*X14
03190 C
03200 C***********************************************************
03210 C
03220 C     CHECK TO SEE IF A STORM IS TO BE SIMULATED.
03230 C
03240      IF(ICOUNT-STORM1) 47,44,41
03250   41 IF(ICOUNT-STORM2) 47,44,42
03260   42 IF(ICOUNT-STORM3) 47,45,43
03270   43 IF(ICOUNT-STORM4) 47,45,48
03280   48 IF(ICOUNT-STORM5) 47,45,49
03290   49 IF(ICOUNT-STORM6) 47,44,47
03300   44 PUCX=PUCX-30
03310      FC900=30.
03320      GO TO 46
03330 C
03340 C     IF ICOUNT EGUALS THE TIME OF A STRONG STORM, ADD AN ADDITIONAL
03350 C
03360 C     30 GC TO FC900 AND REMOVE IT FROM X13.
03370   45 PUCX=PUCX-60.
03380      FC900=60.
03390   46 FC913=DIM(PUCX,.01)
03400      F1309=DIM(0.,PUCX)
03410      FC714=.01*F0913
03420      F1200=.47*X2-F0210-F0207+F0206+F1C02
03430      FL400=.95*X6+F0306+F0506+F0206+F1106-F0607+F0726+F0406
03440      FC700=.33*X7+F0207+F0507+F0907+F0607-F0701-F0726-FU7C9
03460      1-F0710
03470   47 CONTINUE
03480 C
03490 C***********************************************************
03500 C
03510 C
03520 C     COMPUTE THE DIEL CHANGE IN COMPARTMENT STANDING CROP
03530 C
03540    1 CONTINUE
03550      XX1=F07C1-FU110+F1001+FJ401+F0301-FC1C3
03560 C
03570      FC001=DIM(0.,XX1)
03580      FC100=DIF(XX1,0.)
03590 C
03600      XX2=-FJ210+F1002-F0207-F0206-F02CC
03610      XX3=FJ103-F03C1-FU3C4-FU305-F0306-F0312+F1203
03620      XX4=-F0401-F0404+F0304-F0406
03630      XX5=-F0305-F0509-FU506-F0507
03640      XX6=F0305+FU506+F0206+F1106-F0607-F0614+FC706+F0406-FJ60J
03650      XX7=F0207+F0507+F0907+F0607-F0701-F0706-F0709-FJ713-
03660      1FC7C0+F1407-FU714
03670      XX8=F0810+F1308+F110d+F0811-F0813-F0800
03680      XX9=F0509+F0409+F1309-F0913+F0709-F0907-F0900
03690      XX1C=-F0301+F0110+F0710+F0810-F10C2+F0210+F1213+F1410
03700      XX11=-F1106+F0811-F110d+F1211+F1411
03710      XX12=F0312-F1211-F1213+F1210-F1200
03720      XX13=F0913-F1308+F1213-F1314+F1413-F1309-F1300-F1309
03730      XX14=F0614+F1314+F0714-F1407+F1410-F1411-F1413-F1400
03740 C
03750 C     MULTIPLY DIEL CHANGE BY ITERATION INTERVAL(DELT)
03760 C
03770      XX1=XX1*DELT
03780      XX2=XX2*DELT
03790      XX3=XX3*DELT
03800      XX4=XX4*DELT
03810      XX5=XX5*DELT
03820      XX6=XX6*DELT
03830      XX7=XX7*DELT
03840      XX8=XX8*DELT
03850      XX9=XX9*DELT
03860      XX10=XX10*DELT
03870      XX11=XX11*DELT
03880      XX12=XX12*DELT
03890      XX13=XX13*DELT
03900      XX14=XX14*DELT
03910 C
03920 C     INCREMENT COMPARTMENTS.  NOTE** IF CLOSED IS ZERO,
03930 C     CUMPARTMENT 1(CO2 IN THE AIR) IS HELD CONSTANT.  IF
03940 C     CLOSED SYSTEM IS TO BE SIMULATED, CLOSED IS GIVEN THE
03950 C     VALUE OF 1.
03960 C
03970      X1=X1+XX1*CLOSED
03980      X2=X2+XX2
03990      X3=X3+XX3
04000      X4=X4+XX4
04010      X5=X5+XX5
04020      X6=X6+XX6
04030      X7=X7+XX7
04040      X8=X8+XX8
04050      X9=X9+XX9
04060      X10=X10+XX10
04070      X11=X11+XX11
04080      X12=X12+XX12
04090      X13=X13+XX13
04100      X14=X14+XX14
04110 C
04120 C     INCREMENT THE CUMULATIVE FLUX STORAGE ARRAY
04130 C
04140      DO 18 J=1,52
04150      TFLUX(J)=TFLUX(J)+OUTP(J+31)*DELT
04160   18 FLUX(J)=FLUX(J)+OUTP(J+31)*DELT
04170 C
04180 C     INCREMENT THE COUNTERS.  CHECK WHETHER TIMES FOR PLOTTER
04190 C     STORAGE, PRINTING OR END OF YEAR HAVE BEEN REACHED.
04200 C
04210      COUNT=COUNT+DELT
04220      ICOUNT=COUNT+.001
```

```
04230        IPRINT=IPRINT+1
04240        IPLOT=IPLOT+1
04250        ITIME=ITIME+1
04260        IF(JPLOT-IPLOT) 201,200,201
04270   200  DO 19 J=1,91
04280    19  SAVE(J,L)=OUTP(J)
04290        TTI(L)=COUNT
04300        L=L+1
04310        IPLOT=0
04320   201  CONTINUE
04330        IF(JPRINT-IPRINT) 301,300,301
04340   300  CONTINUE
04350        CALL WRITER(0,1,0,0)
04360        IPRINT=0
04370   301  CONTINUE
04380        IF(YEAR-ITIME) 17,16,17
04390    16  CONTINUE
04400        CALL WRITER(0,1,1,0)
04410        YR=YR+1
04420        ITIME=0
04430        IPLOT=0
04440        IPRINT=0
04450        COUNT=0
04460        ICOUNT=0
04470        IF(YR-STPLOT) 320,305,310
04480   305  JPLOT=PLOT/DELT+.5
04490        DO 306 J=1,91
04500        SAVE(J,1)=OUTP(J)
04510   306  CONTINUE
04520        TTI(L)=COUNT
04530        L=L+1
04540   310  IF(ENDPLOT-YR) 320,315,320
04550   315  JPLOT=0
04560   320  CONTINUE
04570        CALL WRITER(1,1,0,0)
04580   C
04590   C     LIST THE ACCUMULATED FLUXES AND THE OVERALL CHANGE IN
04600   C     COMPARTMENT SIZES
04610   C
04620   C
04630    17  CONTINUE
04640   900  FORMAT(1H ,10F12.3)

04650   991  FORMAT(1H ,I12)
04660        WRITE(7,991)INDEX
04670        DO 950 K=1,100
04680        WRITE(7,900) (SAVE(J,K),J=1,91)
04690   950  CONTINUE
04700        WRITE(7,900) (TTI(J),J=1,100)
04710   C
04720   C     CALL SUBROUTINE PLOTER(ISLOG,PRESET,NS,TV,NV,N,MIN,VMIN,MAX,X,Y)
04730   C     ALL ARGUMENTS ARE INTEGER EXCEPT VMIN
04740   C
04750        CALL PLOTER(1,0,1,1,91,49,-4,10.,6,49,101)
04760        STOP
04770        END
04780   C
04790        SUBROUTINE DATA1
04800   C
04810   C     SUBROUTINE DATA1 SETS THE VALUES OF ALL PARAMETERS THAT REMAIN
04820   C     CONSTANT DURING THE ENTIRE SIMULATION RUN
04830   C
04840        COMMON DELT,TIME,L,PLOT,PRINT,N,PP0207,PP0507,PP0607,PPC3C7,
04850       1PP1108,PP0614,PP1308,PP1314,X1A,X2A,X3A,X4A,X5A,X6A,X7A,X8A,
04860       2X9A,X10A,X11A,X12A,X13A,X14A,T0110,T0207,T0304,T0507,T0663,T0907,
04870       3T0913,T1001,T1106,T110d,T0614,T1314,T1308,R0401,R0710,R081C,
04880       4R1410,AE0406,AE0706,AE1411,ACI03,A0202,A0207,A0303,A0304,A0406,AU507,
04890       5EMU0709,EMU0813,EMU1413,ACI03,A0202,A0207,A0303,AU304,A0406,AU507,
04900       6AC607,AU707,A0808,A0907,AIOC2,A1108,AC614,A1314,A1414,
04910       7G0103,G0202,G0207,G0303,G0304,G0404,G0507,G0607,G0707,G0808,
04920       8G0907,G1002,G1008,G1108,G0614,G1308,G1314,G1414,G0110,START,
04930       9EPS0207,EPS0304,EPS0507,EPS0907,EPS1108,EPS1308,EPS1314
04940        DELT=1
04950        TIME=1825
04960        TIDE=.16
04970        BENTH=.5
04980        L=1
04990        PLOT=7.5
05000        PRINT=10
05010        STPLOT=5
05020        ENDPLOT=6.
05030        PP0207=.3
05040        PPC507=.05
05050        PP0607=.15
05060        PP0907=.5
05070        PP1108=.5
05080        PP0614=.5
05090        PP1308=.2
05100        PP1314=.5
05110        X1A=875.
05120        X2A=.2
05130        X3A=53.8
05150        X4A=.101
05160        X5A=116.5
05170        X6A=2.8
05180        X7A=13.9
05190        X8A=43.4
05200        X9A=9.5
05210        X10A=34.2
05220        X11A=34.5
05230        X12A=34.3
05240        X13A=17608.4
05250        X14A=8.8
05260        T0110=.7
05270        T0207=.8
05280        T0304=.25
05290        T0507=.2
05300        T0667=.64
05300        T0907=.4
05310        T0913=.01
05320        T1001=.7
05330        T1106=.008
05340        T1108=.02
05350        T0614=.4
05360        T1308=.02
05370        T1314=.4
05380        R0401=.1
05390        R0710=.04
05400        R0816=.25
05410        R1410=.40
05420        AE0406=.01
05430        AE0706=.01
05440        AE1411=.005
05450        EMU0709=.05
05460        EMU0813=.005
05470        EMU1413=.025
05480        AC0304=0.
05490        A0202=1.
05500        A0207=1.4
05510        A0303=150.
05520        A0304=.25
05530        A0507=125.
05540        A0607=6.
05550        A0707=6.7
05560        A0808=30.
05570        A0907=10.
05580        A1002=7.8
05590        A1108=10.1
05600        A0614=20.
05610        A1308=415.
05620        A1314=200.
05630        A1414=5.
05640        G0103=20
05650        G0110=30
05660        G0202=2.
05680        G0207=.8
```

```
05690        G03C3=186.
05700        G03C4=0
05710        G0404=1.25
05720        G0507=50.
05730        G0607=4
05740        G0707=21.3
05750        G0808=60
05760        G0907=0.
05770        G1001=30
05780        G1002=.1
05790        G11C6=.1
05800        G0614=6.2
05810        G1306=3019
05820        G1314=0.
05830        G1414=20.
05840        EPS0207=.5
05850        EPS0304=.25
05860        EPS0507=.25
05870        EPS0907=.05
05880        EPS1108=.31
05890        EPS1308=.31
05900        RETURN
05910        END
05920 C
05930 C      SUBROUTINE DATA2
05940 C
05950 C      SUBROUTINES DATA2,DATA3,DATA4 AND DATA5 SET THE VALUES OF
05960 C      PARAMETERS THAT CHANGE SEASONALLY DURING WINTER,SPRING,SUMMER
05970 C      AND FALL RESPECTIVELY.   WINTER (AND ALL SIMULATION RUNS)START
05980 C      ON JAN. 1.  ALL SEASONS ARE 91 DAYS LONG
05990 C
06000   92 FORMAT(1H ,*WINTER*)
06010        COMMON/CTU/ T0103,T0312,T0506,T0509,T1002,T1203,R0210,R0301,
06020       1R0701,R1210,AE0206,AE0306,AE1211,EMU0305,EMU1213,EMU0409
06030        T0103=.062
06040        T0312=.03
06050        T0506=.001
06060        T0509=.008
06070        T1002=.071
06080        T1203=.0001
06090        R0210=.013
06100        R0301=.0035
06110        R0701=.004
06120        R1210=.00072
06130        AE0206=.007
06140        AE0306=.002
06150        AE1211=.000053
06160        EMU0305=.023
06170        EMU0409=.23
06180        EMU1213=.5*.0052
06190        WRITE(6,92)
06200        RETURN
06210        END
06220 C
06230 C      SUBROUTINE DATA3
06240 C
06250   93 FORMAT(1H ,*SPRING*)
06260        COMMON/CTU/ T0103,T0312,T0506,T0509,T1002,T1203,R0210,R0301,
06270       1R0701,R1210,AE0206,AE0306,AE1211,EMU0305,EMU1213,EMU0409
06280        T0103=.106
06290        T0312=.071
06300        T0506=.002
06310        T0509=.024
06320        T1002=.3
06330        T1203=.003
06340        R0210=.041
06350        R0301=.0046
06360        R0701=.01
06370        R1210=.0094
06380        AE0206=.021
06390        AE0306=.003
06400        EMU0409=.01
06410        AE1211=.00008
06420        EMU0305=.015
06430        EMU1213=.5*.0056
06440        WRITE(6,93)
06450        RETURN
06460        END
06470 C
06480 C      SUBROUTINE DATA4
06490 C
06500   94 FORMAT(1H ,*SUMMER*)
06510        COMMON/CTU/ T0103,T0312,T0506,T0509,T1002,T1203,R0210,R0301,
06520       1R0701,R1210,AE0206,AE0306,AE1211,EMU0305,EMU1213,EMU0409
06530        T0103=.058
06540        T0312=.024
06550        T0509=.016
06560        T1002=1.3
06570        T1203=.0001
06580        R0210=.24
06590        R0301=.0026
06600        R1210=.0026
06610        AE0206=.125
06620        AE0306=.002
06630        AE1211=.000053
06640        EMU0305=.012
06650        EMU0409=.01
06660        EMU1213=.5*.0069
06670        WRITE(6,94)
06680        RETURN
06690        END
06700 C
06710 C      SUBROUTINE DATA5
06720 C
06730   95 FORMAT(1H ,*FALL*)
06740        COMMON/CTU/ T0103,T0312,T0506,T0509,T1002,T1203,R0210,R0301,
06750       1R0701,R1210,AE0206,AE0306,AE1211,EMU0305,EMU1213,EMU0409
06760        T0103=.034
06770        T0312=.01
06780        T0509=.022
06790        T1002=.145
06800        T1203=.001
06810        R0210=.027
06820        R0301=.0019
06830        R0701=.004
06840        R1210=.0043
06850        AE0206=.014
06860        AE0306=.001
06870        AE1211=.000027
06880        EMU0305=.039
06890        EMU0409=.23
06900        EMU1213=.5*.0045
06910        WRITE(6,95)
06920        RETURN
06930        END
06940 C
06950        SUBROUTINE WRITER(A,B,C,D)
06960 C
06970 C      SUBROUTINE WRITER FORMATS AND PRINTS ALL OUTPUT
06980 C      CALL ARGS A,B,C,AND D ARE ALL INTEGERS
06990 C
07000   20 FORMAT(1H0,5HYEAR=,I2)
07010   25 FORMAT(1H0,4HDAY=,I3)
07020   30 FORMAT(1H ,12HCOMPARTMENTS)
07030   35 FORMAT(1H ,17HFEEDBACK CONTROLS)
07040   40 FORMAT(1H ,6HFLUXES)
07050   45 FORMAT(1H ,19HFEEDING PREFERENCES)
07060   50 FORMAT(1H ,7HDELTAX=,F7.2)
07070   55 FORMAT(1H ,7HIMPORT=,F7.2)
07080   60 FORMAT(1H ,34HPRINT CHANGES IN COMPARTMENTS 1-14)
07090   65 FORMAT(1H ,29HFLUXES 39,32,36,39,45,47,49=)
07100   68 FORMAT(1H ,37HEXPORT FROM X1,X2,X6,X7,X8,X3,(13,X14)
07110   70 FORMAT(1H ,29HCUMULATIVE FLUXES TO THIS DAY)
07120   72 FORMAT(1H ,5X,5HFLUX(,I2,2H)=,F9.3)
07130   75 FORMAT(1H ,14F9.3)
07140        INTEGER YR,A,B,C,D
07150        DIMENSION OUTP(91)
```

```
C7160          COMMON DELT,TIME,L,PLOT,PRINT,N,PP0207,PP0507,PP0607,PP0907,
07170         1PP1109,PP0614,PP1308,PP1314,X1A,X2A,X3A,X4A,X5A,X6A,X7A,X8A,
07180         2X9A,X10A,X11A,X12A,X13A,X14A,T011C,T0207,T3304,T0507,T0607,TV907,
07190         3TV913,T1C01,T113B,T1108,T0614,T1314,T1308,R0401,R071C,R0810,
07200         4R1410,C4C46,AE0706,AE1411,ICOUNT,YK,TIDE,RENTH,STPLCT,ENCPLOT,
07210         5EMU0709,EMU0813,EMU0413,A0103,A0202,A0207,A0303,A0304,A0507,
07220         6AC607,A07C7,A0808,A0907,A1002,A1108,A0514,A1308,A1314,A1414,
07230         7GU163,GE2C6,GU237,GU333,GC3C4,G04C4,G0507,G0607,G0707,G0808,
07240         8G0907,G1002,G1001,G11J3,G0614,G13C8,G1314,G1414,G0110,START,
```

```
07250 C        9EPSC227,EPS0304,EPS0507,EPS0927,EPS1109,EPS1308,EPS1314
07260         COMMON/FKJ/X1,X2,X3,X4,X5,X6,X7,X8,X9,X10,X11,X12,X13,X14
07270         1,FF0103,FF0202,FF0207,FF0303,FF03C4,FF0404,FF0507,FF0507,FFL607
07280         3,FFC707,FF0808,FFU907,FF1002,FF1108,FF0514,FF3507,FFL607
07290         3,FF0001,FC2C6,FU610,FF03J4,F0305,FC306,F0312,F0401,F04C6,F0409,F05C6
07300         4,F0509,F0706,F0706,F0709,F0710,F0810,F0811,F3913,F0913,F1001,F1106
07310         5,F1203,F1210,F1211,F1213,F1410,F1411,F1413,F13J7,F0913,F1001,F1106
07320         5,F0304,FC5C7,FC6J3,F3503,F09C0,F0907,F1002,F1108,F0614,F1308,F1314
07330         7,F0730,F0714,F0800,F130J,F1400,F14C7,F1103,F1309
07340         8,P0207,P0507,P06C7,P0907,PC614,P1314,P1108,P1308
07350         9,FLUX(52),TFLUX(52)
```

```
07360         EQUIVALENCE (X1,OUTP(1))
07370       1 IF(A.LE.C.)GO TO 2
07380         WRITE(6,20) YK
07390       2 IF(B.LE.C.)GO TO 3
07400         WRITE(6,25) ICOUNT
07410         WRITE(6,33)
07420         WRITE(6,75)(OUTP(J),J=1,14)
07430         WRITE(6,35)
07440         WRITE(6,75)(OUTP(J),J=15,31)
07450         WRITE(6,40)
07460         WRITE(6,75)(OUTP(J),J=32,83)
07470         WRITE(6,42)
07480         WRITE(6,75)(OUTP(J),J=84,91)
07490       3 IF(C.LE.0.)GO TO 4
07500         TMPURT=FLUX(1)-FLUX(32)-FLUX(36)-FLUX(3P)-FLUX(45)-FLUX(47)-
07510         1FLUX(48)-FLUX(49)-FLUX(51)
07520         WRITE(6,20) YK
07530         WRITE(6,25) ICOUNT
07540         WRITE(6,70)
07550         DO 100 J=1,52
07560         WRITE(6,72) J,FLUX(J)
07570     100 FLUX(J)=C.
07580         WRITE(6,65)
07590         WRITE(6,66)
07600         WRITE(6,80)
07610         Z1=X1-X1A
07620         Z2=X2-X2A
07630         Z3=X3-X3A
07640         Z4=X4-X4A
07650         Z5=X5-X5A
07660         Z6=X6-X6A
07670         Z7=X7-X7A
07680         Z8=X8-X8A
07690         Z9=X9-X9A
07700         Z10=X10-X10A
07710         Z11=X11-X11A
07720         Z12=X12-X12A
07730         Z13=X13-X13A
07740         Z14=X14-X14A
07750         SUMZ=Z1+Z2+Z3+Z4+Z5+Z6+Z7+Z8+Z9+Z10+Z11+Z12+Z13+Z14
07760         DELTAX=SUMZ-START
07770         START=SUMZ
07780         WRITE(6,73) Z1,Z2,Z3,Z4,Z5,Z6,Z7,Z8,Z9,Z10,Z11,Z12,Z13,Z14
07790         WRITE(6,50) TMPURT
07800         WRITE(6,55) DELTAX
07810       4 IF(D.LE.C.) GO TO 400
07820         WRITE(6,25) ICOUNT
07830         WRITE(6,70)
07840         DO 200 J=1,52
07850         WRITE(6,75) J,TFLUX(J)
07860     200 TFLUX(J)=C.
07870         WRITE(6,65)
07880         WRITE(6,66)
07890     400 CONTINUE
07900         RETURN
07910         END
07920         SUBROUTINE PLOTER(SLOG,PRESETM,NS,IV,NV,N,MIN,VMIN,MAX,X,Y)
07930 C
07940 C       SLOG GREATER THAN ZERO--LOG 10 OF DATA PLOTTED
07950 C       PRESETM GREATER THAN ZERO--PRESET MAX AND MIN IN USE)
07960 C       NS LESS THAN OR EQUAL TO ZERO--MAX AND MIN FOUND BY
07970 C         SEARCHING,THEN MAX SET TO NEXT HIGHER POWER OF 10
07980 C         AND MIN SET TO NEXT LOWER POWER OF 10(CONSTRAINED
07990 C         BY THE VALUE OF VMIN)
08000 C       NS GREATER THAN ZERO-MAX AND MIN FOUND BY SEARCHING
08010 C       AND THEN MAX SET TO NEXT HIGHER INTEGER MULTIPLE OF 10
08020 C         AND RMIN SET TO NEXT LOWER MULTIPLE OF 10
08030 C       IV IS FIRST VARIABLE PLOTTED
08040 C       NV IS LAST VARIABLE PLOTTED
08050 C       N IS TOTAL NUMBER OF TIME INTERVALS(PLOT IN MAIN PROG)
08060 C         PLOTTED ON THE GRAPH
08070 C       MIN IS THE PRESET MINIMUM VALUE OF EACH VARIABLE
08080 C       VMIN SPECIFIES THE UPPER LIMIT OF MIN WHEN VARIABLE
08090 C         SCALING IS EMPLOYED,IF VMIN LESS THAN ZERO IT IS
08100 C         IGNORED
08110 C       MAX IS THE PRESET MAXIMUM VALUE OF EACH VARIABLE
08120 C       X IS THE MAXIMUM NUMBER OF TIME(PLOT)INTERVALS GRAPHED
08130 C         ON A SINGLE PAGE OF OUTPUT
08140 C       Y IS THE TOTAL NUMBER OF PRINTER SPACES SEARCHED FOR A
08150 C         VERTICAL LINE OF DASHES OR PLUS IN THE GRAPH(THIS
08160 C         IS THE HORIZONTAL LINE OF PRINT ON THE OUTPUT
08170 C       Y AND X ALONG WITH THE VARIABLE PLOT IN THE MAIN PROG
08180 C         ARE VARIED TO SCALE THE OUTPUT TO FIT EASILY ON
08190 C         ANY PARTICULAR GRAPH PAPER
08200 C
08210 C
08220   94 FORMAT(1H1,*YEAR=*,12)
08230   96 FORMAT(1H0,*LOGARITHM*)
08240   97 FORMAT(1H0,*RMAX=*,F15.5,5X,*RMIN=*,F15.5)
08250   95 FORMAT(1H,*BMAX=*,F15.5,5X,*BMIN=*,F15.5)
08260   98 FORMAT(1H0,*VARIABLE=*,I5)
08270 C
08280 C       PICT AND FORMAT 99 MUST BE DIMENSIONED GE. Y
08290 C
08300   99 FORMAT(1H ,F7.2,6X,F15.5,5X,101A1)
08310         INTEGER X,Y,SLOG,PRESETM
08320         DIMENSION PICT(101)
08330         COMMON/PLOT/ VAR(9,100),TIME(100),INDEX
08340         DATA DASH,PLUS/1H-,1H+/,BLANK/1H /
08350         JJ=-6
08360         JK=6
08370         RMIN=MIN
08380         RMAX=MAX
08390         BMAX=MAX
08400         DO 753 I=1,Y
08410     753 PICT(I)=BLANK
08420         DO 500 JM=IV,NV
08430         BMIN=MIN
08440         WRITE(6,94) INDEX
08450         WRITE(6,98)JM
08460         JCOUNT=C
08470         IF(SLOG)299,300,299
08480     299 CONTINUE
08490         WRITE(6,96)
08500         DO 250 I=1,N
08510         IF(VAR(JM,I))101,101,100
08520     101 VAR(JM,I)=ABS(VAR(JM,I))+.00001
08530     100 CONTINUE
08540         VAR(JM,I)=ALOG10(VAR(JM,I))
08550     250 CONTINUE
08560     300 CONTINUE
08570         RMAX=0.
08580         RMIN=VMIN
08590         DO 550 I=1,N
08600         IF(RMAX.LT.VAR(JM,I))RMAX=VAR(JM,I)
```

```
08600          IF(RMIN.GT.VAR(JM,I))RMIN=VAR(JM,I)
08610    550  CONTINUE
08620          IF(PRESETM)560,599,560
08630    599  CONTINUE
08640          BMAX=0.
08650          BMIN=0.
08660          IF(SLOG)553,554,553
08670    553  CONTINUE
08680          BMAX=AINT(RMAX)+1.
08690          BMIN=AINT(RMIN)
08700          GO TO 560
08710    554  IF(NS)555,555,605
08720    555  IF(RMAX)552,561,552
08730    552  SMAX=ALOG10(RMAX)
08740          IF(SMAX)556,526,557
08750    556  BMAX=10.**AINT(SMAX)
08760          GO TO 561
08770    557  BMAX=10.**(AINT(SMAX)+1)
08780    561  IF(RMIN)562,560,562
08790    562  SMIN=ALOG10(RMIN)
08800          IF(SMIN)558,559,559
08810    558  BMIN=10.**(AINT(SMIN)-1)
08820          GO TO 560
08830    559  BMIN=10.**AINT(SMIN)
08840          GO TO 560
08850    605  IF(RMAX)620,620,607
08860    607  DO 615 K=JJ,JK
08870          BMAX=AINT((RMAX)/10.**(-K))
08880          IF(BMAX)610,615,610
08890    610  BMAX=(BMAX+1.)*10.**(-K)
08900          GO TO 620
08910    615  CONTINUE
08920    620  IF(RMIN)560,560,625
08930    625  DO 640 K=JJ,JK
08940          BMIN=AINT(RMIN/10.**(-K))
08950          IF(BMIN)630,640,630
08960    630  BMIN=(BMIN-1.)*10.**(-K)
08970          GO TO 560
08980    640  CONTINUE
08990    560  CONTINUE
09000  C***************************
09010  C
09020  C         SPECIAL CONSTRAINTS ON BMAX AND BMIN
09030  C
09040          IF(BMAX-1.)650,650,660
09050    650  BMAX=1.
09060    660  IF(SLOG)555,680,665
09070    665  IF(PRESETM)680,670,680
09080    670  BMIN=BMAX-5.
09090    680  CONTINUE
09100  C
09110  C***************************
09120          WRITE(6,97)RMAX,RMIN
09130          WRITE(6,95)BMAX,BMIN
09140          RANGE=BMAX-BMIN
09150          YY=Y
09160          DIST=RANGE/YY
09170          DO 500 I=1,N
09180          DO 400 J=1,Y
09190          RJ=J
09200          PICT(J)=LAST
09210          IF(VAR(JM,I)-(BMIN+DIST*RJ))800,800,400
09220    400  CONTINUE
09230    800  PICT(J)=PLUS
09240          IF(SLOG)350,350,349
09250    349  VAR(JM,I)=10.**(VAR(JM,I))
09260    350  CONTINUE
09270          WRITE(6,99) TIME(I),VAR(JM,I),(PICT(K),K=1,Y)
09280          JCOUNT=JCOUNT+1
09290          IF(JCOUNT-X)200,200,199
09300    199  CONTINUE
09310          WRITE(6,98) JM
09320          WRITE(6,99) TIME(I),VAR(JM,I),(PICT(K),K=1,Y)
09330          JCOUNT=1
09340    200  CONTINUE
09350          DO 754 J=1,Y
09360    754  PICT(J)=BLANK
09370          JCOUNT=C
09380    500  CONTINUE
09390          RETURN
09400          END
```

## Appendix 2

The sources of data and the general methods of computation
employed in obtaining the new parameters and values of Table 1 rely
on both unpublished and published material.  A brief outline of
these sources follows:

(1)  Information available on the microbial interactions of the
     marsh sediment, both anaerobic and the aerobic transformations
     in and around *Spartina* roots, were provided by Robert Christian,
     Dept. of Biology, Drexel Instit., Philadelphia and
     William Wiebe, Dept. of Microbiology, UGA.  The relevant papers
     are Christian and Hall (1977), Christian *et al.* (1975),
     Sottile (1973), Payne (1970), Gallagher (1975), Christian (1976).
     In addition, unpublished data of Christian were used in the
     evaluation of T1114 and T1314 was determined from unpublished
     data by Wetzel and Hanson.

A major result of the newest studies of the sediment was to change the method of representing the metabolic loss of carbon from a specific loss rate to a proportion of all carbon ingested. In the case of the anaerobes this loss is represented as both a proportion of carbon ingested and a small specific loss rate multiplied times the standing crop of carbon (see Appendix 1, F0810 equation) on the basis that the population will have a small loss of carbon no matter what its rate of ingestion. This point, however, represents a considered judgment on our part and remains to be justified by experiment.

A perusal of Table 1 shows a decrease in the maximum rate of utilization of dissolved organic material (T1108 and T1308 decreased) and little change or an actual increase in the growth rate possible via utilization of particulate organic carbon of the sediments. The decrease in the maximum permissible standing crop of anaerobes (G0808) reflects the improved knowledge of the maximum standing crop in the field.

(2)  New data on the carbon fluxes associated with *Spartina alterniflora* were summarized by Jack Gallagher. Sources of information were Gallagher and Pfeitter (1977), Valiela *et al.* (1976), Gallagher *et al* (1976), Guirgevich and Dunn (1977), Gallagher *et al.* (1972), and unpublished data of Gallagher and of Lloyd Dunn

The most important changes in the model parameters dealing with carbon flows through the *Spartina* compartments were in the maximum gross photosynthesis rate (T0103), the transfer from shoots to roots-rhizomes (T0312), the respiration losses of shoots (R0301) and of roots (R1210), and the seasonal mortality of shoots (EMU0305). The specific rates of photosynthesis, shoot-root transfer and respiration were all lowered by a large percentage. Mortality of shoots were on the average increased and the seasonal pattern was changed as a result of the newest data.

(3)  Roger Hanson provided unpublished data that allowed a more accurate estimate of the parameters representing ingestion by aerobic heterotrophs in the water (T0507, T0607, and T0907). Although this is a large heterogenous group, containing vertebrates, invertebrates and bacteria, we assume in this model that the specific rate of carbon flow is dominated by the microbial utilization values. Hence our reliance on the microbial degradation data of Hanson.

(4)  One parameter, the maximum maintainable density of aerobic heterotrophs in water (G0707), was revised upward with the help of data collected by Michael Pace.

(5)  Lawrence Pomeroy provided a summary of data used to revise the parameters dealing with flows of carbon through the algae. These data dealt with the phytoplankton part of the component and are drawn from unpublished data of Pomeroy. The only major

changes were the reductions in maximum standing crop (G0202) and in threshold response density (A0202).

(6)   Data provided by Wetzel were used to change the respiration rates of aerobic heterotrophs in water (R0701, R0710), the maintainable density of heterotrophs (G0707), and the refuge value of algae (G0207) when used as a food source by this group (Wetzel, 1977).  Loss of $CO_2$ directly to air (F0701) is an addition to the model.  The data are derived from Wetzel (1975) and unpublished data of John Hall.

(7)   David Whitney and Marshall Darley used their unpublished data to provide the benthic algal contribution to the rates discussed in (5) above.  The two sets or parameter estimates (benthic and phytoplankton) were simply averaged because the algal compartment is estimated to comprise one-half phytoplankton and one-half benthic diatoms.

The following sources come from ongoing or recently completed studies that are unpublished:

(8)   Dirk Frankenberg, Larry Pomeroy - pers. comm.

(9)   William Pfeiffer, pers. comm.

# A Mathematical Model of Trophic Dynamics in Estuarine Seagrass Communities

**Randolph L. Ferguson**

National Marine Fisheries Service
Beaufort, North Carolina 28516

**S. Marshall Adams**

Oak Ridge National Laboratory
Environmental Sciences Division
Oak Ridge, Tennessee 37830

A mathematical model of the trophic dynamics of epifauna and omnivorous and carnivorous juvenile fish is used to simulate the response of these organisms to temperature and the biotic factors within an eelgrass community. Optimal feeding and respiration rates are functions of temperature. Actual feeding rate is limited by habitat and preference-scaled prey abundance factors. Omnivorous and carnivorous fish compete for available habitat and prey. Omnivorous fish respond to reduced availability of prey by consuming more detritus. Simulation experiments indicate that the single most important environmental factor is eelgrass which defines the habitat. Principal component analysis suggests that greater than half of the epifaunal and juvenile fishes' response to variation in average annual temperature is indirectly affected by the response of eelgrass to temperature. It is recommended that future models of trophic dynamics in eelgrass beds include a highly resolved eelgrass component.

## Introduction

Seagrass communities are extremely important habitats in marine and estuarine systems (Phillips, 1969; Wood *et al.*, 1969; den Hartog, 1970; Ferguson *et al.*, 1977). They are nurseries for juvenile fish and invertebrates (Blegvad, 1916; Ried, 1954; Kikuchi, 1961; Hoese and Jones, 1963; Adams, 1976a), and secondary productivity is high because of the favorable conditions of food and cover. In the shallow estuarines near Beaufort, N.C., eelgrass, *Zostera marina*, has been estimated by Williams (1973) to supply as much as 64% of the total primary production occurring there. Since a large amount

of this productivity dies and is flushed by tides to open waters of
the estuary, these beds also help support the detritus-based food
chain of estuaries.

In the past 7 years there have been several studies in the Beaufort, N.C., area that have investigated various aspects of the structure (seasonal biomass and diversity of organisms) and functioning
(flows of organic matter) of local eelgrass communities. Primary
productivity studies have been conducted by Dillon (1971) and Penhale
(1976); invertebrate macrofaunal studies by Godfrey (1970), Thayer
*et al.* (1975), and Stuart (1975); and fish studies by Adams (1976 a,
b, c). Most of the data collected thus far is pertinent to community
structure. Other community aspects such as detritus, microbes and the
relative importance of abiotic factors have been less well studied.
Since the seasonal abundance of standing crops at major trophic
levels of the grass beds in Beaufort are relatively well-known, it is
important at this time to synthesize our knowledge of these systems
to assess our understanding of the functioning of major trophic
groups within these communities. Rates of energy or carbon flow between trophic levels are virtually unmeasured although several ap-
proximations have been derived from annual energy budgets (Thayer
*et al.*, 1975; Adams, 1976 a,b).

Our approach to this synthesis was to develop a mathematical
model of the flow of matter through three of the major trophic
groups of seagrass organisms: epifauna and omnivorous and carnivorous juvenile fish. We prepared structural formulae for assimilation
and respiration that are dependent on time, temperature, standing
crops, and biotic interactions and are analogous to the real community interactions. These formulae are used as descriptors of the
dynamical behavior of modeled trophic groups (compartments) and an
experimental and predictive device for components of seagrass communities in general. We used existing field data collected from an
eelgrass bed in the Newport River estuary in N.C. to assemble a
conceptual model of the system similar to the energy-flow diagram
of Thayer *et al.* (1975). For the present computer simulations we
chose to limit the number of compartments monitored, the sophistica-
tion of many of the flow function computations, and the number of
forcing functions. The resulting model includes temperature and
time of year as forcing functions; illumination, nutrients and salinity, for example, are omitted. The computer model simulates
standing crops and flows on a $m^2$ basis with spatial heterogeneity
omitted. Species are lumped into relatively few compartments according to trophic level with taxonomic distinctions omitted. However, the structure of the model allows sufficient resolution of the
epifauna and carnivorous and omnivorous fish compartments to predict
standing crops, composition of diet, emigration and predation rates
through time. The model also predicts such ecologically meaningful
parameters as ecological growth efficiency (production/consumption),
tissue growth efficiency (productions/assimilation), and turnover
rate (production/average standing crop).

Our objectives were to 1) examine the structure and functioning of seagrass communities, 2) increase our understanding of the relative importance of biotic factors and of temperature, 3) test our understanding of ecological processes based on the performance of modeled theoretical relationships, and 4) determine the potential of eelgrass bed models as research devices. The following sections develop the biology of eelgrass beds in the vicinity of Beaufort, N.C., the mathematical basis of the model, the credibility of model simulations and the utility of the model.

## Biology of Eelgrass Beds

The principal compartments and flows of matter within an eelgrass bed and between the bed and its environment are shown in Figure 1. This flow diagram represents the basic structure of the model as conceived for our simulations of the eelgrass systems near Beaufort, N.C. Temperature and season cues are the main outside forces driving the system, including the dynamics of the eelgrass compartment. Other outside forces that affect the functioning of an eelgrass bed but were not included in the present version of the model include tides, wind-driven currents, and light. Pertinent biotic factors external to the bed include predators such as birds and large nekton. Larval fish and zooplankton also are external biotic factors. The former provides a source of recruitment stock for fish in the bed, and both are sources of food for fish inhabiting the bed. The biomasses of fish larvae and zooplankton are functions of time and duplicate empirical field data (Thayer *et al.*, 1974, 1975).

Eelgrass is the principal biotic component of the bed and provides habitat for the animals. Detritus is produced mainly from dying eelgrass and is consumed by all animals of the bed except carnivorous fish. Fish of the bed consume epifauna, infauna, zooplankton, and larval fish and emigrate from the bed when environmental conditions become unfavorable. Outside predators are not residents of the bed but enter the bed to prey on animals. The dominant fish of the eelgrass beds near Beaufort, N.C., and their diets are listed by Adams (1976 a, c). The omnivore fish are dominated by juvenile pinfish (*Lagodon rhomboides*) and pigfish (*Orthopristis chrysoptera*) which account for 47% and 8% of the annual standing crop of the eelgrass fish community. Detritus makes up 30%; plant material (eelgrass and algae), 18% and animal matter, 50% of the juvenile pinfish diet over the year. For juvenile pigfish, detritus contributed 46%, and animal matter, 54% of the diet. The carnivorous fish, dominated by juvenile silversides (*Menidia menidia*) and silverperch (*Bairdiella chrysura*), account for 4 and 2%, respectively, of the average annual fish standing crop. These fish do not prey upon the omnivorous fish of the bed. The diet of silversides consists of 95% animal matter: epifauna, infauna, zooplankton and larval fish (Adams, 1976c). Dominant epifauna in the eelgrass beds which also are preferred food for the juvenile fishes of the bed include *Palaemonetes, Hippolyte,* and gammarid and caprellid amphipods. Calanoid copepods also are a preferred food item of most of these juvenile

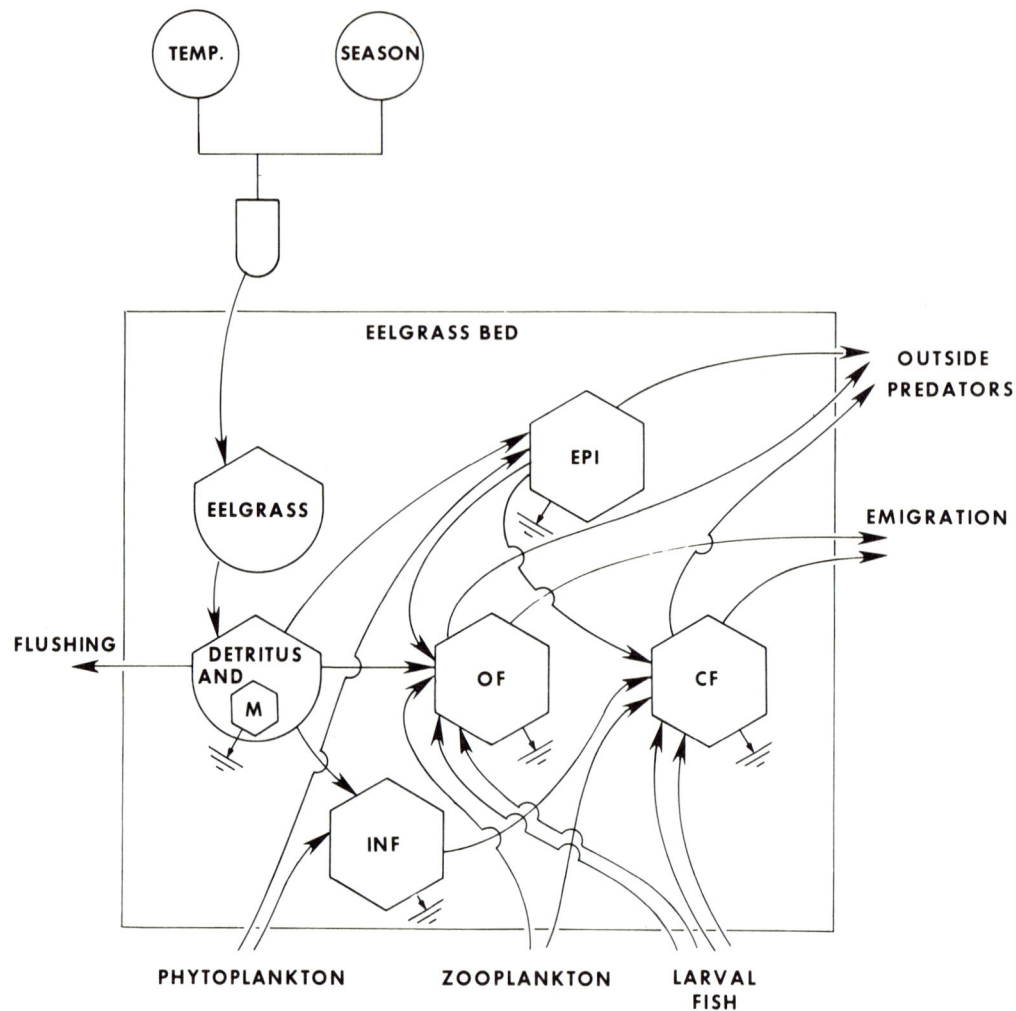

Fig. 1. Carbon flow diagram for eelgrass system. CF, carnivorous fish; INF, infauna; EPI, epifauna; M, heterotrophic microorganisms; OF, omnivorous fish.

fishes.

## Mathematical Description of Trophic Dynamics

The trophic dynamics of omnivorous and carnivorous fish and of epifauna are functions of standing crop (SC), temperature (TDFR), available habitat (AHF), assimilation efficiency(AE), and availability of preferred food items (AFF). Some of the rationale for parameter values and formulations are discussed in the subsequent section.

$$\text{Assimilation} = j \cdot SC \cdot TDFR \cdot AHF \cdot AE \cdot AFF$$

$$\text{Respiration} = k \cdot SC \cdot TDFR$$

For both fish j equals 1.0, and for epifauna j equals 1.5, while k is 0.09, 0.15 and 0.155 for carnivorous fish, omnivorous fish, and epifauna.

$$TDFR = 0.03T-0.0005T^2-0.2$$

where T is the temperature in degrees centigrade. This formulation is based on empirical data presented by Peters *et al.*(1976) for postlarval pinfish. TDFR is in the units fraction of body weight consumed per day. AHF is computed:

$$1 \geq AHF = \left[1 - \left(\frac{SC \ (both \ fish)}{0.02 \ SC \ (eelgrass)} \ or \ \frac{SC \ (epifauna)}{0.4 \ SC \ (eelgrass)}\right)\right] \geq 0$$

AE is constant for carnivorous fish, 0.9, and for epifauna, 0.4. AFF is constant at 1.0 for both omnivorous fish and epifauna. AE for omnivorous fish and AFF for carnivorous fish depend on the relative preference-scaled abundance of meat items present in the habitat (PSAM).

$$0 \leq PSAM = \Sigma \quad \begin{array}{ll} 0.032 & SC \ (infauna) \\ 0.05 & SC \ (epifauna) \\ 1.26 & SC \ (larval \ fish) \\ 1.72 & SC \ (zooplankton) \end{array} \quad \leq 1$$

Thus, for omnivorous fish

$$AE = 0.4 \cdot PSAM + 0.56$$

and for carnivorous fish

$$AFF = PSAM.$$

## Computer Model Description

The computer model consists of 99 interconnected blocks (Fig. 2). Ninety-eight blocks are typed (Table 1) as one of various computational elements whose functions are described in the 1130 Continuous System Modeling Program II, Program Description and Operations Manual (IBM Program Number 5711-X51). The 99th block generates time. Computational elements vary from simple arithmetic functions to numerical approximation of integrals and include both function generators and user definable FORTRAN subroutines called special elements.

The model uses all five special elements available in the CSMP II package. That specified as type 1 is used to provide "random numbers." The jitter element of the package is not used because it generates the same sequence of numbers starting with the same initial number on every simulation run. Those subroutines designated types 2 through 5 are used to perform computations which would have required a larger number of the standard computational element types. This use of subroutines is done to increase the total number of

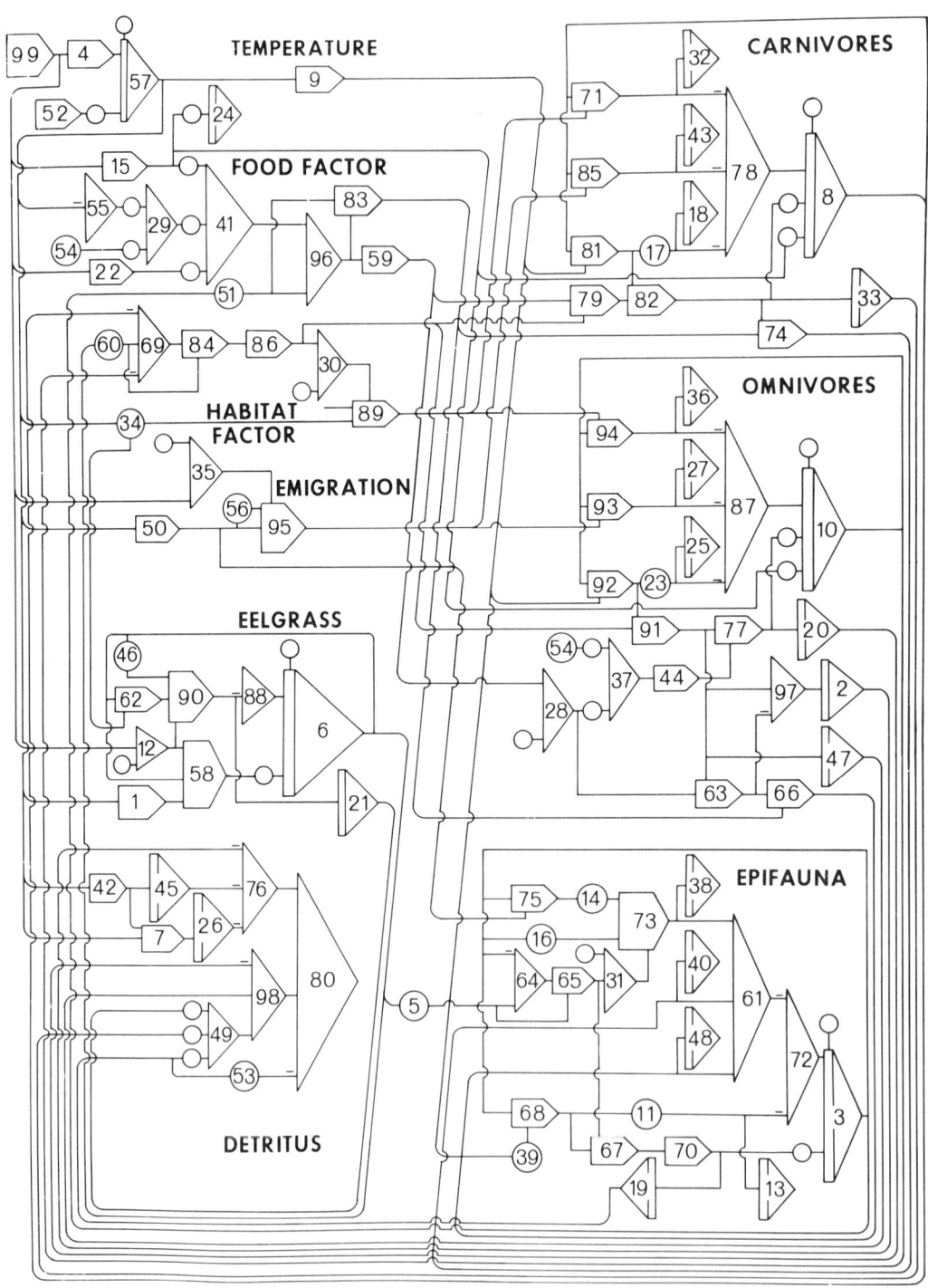

Fig. 2. The eelgrass bed model using modified analogue symbols. For differentiation of multipliers and dividers, etc., refer to Table 1.

Table 1.  Submodel block assignments, inputs and outputs.

| Submodel | Block# | Type* | Inputs 1 | Inputs 2 | Inputs 3 | Output |
|---|---|---|---|---|---|---|
| Temperature | 4 | F | 99 | | | Thermal increment |
| | 9 | 2 | 57 | | | Feeding rate |
| | 52 | 1 | | | | Random deviate |
| | 57 | 1 | 4 | 52 | | Temperature |
| | 99 | | | | | Time |
| | | | | | | |
| Eelgrass | 1 | F | 57 | | | Production rate |
| | 6 | 1 | 88 | 58 | | Standing crop |
| | 12 | 0 | 99 | | | Season cue |
| | 21 | I | 90 | | | Mortality |
| | 46 | G | 6 | | | |
| | 58 | 3 | | 6 | 12 | Net production |
| | 62 | X | 6 | 34 | | |
| | 88 | – | 90 | | | |
| | 90 | R | 12 | 46 | 62 | |
| | | | | | | |
| Epifauna | 3 | I | 72 | 70 | | Standing crop |
| | 5 | G | 6 | | | |
| | 11 | G | 68 | | | |
| | 13 | I | 11 | | | Respiration |
| | 14 | G | 75 | | | |
| | 16 | G | 3 | | | |
| | 19 | I | 70 | | | Consumption |
| | 31 | 0 | 65 | | | Relative standing crop |
| | 38 | I | 73 | | | Cropping by outsiders |
| | 39 | G | 9 | | | Feeding rate |
| | 40 | I | 74 | | | Cropping by carnivores |
| | 48 | I | 66 | | | Cropping by omnivores |
| | 61 | + | 73 | 74 | 66 | |
| | 64 | + | -3 | 5 | | |
| | 65 | / | 64 | 5 | | |
| | 67 | X | 68 | 65 | | |
| | 68 | X | 3 | 39 | | |
| | 70 | N | 67 | | | |
| | 72 | + | -61 | -11 | | |
| | 73 | R | 31 | 16 | 14 | |
| | 75 | X | 3 | 50 | | |
| | | | | | | |
| Omnivores | 2 | I | 97 | | | Detritus consumed |
| | 10 | I | 87 | 77 | 15 | Standing crop |
| | 20 | I | 77 | | | Assimilation |
| | 23 | G | 92 | | | |
| | 25 | I | 23 | | | Respiration |
| | 27 | I | 93 | | | Emigration |
| | 28 | 0 | 59 | | | |
| | 36 | I | 94 | | | Cropping by outsiders |
| | 37 | W | 54 | 28 | | |
| | 44 | N | 37 | | | |
| | 47 | I | 91 | | | Consumption |
| | 54 | K | | | | |
| | 63 | X | 28 | 91 | | |
| | 66 | X | 63 | 83 | | |
| | 77 | X | 91 | 44 | | |
| | 87 | + | -94 | -93 | -23 | |
| | 91 | X | 86 | 92 | | |
| | 92 | X | 10 | 9 | | |
| | 93 | X | 10 | 95 | | |

Table 1 continued

| Submodel | Block# | Type* | Inputs 1 | Inputs 2 | Inputs 3 | Output |
|---|---|---|---|---|---|---|
| | 94 | X | 10 | 89 | | |
| | 97 | + | 91 | -63 | | |
| Carnivores | 8 | I | 78 | 82 | 15 | Standing crop |
| | 17 | G | 81 | | | |
| | 18 | I | 17 | | | Respiration |
| | 32 | I | 71 | | | Cropping by outsiders |
| | 33 | I | 82 | | | Consumption |
| | 43 | I | 85 | | | Emigration |
| | 71 | X | 8 | 89 | | |
| | 74 | X | 83 | 82 | | |
| | 78 | + | -71 | | -17 | |
| | 79 | X | 59 | 86 | | |
| | 81 | X | 8 | 9 | | |
| | 82 | X | 79 | 81 | | |
| | 85 | X | 8 | 95 | | |
| Food Factor | 15 | F | 99 | | | Larval fish |
| | 22 | F | 99 | | | Zooplankton |
| | 24 | I | | 15 | | Fish immigration |
| | 29 | W | 55 | 54 | | Infauna |
| | 41 | W | 15 | 29 | 22 | |
| | 51 | G | 3 | | | |
| | 55 | – | 57 | | | |
| | 59 | L | 96 | | | Food factor |
| | 83 | / | 51 | 96 | | Epifauna per ration |
| | 96 | + | 41 | 51 | | |
| Habitat factor | 30 | O | 84 | | | Relative standing crop |
| | 34 | G | 57 | | | cue |
| | 60 | G | 6 | | | |
| | 69 | + | - 8 | 60 | -10 | |
| | 84 | / | 69 | 60 | | |
| | 86 | N | 84 | | | Habitat factor |
| | 89 | R | 30 | | 34 | Cropping rate |
| Emigration | 35 | O | 99 | | | |
| | 50 | F | 57 | | | Cropping rate |
| | 56 | G | 50 | | | |
| | 95 | R | 35 | 50 | 56 | Emigration rate |
| Detritus | 7 | 5 | 42 | 57 | | |
| | 26 | I | 7 | | | Infauna respiration |
| | 42 | 4 | 57 | | | |
| | 45 | I | 42 | | | Microbial respiration |
| | 49 | W | 21 | 33 | 19 | |
| | 53 | G | 19 | | | |
| | 76 | + | - 2 | -45 | -26 | |
| | 80 | + | 76 | 98 | -53 | Detritus net change |
| | 98 | + | -20 | 47 | 49 | |

| * | F | Function generator | N | Negative clipper | / | Divider |
|---|---|---|---|---|---|---|
| | G | Gain | O | Offset | – | Inverter |
| | I | Integrator | R | Relay | + | Summer |
| | K | Constant | W | Weighted summer | 1-5 | FORTRAN Subroutines |
| | L | Limiter | X | Multiplier | | |

computations that can be performed within the limits of the program. The FORTRAN programs for these subroutines are included as an appendix to this manuscript.

Function generators receive an input variable with known maximum and minimum values. Their range is subdivided into deciles by 11 intercepts which define the output at those values of the input variable. Output within deciles is based on linear interpolation. We use five such elements to incorporate rates and empirical standing crops into the model (Table 2). Block 1 receives temperature and outputs unscaled temperature-dependent growth rate of eelgrass to the block specified special element type 3 (See appendix). This rate is ultimately scaled by a constant multiplier when it enters block 6. The estimation of net production of eelgrass is based on discussions of field observations of standing crops and estimated turnover rates with G. W. Thayer (personal communication). Block 4 generates increments of temperature throughout the year. These increments are summed along with the random deviate generated by block 52 to simulate seasonally biased stochastic temperature changes. Blocks 15 and 22 generate standing crops of larval fish and zooplankton as a function of time and are based on empirical data (Thayer *et al.*, 1974). The output of block 50 is a temperature-dependent-fish emigration rate and loss rate of epifauna due to cropping by outside predators.

The model consists of 9 submodels: temperature, the biological compartments, eelgrass, epifauna, omnivorous fish and carnivorous fish, the rate-determining food factor and habitat space factor, and

Table 2. Description of function generators.

| | Functional Elements (Block #) | | | | |
|---|---|---|---|---|---|
| | 1 | 4 | 15 | 22 | 50 |
| Input | Temperature | Time | Time | Time | Temperature |
| Minimum | 5.0 | 0.0 | 0.0 | 0.0 | 5.0 |
| Maximum | 35.0 | 360. | 360. | 360. | 35.0 |
| Intercepts | | | Values of Output | | |
| 1 | 0.01 | 0.0 | 0.011 | 0.115 | 0.2 |
| 2 | 0.02 | 0.071 | 0.037 | 0.138 | 0.1 |
| 3 | 0.04 | 0.162 | 0.07 | 0.12 | 0.05 |
| 4 | 0.05 | 0.128 | 0.043 | 0.101 | 0.01 |
| 5 | 0.05 | 0.15 | 0.0041 | 0.091 | 0.005 |
| 6 | 0.05 | 0.012 | 0.0 | 0.065 | 0.005 |
| 7 | 0.04 | 0.032 | 0.0 | 0.083 | 0.005 |
| 8 | 0.01 | −0.146 | 0.0 | 0.07 | 0.005 |
| 9 | 0.0 | −0.123 | 0.001 | 0.061 | 0.1 |
| 10 | 0.0 | −0.153 | 0.0041 | 0.075 | 0.2 |
| 11 | 0.0 | −0.146 | 0.0 | 0.115 | 0.4 |
| Output | Unscaled Production Rate | Thermal Increment | Larval Fish | Zooplankton | Cropping Rate |

the emigration and detritus submodels (Fig. 2). The highly resolved epifaunal, carnivorous and omnivorous fish submodels all have integrators which generate instantaneous standing crops and which record individual positive and negative inputs to standing crops. All three of these compartments can become habitat- or space-limited. Habitat is a linear function of the standing crop of eelgrass. The habitat factor, block 86 for both fish and block 65 for epifauna, is used in computing consumption rates, and the fish compete with each other for available habitat (block 69). Carnivorous fish also can become food-limited. In the model, it is assumed that omnivores respond to reduced animal matter availability by consuming the same total amount of food, and more detritus, which is assimilated with lower efficiency. Both types of fish eat epifauna and compete for available animal matter. The food factor is computed by scaling and totaling standing crops of consumed and available food items exclusive of detritus. The scaling is similar to a concentration factor (food preference factor) based on empirical relative abundance data of concentrations of food items in the gut of fish collected from eelgrass beds (Adams, 1976c) compared to concentrations of these items in the natural habitat (Thayer *et al.*, 1975). This factor also is scaled so that it ranges from 0 with no food items present to 1 when total preference-scaled food abundance in the model simulation reached the maximum based on empirical maximum standing crops of food items in seagrass beds near Beaufort, N.C. Both the habitat and food factors are restricted to values between 0 and 1. Respiration rate is a fraction of TDFR, the temperature-dependent feeding rate; TDFR is the optimum rate of consumption subject to habitat or food limitation. Thus, respiration rate (fraction SC per day) is not limited, but consumption is limited by factors other than temperature. Fish immigration is a linear function of larval fish standing crop. Fish emigration rates are functions of fish standing crops, temperature, and season. Cropping rates by outside predators are functions of prey standing crops, temperature, and available habitat. Both immigration and emigration rates and cropping rates by outside predators on fish and epifauna of the bed cannot be based upon empirical data because these data do not exist. Parameters were adjusted during submodel development to provide reasonable model behavior.

Infaunal standing crop and respiration are computed in the food factor submodel as a function of temperature based on data in Thayer *et al.* (1975). Detritus is recorded as net change after summing all positive and negative inputs (block 80) which do not reflect levels of detritus present. The net change of detritus is highly dependent on both the fraction of eelgrass mortality retained in the bed and microbial respiration rate. For the present we assume a mortality retention of 50% and a microbial respiration which increases from 0.07 at 10°C to 0.58 gm/m$^2$/day at 30°C which simulates small annual net changes in detritus in the bed. The estimation of microbial respiration rate is based on ATP estimates of average standing crops of microorganisms in sediments of the Newport River estuary

(Ferguson and Murdoch, 1975) and on estimated turnover rates. The dependence of the biological compartments on time (season) and temperature can be seen by tracing the outputs of blocks 99 and 57 in Figure 2.

### THE STANDARD MODEL

Some of the parameters of the model were tuned to simulate, to a reasonable degree, the empirical field data from the Beaufort area. In the model, these include the standing crops of the poorly re-solved biological compartments of infauna, larval fish, zooplankton, and the annual temporal patterns of temperature. The growth of eel-grass approximates the pattern that might be observed in an "optimal" growth year in the Newport River estuary. This was done, not so much to simulate eelgrass growth, as to provide an appropriate "habi-tat" within which the more highly resolved epifauna and juvenile fish compartments could respond. The parameters in Table 3 reflect this tuning. The table also contains parameter values that were chosen and based on preliminary testing of the individual submodels. Some of the parameters associated, for example, with cropping by outside predators and fish immigration and emigration rates cannot at present be based on or tested against field data. This type of data does not exist because of the extreme difficulty in quantifying these processes in open systems. The various responses of the high-ly resolved components in the model could, in fact, be fine-tuned to simulate more faithfully the extant laboratory and field data, but this fine-tuning has not been done because the major objective of this study was not limited to faithful reproduction of observed data. Initial conditions were set to approximate those occurring at the end of a year's simulation of the standard model. The stochastic temperature option was not utilized in any of the simu-lations described in this paper.

### PRINCIPAL COMPONENT ANALYSIS OF THE MODEL OUTPUTS

Principal component analysis (Service, 1972) was used as a means of condensing the information generated in simulation experi-ments as outputs of monitored dependent variables. That is, it was used to resolve the information of up to 28 intercorrelated depen-dent variables into one or more independent, orthogonal, composite variables. An early version of the model was subjected to fluctua-tions in eelgrass growth rate in a first and average annual tempera-ture in a second series of year-long simulations. The ranges ob-served indicate the sensitivity of each of the dependent variables to these factors (Table 4).

The 24 dependent variables observed in simulation experiment 1 (eelgrass growth rate) resolve into a single principal component (with a characteristic root greater than 1.0) which accounts for 95.8% of the total variation in the data (Table 5). All of the ori-ginal 24 variables including average eelgrass standing crop are posi-tively correlated with this single principal component (Table 6 and Figure 3). This component increases with eelgrass growth rate but

Table 3.  Non-zero initial conditions and parameter assignments for standard model.

| Block # | Parameters | | |
|---|---|---|---|
| | IC*/1 | 2 | 3 |
| 1 | 35. | 5. | |
| 3 | 0.2 | 0.5 | |
| 4 | 360. | | |
| 5 | .04 | | |
| 6 | 1.0 | 5.0 | |
| 7 | -0.018 | 0.63 | |
| 8 | 0.01 | 0.9 | 0.15 |
| 9 | 0.00025 | 0.015 | 0.1 |
| 10 | 0.02 | 1.0 | 0.15 |
| 11 | 0.155 | | |
| 12 | -275. | | |
| 14 | 1.0 | | |
| 15 | 360. | | |
| 16 | 0.001 | | |
| 17 | 0.09 | | |
| 22 | 360. | | |
| 23 | 0.15 | | |
| 24 | | 0.15 | |
| 28 | -0.1 | | |
| 29 | 0.2 | 7.0 | |
| 30 | -0.5 | | |
| 31 | -0.5 | | |
| 34 | 0.0003 | | |
| 35 | -165. | | |
| 37 | 0.6 | 0.4 | |
| 39 | 1.5 | | |
| 41 | 1.26 | 0.032 | 1.72 |
| 42 | 16. | 400. | 25000. |
| 46 | 0.04 | | |
| 49 | 0.8 | 0.1 | 0.6 |
| 50 | 35. | 5. | |
| 51 | .05 | | |
| 52 | 1. | | |
| 53 | 0.5 | | |
| 54 | 1. | | |
| 56 | 0.3 | | |
| 57 | 9.1 | | |
| 58 | 0.1 | 0.01 | 500. |
| 59 | 1.0 | 0.1 | |
| 60 | 0.02 | | |

*This assigns initial condition for intergrators for which parameter
1 is assumed 1.0.

Table 4.  Selected dependent variable responses observed during 19 year-long simulations.  In experiment 1 the average annual growth rate of eelgrass ranged from 8 to 18% of the standing crop per day in 8 simulations.  In experiment 2 the average annual temperature ranged from 16.2 to 26.3°C in 11 simulations.

| Dependent variable | Exp 1 | | | Exp 2 | | |
|---|---|---|---|---|---|---|
| | Low | High | Mean | Low | High | Mean |

$$\text{------------------------------ g dry/m}^2 \text{ ------------------------------}$$

| | Low | High | Mean | Low | High | Mean |
|---|---|---|---|---|---|---|
| **Standing Crop**** | | | | | | |
| Eelgrass (EE) | 22.8 | 327. | 109. | 106. | 209. | 176. |
| Epifauna (EP) | 0.41 | 2.83 | 1.93 | 1.40 | 2.30 | 1.99 |
| Carnivorous Fish (CF) | 0.15 | 0.53 | 0.38 | 0.28 | 0.45 | 0.37 |
| Omnivorous Fish (OF) | 0.24 | 3.05 | 1.78 | 1.00 | 1.99 | 1.66 |
| Infauna (IN) | | | | 1.77 | 4.26 | 2.82 |
| Detritus (DE) | -256. | 122. | - 39.9 | -237. | 17.0 | -64.4 |

$$\text{------------------------------ g dry/m}^2\text{/yr ------------------------------}$$

| | Low | High | Mean | Low | High | Mean |
|---|---|---|---|---|---|---|
| **Production** | | | | | | |
| EE | 76.0 | 1330. | 770. | 371. | 1040. | 726. |
| EP | 0.20 | 9.70 | 5.06 | 1.50 | 8.20 | 4.56 |
| CF | - 0.34 | 0.67 | 0.22 | 0.02 | 0.24 | 0.18 |
| OF | 0.00 | 10.8 | 5.72 | 2.20 | 8.10 | 5.13 |
| DE | 47.0 | 724.0 | 426. | 216. | 421. | 338. |
| **Respiration** | | | | | | |
| EP | 3.90 | 24.3 | 17.1 | 13.7 | 19.6 | 17.3 |
| CF | 0.59 | 2.27 | 1.63 | 1.04 | 2.07 | 1.64 |
| OF | 1.90 | 16.6 | 9.84 | 5.8 | 10.4 | 9.04 |
| IN | | | | 7.89 | 10.0 | 9.48 |
| **Consumption** | | | | | | |
| EP | 10.2 | 85.0 | 55.4 | 39.0 | 61.8 | 54.7 |
| CF | 0.28 | 3.27 | 2.05 | 1.38 | 2.55 | 2.03 |
| **Predation Loss** | | | | | | |
| EP to CF | 0.04 | 0.84 | 0.51 | 0.38 | 0.63 | 0.51 |
| EP to OF | 0.05 | 5.65 | 2.71 | 0.82 | 3.74 | 2.23 |
| EP to Outsiders (OU) | 0.04 | 0.60 | 0.34 | 0.16 | 0.45 | 0.31 |
| CF to OU | 0.35 | 0.93 | 0.66 | 0.30 | 0.92 | 0.66 |
| OF to OU | 0.58 | 7.17 | 4.07 | 2.59 | 4.52 | 3.64 |

$$\text{------------------------------ No Units ------------------------------}$$

| | Low | High | Mean | Low | High | Mean |
|---|---|---|---|---|---|---|
| **Consumption Rate Multipliers*** | | | | | | |
| EP habitat | 0.17 | 0.69 | 0.42 | 0.32 | 0.45 | 0.40 |
| CF and OF habitat | -0.23 [0] | 0.37 | 0.16 | -0.10 [0] | 0.18 | 0.09 |
| CF food | 0.29 | 0.41 | 0.37 | 0.30 | 0.41 | 0.40 |

$$\text{------------------------------\%------------------------------}$$

| | Low | High | Mean | Low | High | Mean |
|---|---|---|---|---|---|---|
| **Miscellaneous** | | | | | | |
| OF assimilation efficiency | 57.7 | 62.6 | 60.8 | 58.2 | 62.5 | 60.6 |

$$\text{------------------------------ \% SC Consumed/Day ------------------------------}$$

| | Low | High | Mean | Low | High | Mean |
|---|---|---|---|---|---|---|
| Temperature Dependent Feeding Rate (TDFR)** | | | | 6.50 | 11.1 | 9.24 |
| Average temperature | | | | 16.2 | 26.2 | 21.2 |

* See text
** Annual mean except detritus which is net change over the year.

Table 5.   Statistics derived from principal component analysis
of Experiments 1 and 2.

| Experiment | Independent variable levels | Dependent variables | Characteristic roots ( >1.0) | Cumulative variation accounted for |
|---|---|---|---|---|
| | --------Number--------- | | | ----------%---------- |
| 1 | 8 | 24 | 23.48 | 97.83 |
| 2 | 11 | 28 | 16.99 | 60.68 |
| | | | 8.32 | 90.41 |
| | | | 1.38 | 95.32 |

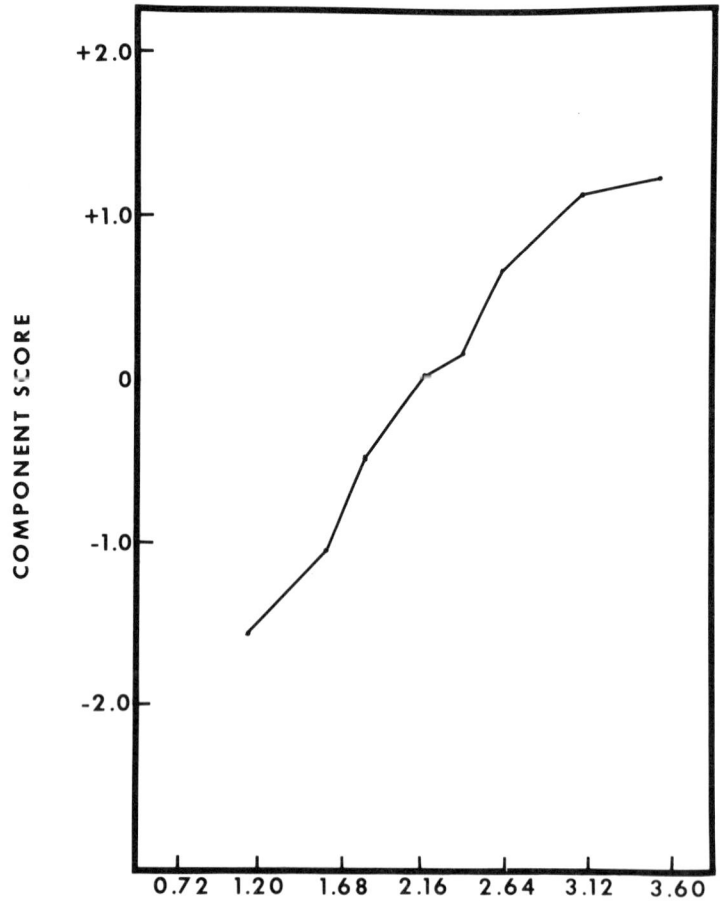

EELGRASS GROWTH RATE

(% SC/day)

Fig. 3.   Principal component of the eelgrass growth rate
simulations where the rate is the average output of block one
(over the simulated temperature range) and is subject to scaling
by a factor of 5 throughout the year and also by 0.1 or by 0.01,
according to season or maximum standing crop cues.

Table 6.  Orthogonal factor matrices (Varimax) for Experiments 1 and 2.

| Dependent Variable[1] | Principal Components | | | |
|---|---|---|---|---|
| | Exp 1 | Exp 2 | | |
| | I | I | II | III |
| **Standing Crop** | | | | |
| EE | 0.997 | 0.91 | 0.36 | -0.08 |
| EP | 0.991 | 0.97 | -0.04 | -0.19 |
| CF | 0.995 | -0.22 | 0.90 | -0.16 |
| OF | 0.998 | 0.91 | 0.39 | -0.10 |
| IN | | 0.86 | -0.43 | -0.12 |
| DE* | 0.998 | 0.96 | 0.00 | -0.21 |
| **Production** | | | | |
| EE | 0.998 | 0.95 | -0.26 | -0.13 |
| EP | 0.988 | 0.92 | -0.38 | -0.03 |
| CF | 0.998 | 0.80 | 0.54 | 0.11 |
| OF | 0.996 | 0.97 | -0.24 | -0.01 |
| DE | 0.998 | 0.57 | 0.07 | 0.68 |
| **Respiration** | | | | |
| EP | 0.985 | 0.00 | 0.91 | -0.34 |
| CF | 0.993 | -0.49 | 0.87 | -0.01 |
| OF | 0.997 | 0.57 | 0.79 | -0.16 |
| IN | | 0.81 | 0.04 | -0.56 |
| **Consumption** | | | | |
| EP | 0.996 | 0.81 | 0.43 | -0.32 |
| CF | 0.998 | -0.34 | 0.94 | -0.01 |
| **Predation Loss** | | | | |
| EP to CF | 0.997 | 0.12 | 0.97 | 0.00 |
| EP to OF | 0.978 | 0.98 | -0.13 | -0.02 |
| EP to OU | 0.975 | 0.84 | -0.40 | -0.23 |
| CF to OU | 0.994 | -0.68 | 0.72 | 0.15 |
| OF to OU | 0.996 | -0.08 | 0.99 | 0.00 |
| **Consumption rate multipliers** | | | | |
| EP habitat | 0.947 | 0.45 | 0.31 | -0.73 |
| CF and OF habitat | 0.951 | -0.48 | 0.57 | -0.53 |
| CF available food | 0.980 | 0.92 | -0.30 | -0.14 |
| **Miscellaneous** | | | | |
| OF assimilation efficiency | 0.991 | 0.93 | -0.28 | -0.22 |
| TDFR | | -0.84 | 0.51 | 0.11 |
| Average Temperature | | -0.88 | 0.40 | 0.23 |

[1]See Table 4        *Net Change

levels off at average growth rates of 16 to 18% standing crop/day due to a limit on standing crop based on an empirical maximum; i.e., marked reduction of growth rate occurs when the instantaneous standing crop of eelgrass exceeds 500 $g/m^2$.

In the second simulation experiment (average annual temperature), the 28 dependent variables resolve into 3 principal components (with characteristic roots greater than 1.0), which together account for 95.3% of the total variation in the data (Table 5). Seventeen of the 28 variables are highly correlated with component 1; 15 of these including average eelgrass standing crop are positive correlations (Table 6). Seven of the remaining variables are highly and positively correlated with the second component. None of the variables is highly correlated with the third principal component which we feel is unimportant and accounts for less than 5% of the total variation. The first component is negatively related to average annual temperatures above 17.2°C (Fig. 4). In the average annual temperature range of 18.2 to 24.2°C, the relation has a very small slope. Apparently, at intermediate temperatures there are compensatory mechanisms in the model which moderate the effects of small changes in average temperature. The second component parallels component I at the

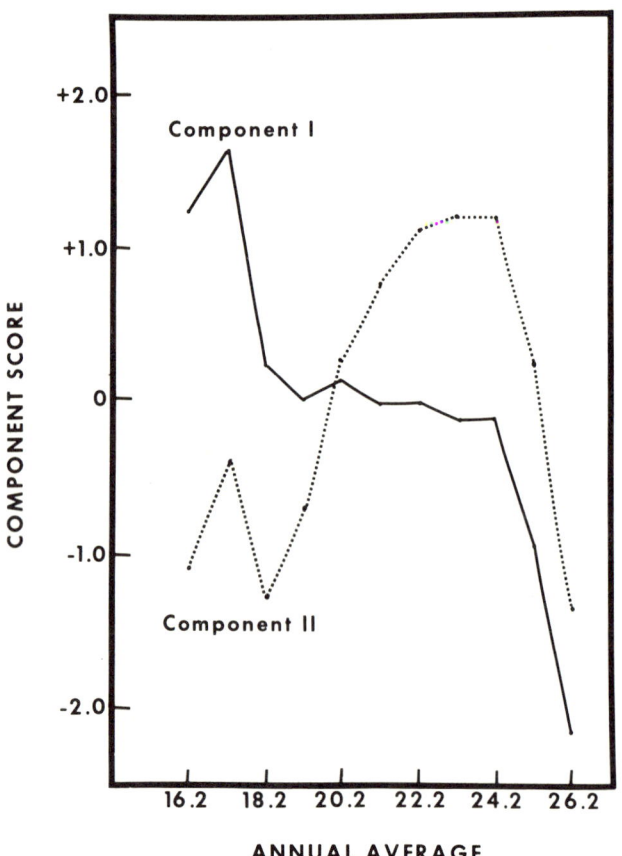

ANNUAL AVERAGE
TEMPERATURE (C)

Fig. 4. Principal components I and II of the average annual temperature simulations.

extremes of average annual temperatures but also is related positively to temperature, resulting in the distorted cruciform shape of
this figure. The positive correlation to temperature is most evident
at intermediate average annual temperatures where component I changes
slowly with temperature, but component II increases rapidly.

All of the monitored model responses in experiment 1 are
tightly coupled to eelgrass productivity via its effect on standing
crop throughout the year. Eelgrass is the habitat. The importance
of eelgrass is also evident in the second experiment in which the
two components can be described as representing those variables which
respond to temperature indirectly via the temperature effect on eelgrass standing crop and those variables which also respond to temperature directly as follows:

*Component I (Standing Crop and Productivity).* Those dependent
variables, predominantly standing crops and productions, which respond to average annual temperature indirectly through the response
of eelgrass standing crop are positively correlated with this component. Infaunal standing crop is generated inversely from the
temperature; thus, it and infaunal respiration correlate negatively
with this component. Average temperature and the temperature-
dependent feeding rate are, of course, both negatively correlated
to it.

*Component II (Respirations and Carnivorous Fish).* Respirations
of epifauna, carnivorous fish, and omnivorous fish respond indirectly to average annual temperature because of their positive correlation with their own standing crops (Component I); however, they
also show a strong, direct, positive correlation with average annual
temperature. This is best observed at intermediate average annual
temperatures where standing crop changes seem to be moderated by
compensatory mechanisms but where component II continues to increase.
The standing crop, total consumption and predation on epifauna of
carnivorous fish, and cropping by outside predators on omnivorous
fish also are correlated with this component. The leveling of component II between 22.2 and 24.2°C is due to the stability of average
standing crops in this average annual temperature range combined
with an inflection in the temperature-dependent feeding rate (TDFR)
which is used to estimate respiration rates of consumer organisms.

Those variables not highly correlated with either component I
or II include the habitat factors for both epifauna and fish consumption rate computations. Detritus net change and croppings by
outside carnivores on the carnivorous fish of the bed also show low
correlations.

## Credibility and Utility of the Model

According to Patten *et al.* (1975), there are three main criteria
that can be employed to determine the credibility or validity of a
model: 1) the compartmental behavior of the model, if it lies within
realistic ranges of measured initial states and if temporal sequences
are reasonable, 2) the postulated internal organization, if it is
consistent as evidenced by the model's ability to generate reasonable

output dynamics not expressly incorporated into the model, and
3) the responses to small perturbations, if they are believable.

## COMPARTMENTAL BEHAVIOR AND TEMPORAL SEQUENCES

The temporal outputs of temperature, of eelgrass standing crop,
and of the summed standing crops of omnivorous and carnivorous fish
correspond to field data reasonably well (Figs. 5-7). Field data are
from Adams (1976a). The average annual temperature in the simulation
was 21.2, and that reported from the field was 20.1°C, but the temporal sequences were similar (Fig. 5). The bias in the simulation
curves may occur because the observed temperatures were taken during
high tides and because the simulation is designed to predict temperatures that approximate an average of those that might be observed at
low and high tide.

The simulated values of the eelgrass standing crop compartment
generally are much higher than the field observations of standing
crop, but temporal patterns are similar (Fig. 6). The mean standing
crop in the simulation is 201, and in the observations it was 45 in
1971-1972 and 173 gms/m$^2$ for 1975. Standing crop is low in winter,
increases rapidly through the spring, and decreases in the summer
and fall. There are two main sources of variation between simulated
and observed standing crops. The first results from the evolutionary nature of eelgrass systems in the Beaufort area. The eelgrass
bed upon which we have based our model design has progressed from
an immature to a more mature community. Annual mean eelgrass (SC)
has increased from 18 in 1969 when the bed was first observed to
45 in 1971-1972, 105 in 1974, and finally 173 gms/m$^2$ in 1975 (Thayer
*et al.*, 1977). The bed thus seems to be evolving towards progressively higher biomass, and the predicted annual pattern of eelgrass
standing crop, therefore, reflects our understanding of the annual
temporal sequence, while the higher average standing crop more accurately predicts densities that may be obtained in the next few
years. Secondly, simulated and observed values also may differ due
to the effects of unmodeled parameters such as the interaction of
tide level and diurnal temperature of water over the bed. As the
eelgrass community has become more mature, the bed has built up
sediment, and therefore the average depth of the water overlying
the bed is becoming progressively less (Thayer *et al.*,1975). The
densitiy of the eelgrass standing crops and mortality of grass during
the warmer periods of the year is highly dependent not on the average
water temperature over the bed but on the timing of low tides during
periods of temperature extremes. In general, the shallower the water
over a bed, the higher the mortality of grass blades due to desiccation. This is especially true in the summer when temperatures typically reach 28°C and higher in the bed. Dieback of grass blades
has been observed during these conditions, particularly during extreme low tides when beds become exposed. For example, in the summer of 1974, extreme low tides coincided with the midafternoon
temperature maximum, and desiccation occurred. Our model does not
consider mass mortality of grass under these severe conditions;

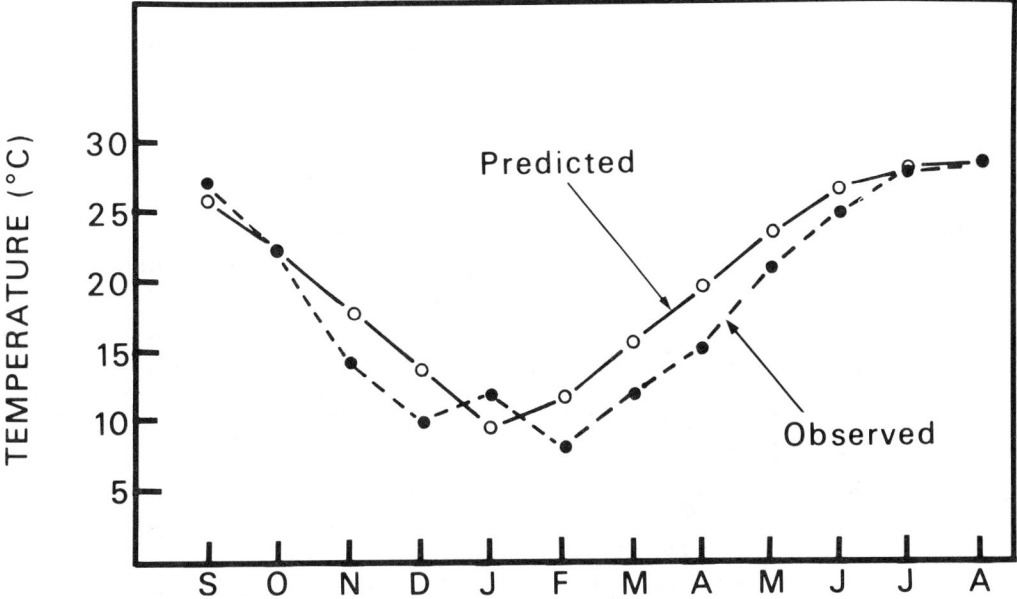

Fig. 5. Simulated and observed temporal sequence of temperature.

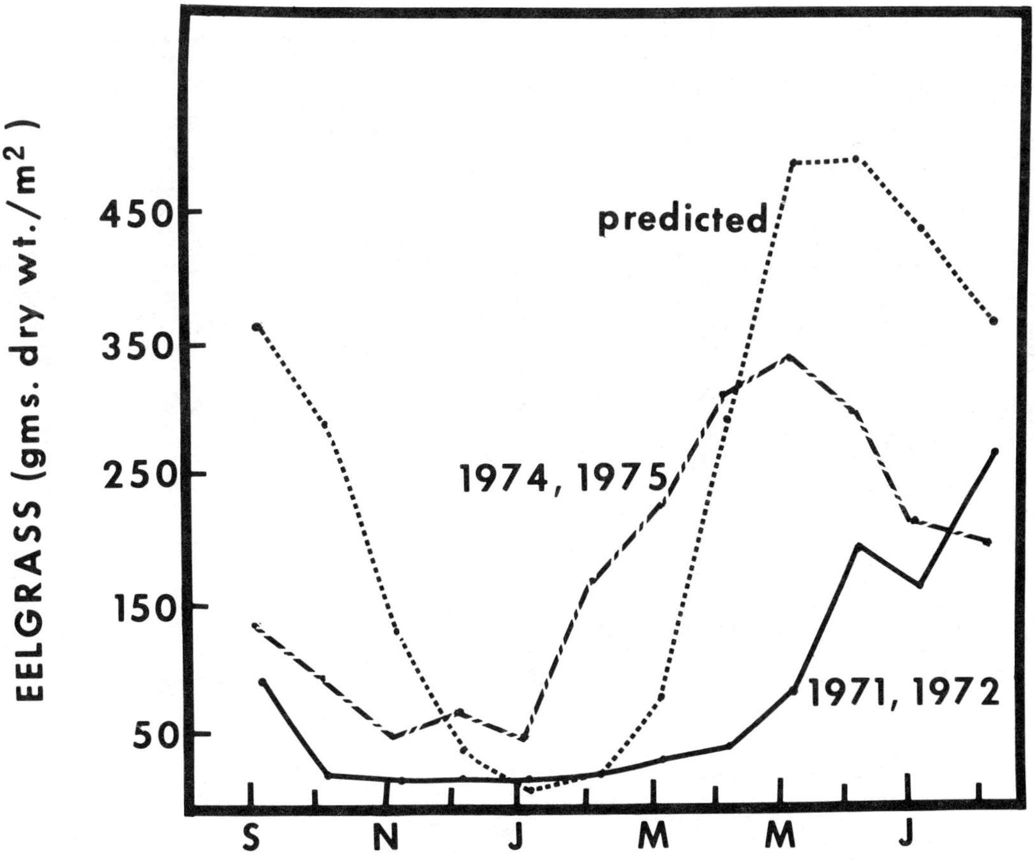

Fig. 6. Simulated and observed temporal sequences in eelgrass standing crop.

rather it simulates increasingly larger percentages of grass mortality both with temperature increases and during the fall and winter seasons.

The simulated total fish standing crop compares favorably with field data (Fig. 7), with the simulated annual mean, 1.63 gms/m$^2$, only somewhat higher than the observed 1.50 gms/m$^2$. The major discrepancy in temporal behavior occurs in May and June when the simulated standing crop of fish was about 50% higher than the observed. Also, the simulated standing crop reached its maximum about the first of June, but the observed values peaked about 7 weeks later. These variations between the model simulation and the field measurements may be due, in part, to the higher grass and epifaunal standing crops in the simulations. It should also be noted, however, that the drop net used by Adams (1976a) to capture fish for biomass estimation is not 100% efficient and leads to underestimation of the biomass of fish present (Kjelson and Johnson, 1973). A suitable correction factor may be as high as 1.76 (*ibid.*). The rapid decline in the simulated standing crop of fish after June is due to the space-limitation factor, emigration, and increased predation. In the model, the fish reached their critical space-limitation and declined because of increased predation and movement out of the bed. In the field data, the increase in standing crop was moderated perhaps by these same factors.

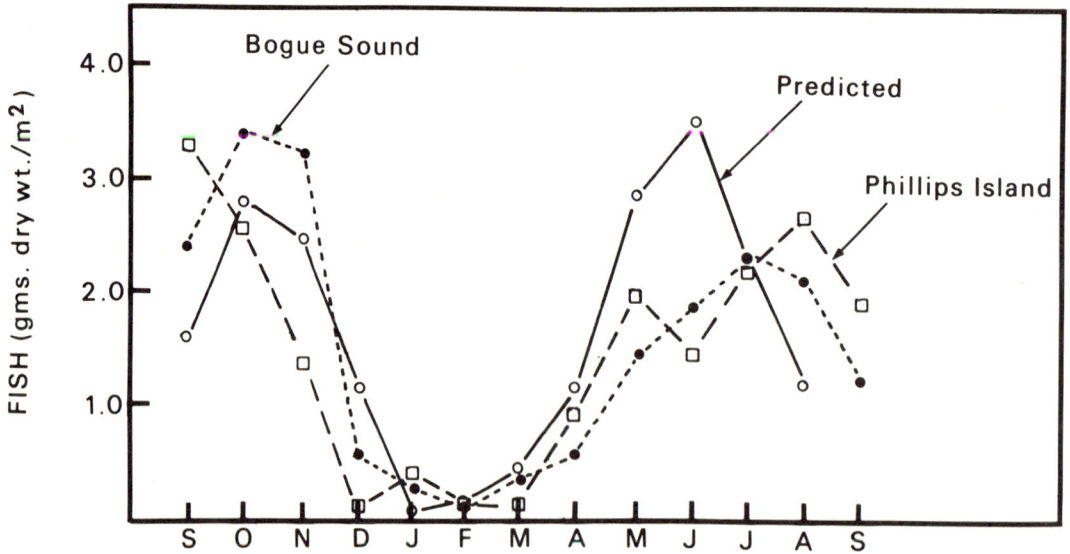

Fig. 7.  Simulated and observed temporal sequences in standing crops of juvenile fish.

Factors that were incorporated into the model, for which comparative data are not available, include immigration, emigration and cropping by outside predators. Empirical data on these flux rates are scarce or nonexistent mainly because of the difficulty involved in the quantification of these factors. In our model simulations, immigration is a linear function of larval fish standing crop. Emigration is a function of temperature, season, and standing crop. Previous field studies have found that fish emigrate out of areas that are thermally stressed (Cameron, 1969; Adams, 1976), and in our model, fish emigration out of the bed increases when the temperature reaches about 27°C in mid-June. The fish also leave the bed during the fall when temperatures are lower than the preferred range of temperatures. Cropping of fish and epifauna by larger outside predators is a function of the habitat factor and respective standing crops of the prey. As fish density increases relative to grass biomass, predation increases, a simulation of the increased chance of a predator finding its prey.

### POSTULATED INTERNAL ORGANIZATION

A second approach to examine the credibility of the model is to examine the output characteristics of the ecosystem as generated by the model, i.e., characteristics that were not specifically introduced in the initial modeling process but are emergent from it. Such properties can be evaluated for reasonableness as an indication of the model's strengths and weaknesses (Patten *et al.*, 1975). Some of the properties that can be examined include ecological efficiency, secondary production, metabolism and respiration.

One of the most important factors in assessing the flow of energy through a trophic level is the ecological efficiency, the ratio of the steady-state yield to energy-consumed (Slobodkin, 1960). The production/consumption (P/C) ratio can be used as a measure of the ecological efficiency assuming, as Mann (1965) and Thayer *et al.* (1975) have done, that all of the production is utilized by predators.

In our model, production of a compartment is calculated as the sum of cropping, emigration, and change in the standing crop minus immigration. For epifauna, the annual cropping by outside predators is 4.05 gms/m$^2$. Thus, adding cropping by carnivorous fish within the bed, 0.59, and by omnivorous fish within the bed, 4.72, the total is 9.36 gms/m$^2$. The net change in standing crop was 0.86, giving a total annual production of 10.22 gms/m$^2$ (Table 7). Consumption over the year was 91.1 and respiration was 26.2 gm/m$^2$. The ecological efficiency for the epifauna in our model is thus 10.22/91.1 = 0.11. This ratio compares well with 0.10 estimated by Thayer *et al.* (1975) for the epifaunal community in the Phillips Island eelgrass bed. With an average caloric content of epifauna in Thayer's study of 5.0 cal/mgDW, the production in the model simulation of 10.22 is reasonably close to the observed value of 9.0 gm/m$^2$. Also, the efficiency of energy dissipation, or the respiration/consumption (R/C) ratio of 0.29 in the model, compares favorably to 0.30 estimated by Thayer *et al.* (1975). The annual production of omnivorous and carnivorous fish compartments in the model is 14.0 and 0.97 gms/m$^2$ (Table 7). Dividing these production values by the total consumption for each fish gave P/C ratios of 0.42 and 0.43 for the omnivorous and carnivorous fish. These efficiencies are much higher than the P/C ratio of 0.24 estimated by Thayer *et al.* (1975) and Adams (1976b). The discrepancy in these results may be due, in part, to the fact that simulated respirations of the fish in the model did not approach the levels which occur

Table 7.  Summary of Integrative Functions for Standard Model.

| Function | Value (gms/m$^2$) |
|---|---|
| Epifaunal Consumption | 91.1 |
| Epifaunal Respiration | 26.2 |
| Epifaunal Predation by Outside Predators | 4.05 |
| Cropping of Epifauna by Omnivores in bed | 4.72 |
| Cropping of Epifauna by Carnivores in bed | 0.59 |
| Infaunal Respiration | 100.07 |
| Microbe Respiration | 463.8 |
| Detritus Consumption by Omnivores | 23.5 |
| Eelgrass Mortality | 1084.0 |
| Omnivorous Fish Consumption | 33.1 |
| Omnivorous Fish Respiration | 9.7 |
| Omnivorous Fish Assimilation | 23.7 |
| Omnivorous Fish Emigration | 12.7 |
| Omnivorous Fish Predation by Outsiders | 1.7 |
| Carnivorous Fish Consumption | 2.25 |
| Carnivorous Fish Emigration | 1.60 |
| Carnivorous Fish Respiration | 1.06 |

with fish in the bed during the summer.  As mentioned previously, a principal characteristic of eelgrass systems that controls its dynamics is the extremes in temperature fluctuations that occur within the bed especially during the summer.  When the depth of water over beds in the Beaufort area is very shallow, 0.3m or less, the water temperature frequently exceeds 32°C in the summer (H. Stuart, unpublished data), therefore causing extreme metabolic stress for the fish in the beds (Adams, 1976b).  A more appropriate simulation, therefore, would increase respiration rate of the fish community during the summer when average temperatures approach the upper 20's.  In our model, respiration is calculated as a constant fraction of the temperature-dependent feeding rate times standing crop.  When the standing crop of fish is declining during the latter half of the summer, respiration also is declining due to both lower standing crop and the inflection in the temperature-dependent feeding rate at around 30°C.  In future iterations of the model, the respiration rate of fish will be increased both as a direct function of temperature and as a function of actual feeding rate.  This increase in respiration rate will produce lower P/C ratios during periods of higher temperatures, even if feeding rate is also high. Temperature-dependent respiration then will simulate a basal rate and increased rates during periods of intense feeding (specific dynamic action).

In our simulations, the omnivorous fish compartment had, as part of its dynamics, a variable assimilation efficiency that was dependent on the proportions of animal matter and detritus consumed. The higher the proportion of animal material consumed, the higher the assimilation rate. When prey is scarce in the simulated or in the natural environment, omnivorous fish consume mainly detritus, but in our simulations the assimilation of detritus is too high. For example, consumption of 30% animal matter results in a 68% assimilation efficiency. Production and therefore the ecological efficiency of the omnivorous fish compartment is high because the assimilation efficiency is too high. The true assimilation efficiency of fish consuming mainly detritus may be closer to 30-40%.

The respiration of the microbial and infaunal compartments, which was a function of temperature, also was monitored. The simulated annual respiration of the microorganisms was 464 and of the infauna, 100 gms/m$^2$. Thayer $et$ $al.$ (1975) estimated that the microbial compartment of an eelgrass bed may respire 47% of the total respiration of all of the invertebrate and microbial compartments combined. The total respiration of the infauna, microbes, and epifauna in our model was 590 gms/m$^2$, and microbes contributed 70% of this total.

## RESPONSES TO PERTURBATIONS

The third approach in examining the credibility of the model is to determine if the model simulates believable responses to small perturbations. According to Patten $et$ $al.$ (1975), such disturbances applied to a model should not cause gross structural alteration, and organizational integrity should be maintained. Perturbations that were applied to flows between compartments in the model include reduction of fish and epifaunal growth, increased cropping of fish, and increased mortality of grass. These types of changes in the flow rates help to illustrate the importance of such biotic influences as competition for space, competition for food, and predation. In general, drastic alterations, except for increases in mortality rate of the grass, produce only minor responses in the simulation. The response of the epifauna and omnivorous fish compartments to these perturbations are shown in Figures 8 and 9 and Table 8. Increased cropping by outside predators on the omnivorous fish increases the average epifaunal standing crop by 5% and most of this increase occurs from June-September. The effect of no growth of omnivorous fish causes the average standing crop if epifauna to increase by 10%. When the mortality rate of eelgrass is increased to twice the normal rate, the average standing crop of epifauna and omnivorous fish declines by 82% (Table 8). No growth by the carnivorous fish increases the average standing crop of omnivorous fish

64

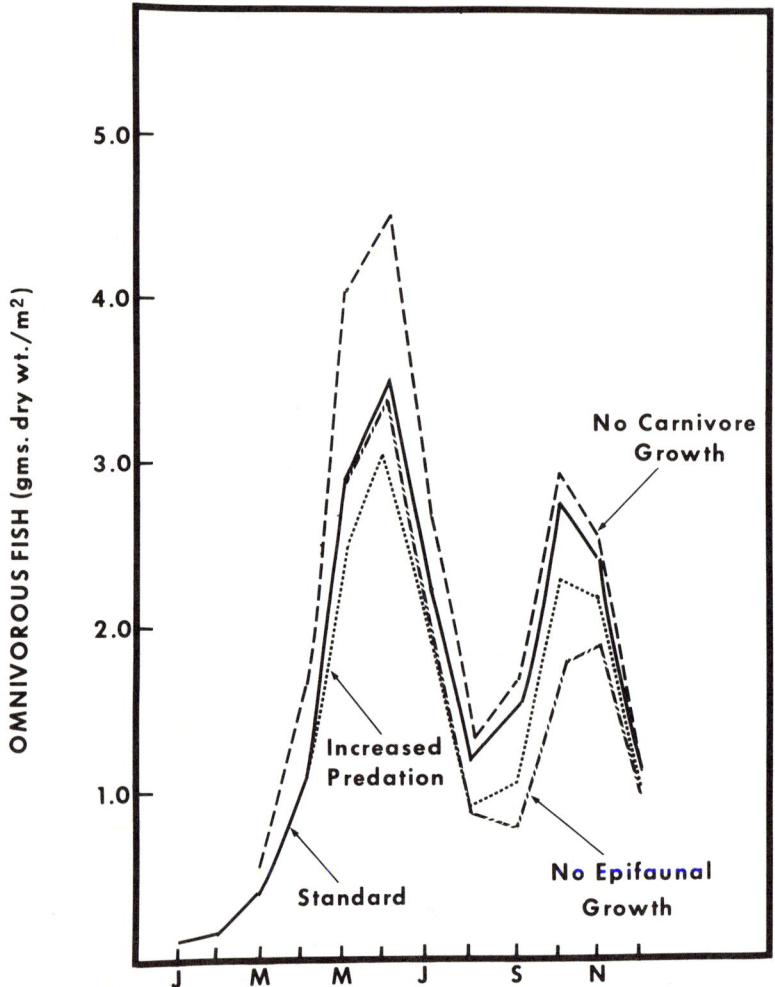

Fig. 8. Simulated temporal sequences in the standing crop of
omnivorous fish in response to various perturbations.

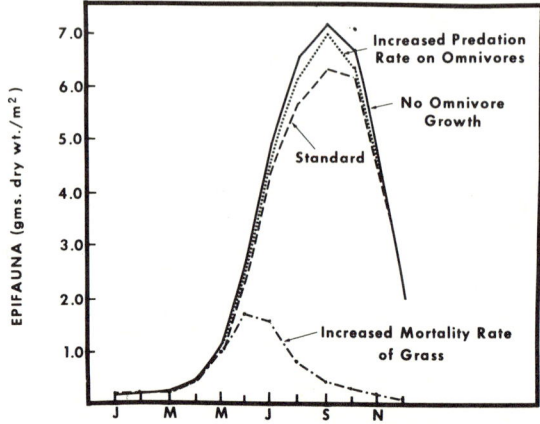

Fig. 9. Simulated temporal sequences in standing crop of
epifauna in response to various perturbations.

Table 8. Responses of model ecological parameters of the eelgrass community to various perturbations.

| Perturbation | | Responses | | | |
|---|---|---|---|---|---|
| Type | Degree | Epifauna | Omnivorous Fish | | |
| | | SC | SC | Prod | P/C |
| | | ---------------------- % Change --------------- | | | |
| Omnivorous Fish | | | | | |
|   Predation by outsiders | +80 | +5 | -5 | -1 | +7 |
|   None Present | NA | +10 | NA | NA | NA |
| No Carnivorous Fish | NA | +2 | +18 | NO | NO |
| No Epifauna | NA | NA | -15 | -17 | -10 |
| Eelgrass Mortality Rate | +200 | -82 | -82 | -80 | -10 |

NA - Not applicable      NO - No observations

by 18% with the majority of this occurring during the spring. The effect of an 80% increase in cropping of omnivores by the outside predators is to decrease the omnivorous fish standing crop by 5%. Also, the effect of not allowing the epifauna to grow is to reduce the average omnivorous fish standing crop by 15%.

Increased eelgrass mortality, increased cropping of both fish by outside predators, and no epifaunal growth, all have an effect on the P/C ratio of the omnivorous fish (Table 8). When eelgrass mortality increases and less grass is present, epifaunal standing crop is reduced. Omnivorous fish, therefore, consume a greater percentage of detritus which results in a lowered assimilation efficiency and therefore a lowered P/C ratio. When cropping of both fish by outside predators is increased, omnivorous fish standing crop and consumption decreases, and ecological efficiency increases for three reasons. First, lowered standing crops decrease the R/C ratio via the habitat factor's effect on consumption rate. Second, the reduced cropping pressure on animal matter increases standing crops of food items (animal matter) and increases the assimilation relative to respiration via the food factor's effect on assimilation efficiency. Finally, increased cropping pressure on the carnivorous fish increases the consumption relative to respiration of omnivorous fish because of the reduced competition for habitat and animal matter. When there is no epifaunal growth, the omnivorous fish consume a greater percentage of detritus, thus lowering the ecological efficiency.

## DREDGING EXPERIMENTS

We also simulated the potential effects of increased turbidity due to dredging on the growth rate of grass. We did not attempt to simulate other impacts of dredging such as siltation. The growth of grass was decreased by varying its growth rate at different times of the year. Since actual dredging operations in a particular area are usually temporary, our experiments consisted of a series of annual simulations in which grass growth rate was reduced for short time intervals at different times of year. This was done in order to look at relative effects of dredging operations conducted during different seasons. We reduced the rate of grass growth by 25% during one quarter of the year and held the growth rate of grass normal the remaining three quarters. This was done in turn for each quarter, and in a final simulation the growth rate was reduced by 25% for the entire year.

Table 9 summarizes the results of these five simulation experiments in terms of the principal dynamic features of the model. These simulations indicate that the model is sensitive to the timing of perturbations. The largest reductions in all of the monitored compartments occurred when the perturbation was in the first quarter. This is the time of the initial and the most rapid vegetative growth phase of the grass. Perturbations during the remainder of the year, in general, caused only a 1–7% change in all parameters. The effects of perturbations seemed to be the smallest in the fourth quarter when the grass was in a mortality phase. The actual effect of

Table 9. Decreases in selected variables due to simulated dredging effect, eelgrass growth rate reduced 25% in each quarter.

| Dependent variable | Season of Growth Reduction | | | | |
|---|---|---|---|---|---|
| | 1 | 2 | 3 | 4 | All |
| | % decrease | | | | |
| **Epifauna** | | | | | |
| Consumption | 12 | 2 | 5 | 1 | 49 |
| Respiration | 10 | 2 | 5 | 1 | 47 |
| Production | 16 | 3 | 5 | 2 | 53 |
| P/C | 0 | 0 | 0 | 0 | 0 |
| **Omnivorous fish** | | | | | |
| Consumption | 26 | 5 | 3 | 1 | 63 |
| Respiration | 28 | 6 | 4 | 1 | 63 |
| Assimilation | 27 | 5 | 3 | 1 | 65 |
| Emigration | 30 | 5 | 2 | 1 | 70 |
| Production | 36 | 5 | 3 | 7 | 66 |
| P/C | 0 | 0 | 0 | 0 | 7 |
| **Carnivorous fish** | | | | | |
| Consumption | 27 | 3 | 0 | 0 | 52 |
| Respiration | 22 | 3 | 0 | 0 | 38 |
| Assimilation | 27 | 3 | 0 | 0 | 52 |
| Production | 33 | 5 | 1 | 0 | 68 |
| P/C | 9 | 0 | 0 | 0 | 33 |

dredging during the last quarter on the growth of eelgrass the following year (such as production of underground rhizomes) may be significant, and field studies are necessary to determine the extent of such effects. Turbidity had a large effect on the production of epifauna and fish but little effect on the P/C ratio because when production decreased, consumption also decreased. During the year-long 25% reduction experiment, the ecological efficiency of the carnivorous fish had a much larger decrease than that of the omnivorous fish. When grass and animal matter became scarce, omnivorous fish consumed detritus to compensate in their diet, while the carnivores did not and continued to respire but not to incorporate energy.

## Summary

We believe that our trophic dynamics model satisfied to a large extent the criteria for credibility and utility. There remain, however, major sources of inconsistency in the simulations. First, we know that the computational scheme for respiration must be modified in order to simulate more realistically the relations between respiration, temperature, and consumption rate of the omnivorous and carnivorous fish. Second, the relation between assimilation efficiency and the proportions of animal matter and detritus in the diet of omnivorous fish must be adjusted. Both of these problems can be handled with little difficulty and will be included in subsequent iterations of the model using the IBM 1130 computer and the CSMP II program package. We believe these changes will provide significant improvement in the resolution of the dynamics of the fish of eelgrass communities in response to temperature. The third inconsistency relates to the overriding importance of the eelgrass itself in determining the habitat and thus the dynamics of the remaining trophic levels of the community. The resolution of the eelgrass response to temperature as this response interacts with factors such as diurnal temperature and tidal exposure extremes is essential to an adequate predictive capacity. Such resolution will require a larger computer.

The model implies that eelgrass standing crop is the single most important factor in determining the dynamics of the eelgrass community. The majority of community responses to temperature, for example, can be explained as an indirect response via the temperature effect on eelgrass. State-of-the-art resolution of the response of seagrasses to natural and man-induced fluctuations in temperature and other abiotic factors, such as light and tidal regimes, is essential to predict adequately the fate of seagrass communities in our changing coastal environments.

## Acknowledgments

We thank G. W. Thayer for his helpful discussions and for making unpublished data on the growth of eelgrass available to us. The helpful criticisms of the manuscript by D. R. Colby, M. A. Kjelson, and W. E. Schaaf, the assistance in FORTRAN programming by J.E. Hollingsworth, and the help in preparation of the figures by H. R. Gordy and M. B. Murdoch, all our NMFS colleagues, are gratefully appreciated. This research was supported by National Marine Fisheries Service and U.S. Energy Research and Development Administration Agreement AT(49-7)-5.

## References

Adams, S. M. 1976a. The ecology of eelgrass, *Zostera marina* (L.), fish communities. I. Structural analysis. J. Exp. Mar. Biol. Ecol.: 269-291.

Adams, S. M. 1976b. The ecology of eelgrass, *Zostera marina* (L.), fish communities. II. Functional analysis. J. Exp. Mar. Biol. Ecol.: 293-311.

Adams, S. C. 1976c. Feeding ecology of eelgrass fish communities. Transactions of the American Fisheries Society 4:514-519.

Blegvad, H. 1916. On the food of the fish in the Danish waters within the Skaw. Reports of the Danish Biological Station 23:17-27.

Cameron, J. N. 1969. Growth, respiratory metabolism and seasonal distribution of juvenile pinfish (*Lagodon rhomboides* Linnaeus) in Redfish Bay, Texas: 19-36.

den Hartog, C. 1970. *The Sea-grasses of the World*. North-Holland Publishing Company, London.

Dillon, R. C. 1971. A comparative study of the primary productivity of estuarine phytoplankton and macrobenthic plants. Ph.D. thesis, University of North Carolina, Chapel Hill.

Ferguson, R. L. and M. B. Murdoch. 1975. Microbial ATP and organic carbon in sediments of the Newport River estuary, North Carolina. In: *Estuarine Research* (L. E. Cronin, ed.). Vol. 1, pp. 229-250. Academic Press, New York.

Ferguson, R. L., G. W. Thayer, and T. R. Rice. 1977. Chapter 2. Marine primary producers. In: *Functional Adaptations of Marine Organisms* (F. J. Vernberg and W. Vernberg, eds.). In press.

Godfrey, M. M. 1970. Seasonal changes in the macrofauna of an eelgrass (*Zostera marina*) community near Beaufort, North Carolina. M.S. thesis, Duke University, Durham, North Carolina.

Hoese, H. D. and R. S. Jones. 1963. Seasonality of larger animals in a Texas grass community. Publications of the Institute of Marine Science, University of Texas 9: 37-46.

Kikuchi, T. 1961. An ecological study on animal community of *Zostera* belt, in Tomioka Bay, Amakusa, Kyushu. I. Fish fauna. Records of Oceanographic Works of Japan 5: 211-223.

Kjelson, M. A. and G. N. Johnson. 1973. Description and evaluation of a portable drop-net for sampling nekton populations. Proceedings of the 27th Annual Conference of the Southeastern Association of Game and Fish Commissioners, pp. 653-662.

Mann, K. H. 1965. Energy transformations by a population of fish in the River Thames. Journal of Animal Ecology 34: 253-275.

Patten, B. C., B. A. Egloff, and T. H. Richardson. 1975. Total ecosystem model for a cove in Lake Texoma. In: *Systems Analysis and Simulation in Ecology* (Patten, B. C., ed.).Vol. III, pp. 205-421. Academic Press, New York.

Penhale, P. A. 1976. Primary production, dissolved organic carbon excretion and nutrient transport in an epiphyte-eelgrass (*Zostera marina*) system. Ph.D. thesis, North Carolina State University, Raleigh, North Carolina.

Peters, D. S., M. T. Boyd, and J. C. Devane, Jr. 1976. The effect of temperature, salinity and food availability on the growth-and-food-conversion efficiency of postlarval pinfish. In: *Thermal Ecology* (G. W. Esch and R. W. McFarlane, eds.). II, pp 106-112. ERDA Symposium Series 40, CONF-750425.

Phillips, R. C. 1969. Temperate grass flats. In: *Coastal Ecological Systems of the United States* (H. T. Odum, B. J. Copeland, and E. A McMahan, eds.). pp. 737-773. Vol. 2, Report of the Federal Water Pollution Control Administration, Contract RFP 68-128.

Ried, G. K., Jr. 1954. An ecological study of the Gulf of Mexico fishes, in the vicinity of Cedar Key, Florida. Bull. Mar. Sci. Gulf Carib. 4: 1-94.

Service, J. 1972. *A User's Guide to the Statistical Analysis System*. Student Supply Stores, N.C. State Univ., Raleigh, N.C.

Slobodkin, L. B. 1960. Ecological energy relationships at the population level. Am. Natur. 94: 213-236.

Stuart, H. H. 1975. Distribution and summer energetics of invertebrate epifauna in an eelgrass (*Zostera marina*) bed. M. S. thesis, N.C. State Univ., Raleigh, N.C.

Thayer, G. W., D. E. Hoss, M. A. Kjelson, W. F. Hettler, Jr., and M. W. LaCroix. 1974. Biomass of zooplankton in the Newport River estuary and the influence of postlarval fishes. Chesapeake Sci. 15: 9-16.

Thayer, G. W., S. M. Adams, and M. W. LaCroix. 1975. Structural and functional aspects of a recently established *Zostera marina* community. In: *Estuarine Research* (L. E. Cronin, ed.). Vol. 1, pp. 518-540. Academic Press, New York.

Thayer, G. W., O. W. Engel, and M. W. LaCroix. 1977. Seasonal distribution and changes in the nutritive quality of living, dead and detrital fractions of *Zostera marina*. (in press).

Williams, R. B. 1973. Nutrient levels and phytoplankton productivity in the estuary. In: *Proceedings of the Coastal Marsh and Estuary Management Symposium* (R. A. Chabreck, ed.), pp. 59-89. Louisiana State University, Baton Rouge.

Wood, E. J. F., W. E. Odum, and J. C. Zieman. 1969. Influence of seagrasses on the productivity of coastal lagoons. In: *Coastal Lagoons, a Symposium* (Ayala-Castanares and F. B. Phleger, eds.) pp. 495-502. Universidad Nacrional Autonoma de Mexico, Ciudad Universitaria.

# Appendix

I = BLOCK NUMBER OF COMPUTATIONAL ELEMENT

MTRX2(I) = BLOCK NUMBER OF FIRST INPUT

MTRX3(I) = BLOCK NUMBER OF SECOND INPUT

MTRX4(I) = BLOCK NUMBER OF THIRD INPUT


C(J) CONTAINS THE CURRENT VALUE OF BLOCK J

PAR1(I) CONTAINS PARAMETER 1 OF BLOCK I

PAR2(I) CONTAINS PARAMETER 2 OF BLOCK I

PAR3(I) CONTAINS PARAMETER 3 OF BLOCK I

C(I) CONTAINS THE OUTPUT OF BLOCK I

Subroutine 1 generates a sequence of "random numbers" between plus and minus 1. The user specifies parameter 1 of the block (I) assigned as type 1 prior to each simulation. This allows generation of up to 99 different sequences of "random numbers"

JX = PAR1(I)

J = JX

IRAND = C(J)

IR = 259 * (IR + IRAND)

C(I) = FLOAT (IR)/32767.0

where 32767.0 is the largest integer specified by the IBM 1130 computer.

Subroutine 2 receives temperature and generates temperature-dependent feeding rate. Input 1 of block I, specified type 2, is the block whose output is temperature. The user defines the 3 parameters of block I for the computation:

J = MTRX2(I)

ACUTM = C(J)

TDFR = ACUTM * PAR2(I)-(PAR3(I) + ACUTM**2*PAR1(I))

IF (TDFR) 1,2,2

1   TDFR = 0.

2   C(I) = TDFR

Subroutine 3 receives temperature-dependent growth rate, standing crop of eelgrass, and a season cue from the blocks specified as inputs 1, 2, and 3 of block I specified as type 3 and generates net production of eelgrass. The user specifies the first 2 parameters of block I which are growth-reducing factors for season and for high standing crop:

J = MTRX2(I)

K = MTRX3(I)

L = MTRX4(I)

IF (C(L)) 1,1,2

1   D = 1.

GO TO 3

2   D = PAR1(I)

3   IF (C(K) - PAR3(I)) 4,4,5

```
4  E = 1.

   GO TO 6

5  E = PAR2(I)

6  C(I) = C(J) * C(K) * D * E
```

Subroutine 4 receives temperature, is assigned 3 parameters, and generates microbial respiration rate.

```
   J = MTRX2(I)

   C(I) = (C(J) = PAR1(I)) * (C(J) ** 2 = PAR2(I))/ PAR3(I)
```

Subroutine 5 receives microbial respiration rate and temperature, is assigned 2 parameters, and generates infaunal respiration rate.

```
   J = MTRX2(I)

   K = MTRX3(I)

   C (I) = C(J) * (C(K) * PAR1(I) + PAR2(I))
```

# Hydrodynamic and Water Quality Model of the Scheldt Estuary

Jacques C.J. Nihoul
François C. Ronday
J. Smitz
University of Liege
Belgium

**G. Billen**
University of Brussels
Belgium

The flocculation and sedimentation of the suspended load in the central part of the Scheldt Estuary is explained with the help of a two-dimensional width-integrated hydrodynamic model.
A one-dimensional (cross-section integrated) active dispersion model based on the former is then used to describe the detailed oxydo-reduction kinetics of the two zones, respectively, upstream and downstream of the sedimentation region.

## Introduction

The Scheldt estuary, 120 km in length, is a partially stratified estuary which is severely polluted upstream by urban and industrial discharges amounting to some 150 tons of organic matter per year.

A typical diagram of the longitudinal salinity profile is shown in Figure 1 for a river flow rate characteristic of the summer situation. At higher (lower) flow rates, the diagram is not significantly modified but simply shifted some ten kilometers downstream (upstream). Critical salinity values of 1-2 $^o/oo$ are classically obtained in central sections of the estuary (*i.e.*, between km 40 and km 60). At such values of the salinity, intense flocculation and sedimentation take place, and the suspended load decreases abruptly as shown in Figure 1 by the profiles of turbidity and COD. Variations of the river flow rate and the subsequent displacement of the region of critical salinity reflect in the sedimentation and tend to spread the deposition of suspended matter over the central part of the estuary. It is found, however, that bottom sediments tend to accumulate in a narrower region than one might expect, and this is explained in the following with the help of a two-dimensional width-integrated residual circulation model.

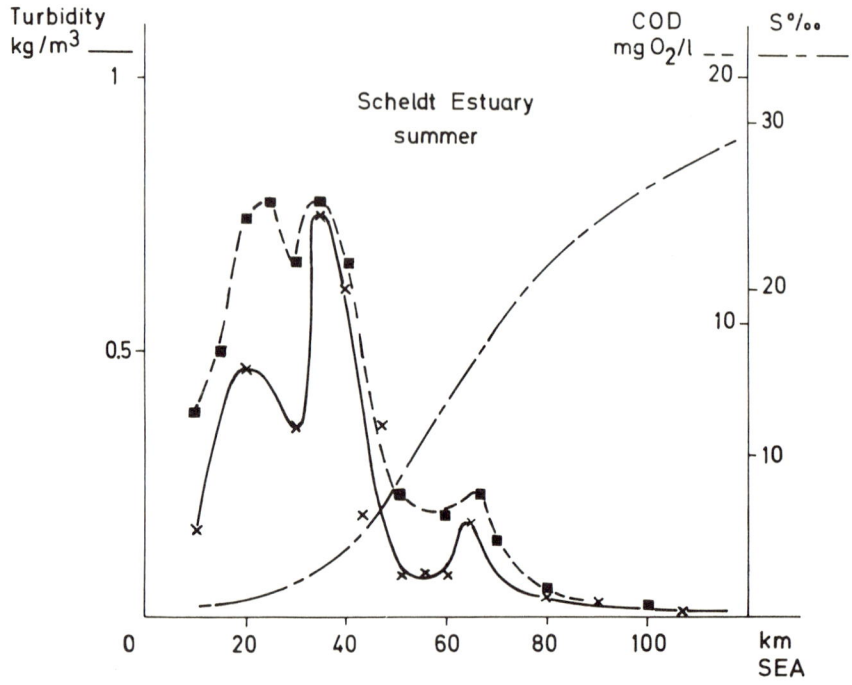

Fig. 1. Longitudinal profiles of salinity, turbidity, and COD in the Scheldt Estuary in a typical summer situation.

Flocculation and sedimentation divide the estuary into two distinct zones. In the upstream part, the organic load induces bacterial proliferation and degradation, resulting in complete exhaustion of the oxygen, followed in anaerobic conditions by successive reduction of other oxidizers, producing toxic elements such as $Mn^{++}$, $NH_4$, $Fe^{++}$ and $H_2S$ and a low redox potential. In the downstream part, the combined effect of reaeration and dilution by sea water allows the progressive reoxidation of the different oxidizers and the reappearance of oxygen. These observations are explained in the following with the help of a one-dimensional (cross-section integrated) "active dispersion" model which takes into account the detailed oxydo-reduction kinetics.

## Residual Circulation in the Scheldt

The residual circulation can be defined as the mean circulation over a period of time sufficiently long to smooth out tidal oscillations and transitory wind currents. It is determined by the river flow rate and the gravity currents associated with the stratification (Nihoul, 1975).

The residual circulation can be described by a two-dimensional model integrated over the estuary's width. If $u_1$ and $u_3$ denote, respectively, the horizontal component of the stream and the vertical component of the velocity vector, if $a$ represents buoyancy (the relative density difference multiplied by gravity), and if $V_1$, $V_3$, and A represent their respective integrals over the width, the basic hydrodynamic equations can be written (Nihoul and Ronday, 1976):

$$\frac{\partial V_1}{\partial x_1} + \frac{\partial V_3}{\partial x_3} = 0 \tag{1}$$

$$\frac{\partial A}{\partial t} + \frac{\partial}{\partial x_1} (L^{-1}V_1 A) + \frac{\partial}{\partial x_3} (L^{-1}V_3 A)$$

$$= \frac{\partial}{\partial x_1} (\Lambda_1 \frac{\partial A}{\partial x_1}) + \frac{\partial}{\partial x_3} (\Lambda_3 \frac{\partial A}{\partial x_3}) \tag{2}$$

$$A = - L \frac{\partial \overline{q}}{\partial x_3} \tag{3}$$

$$\frac{\partial V_1}{\partial t} + \frac{\partial}{\partial x_1} (L^{-1}V_1 V_1) + \frac{\partial}{\partial x_3} (L^{-1}V_1 V_3)$$

$$= - L \frac{\partial \overline{q}}{\partial x_1} + \frac{\partial}{\partial x_1} (N_1 \frac{\partial V_1}{\partial x_1}) + \frac{\partial}{\partial x_3} (N_3 \frac{\partial V_1}{\partial x_3}) \tag{4}$$

where L is the width and $\overline{q}$ the width-averaged of

$$q = \frac{p}{\rho_m} + gx_3 \tag{5}$$

(the difference between the pressure and the hydrodynamic pressure at the constant reference water density $\rho_m$, divided by $\rho_m$). $\Lambda_1$, $\Lambda_3$, $N_1$, and $N_3$ are "shear effect and turbulent" dispersion coefficients. If now < > denotes a time-average over a time $\theta$ covering several tidal periods and if

$$U_i = <V_i> \qquad i = 1,2 \tag{6}$$

$$V_i = U_i + W_i \qquad i = 1,2 \tag{7}$$

$$B = <A> \tag{8}$$

$$A = B + C \tag{9}$$

$$\pi = <\overline{q}> \tag{10}$$

one has, integrating eqs. (1) to (4) over $\theta$ (and assuming $\theta$ sufficiently large to neglect the contribution of the time derivative):

$$\frac{\partial U_1}{\partial x_1} + \frac{\partial U_3}{\partial x_3} = 0 \tag{11}$$

$$\frac{\partial}{\partial x_1} (L^{-1} U_1 B) + \frac{\partial}{\partial x_3} (L^{-1} U_3 B)$$

$$= \frac{\partial}{\partial x_1} \left\{ \Lambda_1 \left| \frac{\partial B}{\partial x_1} + L^{-1} <-W_1 C> \right\} + \frac{\partial}{\partial x_3} \left\{ \Lambda_3 \frac{\partial B}{\partial x_3} + L^{-1} <-W_3 C> \right\} \tag{12}$$

$$\frac{\partial}{\partial x_1} (L^{-1} U_1 U_1) + \frac{\partial}{\partial x_3} (L^{-1} U_1 U_3) = -L \frac{\partial \pi}{\partial x_1}$$

$$+ \frac{\partial}{\partial x_1} \left\{ N_1 \frac{\partial U_1}{\partial x_1} + L^{-1} <-W_1 W_1> \right\} + \frac{\partial}{\partial x_3} \left\{ N_3 \frac{\partial U_1}{\partial x_3} + L^{-1} <-W_3 W_1> \right\} \tag{13}$$

$$B = - L \frac{\partial \pi}{\partial x_3} \tag{14}$$

One can see that the $\theta$ -average of the quadratic terms gives two contributions: the first is related to the product of the means while the second represents the mean product of the fluctuations around the mean. This contribution has a structure which is analogous to a turbulent stress, and it is found to have similarly an effect of dispersion which can be lumped with shear effect and turbulence and parameterized with the help of appropriate new dispersion coefficients $K_1$ , $K_3$ , $M_1$ , and $M_3$ .

One sets

$$\Lambda_1 \frac{\partial B}{\partial x_1} + L^{-1} <-W_1 C> = K_1 \frac{\partial B}{\partial x_1} \tag{15}$$

$$\Lambda_3 \frac{\partial B}{\partial x_3} + L^{-1} <-W_3 C> = K_3 \frac{\partial B}{\partial x_3} \tag{16}$$

$$N_1 \frac{\partial U_1}{\partial x_1} + L^{-1} <-W_1 W_1> = M_1 \frac{\partial U_1}{\partial x_1} \tag{17}$$

$$N_3 \frac{\partial U_1}{\partial x_3} + L^{-1} <-W_3 W_1> = M_3 \frac{\partial U_1}{\partial x_3} \tag{18}$$

with, according to observations MKS (Ronday, 1975),

$$K_1 \sim 10^2 \; ; \; K_3 \sim 10^{-2} \; ; \; M_1 \sim 10^3 \; ; \; M_3 \sim 10^{-1} \; .$$

Clearly, $K_1$, $K_3$, $M_1$, and $M_3$ must be regarded as "control" parameters to be determined by inspection of the data base and sideways modeling investigation. In general, even if they can be taken, in the first instance, as independent of $x_3$, they are functions of $x_1$ (Nihoul and Ronday, 1976). If $\ell_1$ and $\ell_3$ represent characteristic length scales of horizontal and vertical variations of the velocity field, one must have, by eq. (11):

$$\frac{U_1}{\ell_1} \sim \frac{U_3}{\ell_3}$$

In eq. (13), the terms in the left-hand side are of the order

$$\frac{U_1 U_1}{L \, \ell_1}$$

The second term in the right-hand side, using (17), is of the order of

$$\frac{M_1 U_1}{\ell_1}$$

while the last term in the right-hand side, using (18), is of the order of

$$\frac{M_3 U_1}{\ell_3}$$

Taking (Ronday, 1975), in MKS units

$$\ell_3 \sim 10 \; ; \; \ell_1 \sim 10^4 \; ; \; \left| L^{-1} U_1 \right| \sim 10^{-2}$$

one can see that vertical dispersion is $10^2$ to $10^3$ times larger than advection and horizontal dispersion and must be consequently balanced by

$$-L \frac{\partial \pi}{\partial x_1}$$

Such a simplification is, in general, not possible in eq. (12), which contains no term analogous to the pressure gradient and where the vertical dispersion term is comparatively much less important, vertical gradients of salinity being considerably less marked than vertical gradients of velocity.

Eq. (11) suggests the introduction of a residual stream function $\psi_o$ so that

$$U_1 = \frac{\partial \psi_o}{\partial x_3} \tag{19}$$

$$U_3 = - \frac{\partial \psi_o}{\partial x_1} \tag{20}$$

Eliminating $U_1$, and $U_3$, and B, one obtains

$$\frac{\partial \psi_o}{\partial x_3} \frac{\partial^2 \pi}{\partial x_1 \partial x_3} - \frac{\partial \psi_o}{\partial x_1} \frac{\partial^2 \pi}{\partial x_3^2} = \frac{\partial}{\partial x_1}\left\{ K_1 \frac{\partial}{\partial x_1} (L \frac{\partial \pi}{\partial x_3}) \right\} \tag{21}$$

$$+ \frac{\partial}{\partial x_3}\left\{ K_3 \frac{\partial}{\partial x_3} (L \frac{\partial \pi}{\partial x_3}) \right\}$$

$$\frac{\partial}{\partial x_3} (M_3 \frac{\partial^2 \psi_o}{\partial x_3^2}) = L \frac{\partial \pi}{\partial x_1} \tag{22}$$

These equations constitute a complete system for the two variables $\psi_o$ and $\pi$ . They have been applied to the Scheldt estuary taking into account the special geometry of the basin (for instance, the function $L(x_1 , x_3)...$) and the appropriate boundary conditions.
    Figure 2 represents, for a section of the Scheldt going from the Ruppel River (approximately km 30 on Figure 1) to the sea, the lines of equal horizontal velocity calculated by the model for a flow rate equal to four times its low water value. Currents are expressed in m/sec. The curve in dot and dashes (-·-·-), noted $n = 4$, indicates the line of zero residual current. It is readily seen on the salinity scale printed at the top of the diagram (and corresponding to the same flow rate) that the residual current goes to zero at the bottom of the very place where critical salinity values and flocculation are observed. Sedimentation thus occurs in a region where bedload transport is negligible and sediments accumulate on the bottom.
    For reference, two curves in dashes (---), noted $n = 1$ and $n = 10$, respectively, indicate the lines of zero residual current for flow rates equal to the low water value and 10 times that value. The salinity scale is, of course, shifted accordingly, and the zone of precipitation appears to coincide in most cases with the region of zero bottom current. However, when the flow rate increases or decreases and comes back to typical values ($n \sim 4$), bottom currents appear in the regions where they were zero for lower and higher

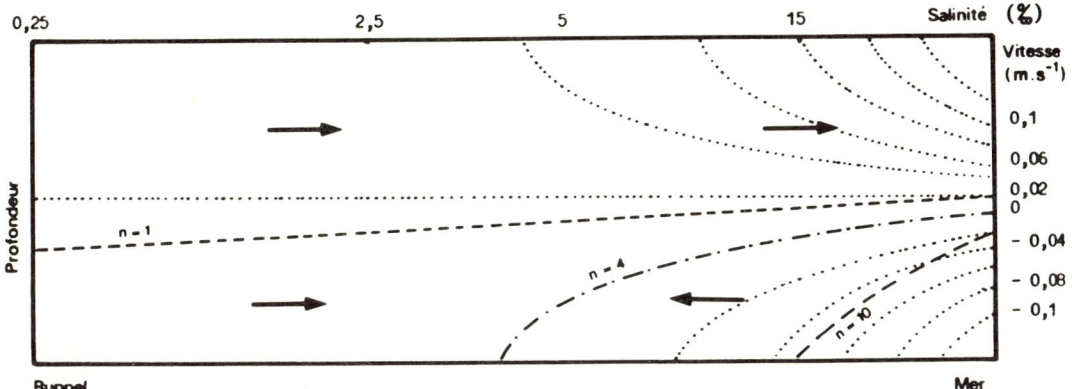

Fig. 2. Lines of equal width-averaged horizontal velocity calculated by the model for a typical river flow rate (4 times the low water value).

flow rates, and their directions are such that, as indicated by the arrows on Figure 2, they tend to carry back the freshly deposited sediments to the median zone. This explains the accumulation of muddy sediments in a relatively limited section of the Scheldt estuary.

## Oxidation--Reduction Processes in the Scheldt

Owing to the importance of the organic load, the dissolved oxygen concentration is not sufficient to describe the quality of the water of the Scheldt, as it is done in most classical models of river pollution. Indeed, other oxidants than oxygen are used by heterotrophic metabolisms ($MnO_2$, $NO_3^-$, $Fe(OH)_3$, $SO_4^{--}$), and their related reductors ($Mn^{++}$, $NH_4^+$, $Fe^{++}$, $HS^-$) are produced in the upstream zone of the estuary and reoxidized in the downstream zone. A typical longitudinal profile of the concentration of these redox substances is shown as an example in Figure 3.

The bacterial heterotrophic activity controls these oxidation-reduction processes by imposing a flux of electrons ($H(x, t)$) on the system, according to the equation

$$CH_2O + H_2O \xrightarrow{\text{bact.}} CO_2 + 4H^+ + 4\ e^- \tag{23}$$

The electron flux induces one or several of the following relations:

a) $\quad 4\ e^- + O_2 + 4H^+ \rightleftharpoons 2H_2O \qquad\qquad X_1 \tag{24}$

b) $\quad 8\ e^- + NO_3^- + 10H^+ \rightleftharpoons NH_4^+ + 3H_2O \qquad X_2 \rightleftharpoons Y_2 \tag{25}$

c) $\quad 2\ e^- + MnO_2 + 4H^+ \rightleftharpoons Mn^{++} + 2H_2O \qquad X_3 \rightleftharpoons Y_3 \tag{26}$

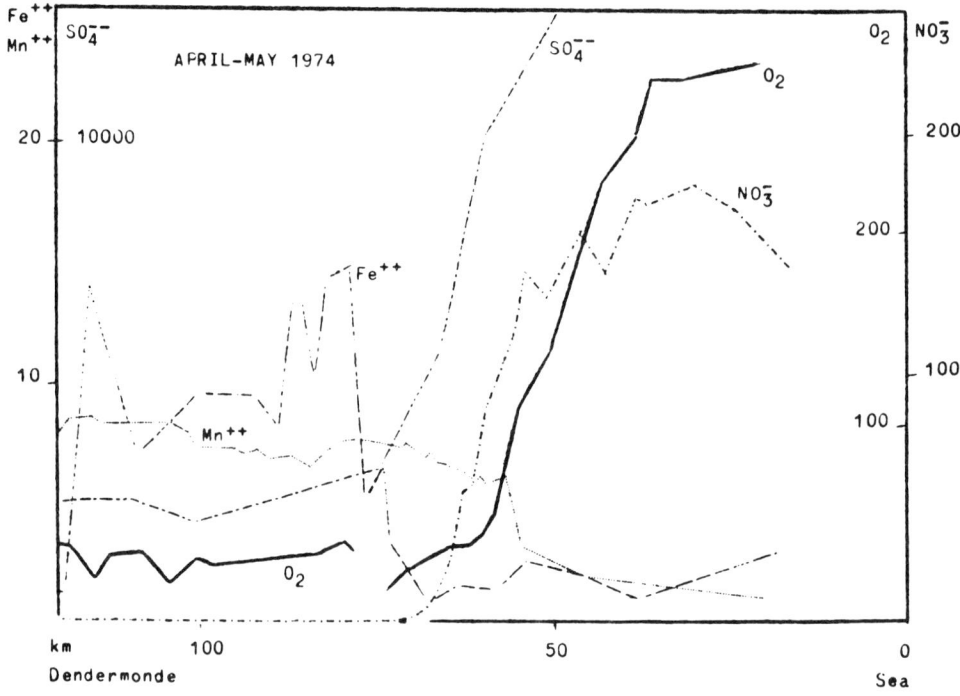

Fig. 3. Longitudinal profiles of concentrations of redox substances measured in the estuary.

d)  $1 \ e^- + Fe(OH)_3 + 3H^+ \rightleftharpoons Fe^{++} + 3H_2O$  $\quad X_4 \rightleftharpoons Y_4$ (27)

d')  $Fe^{++} + HCO_3^- \rightleftharpoons FeCO_3 + H^+$  $\quad X_4' \rightleftharpoons Y_4'$ (28)

e)  $14 \ e^- + 2SO_4^{--} + Fe^{++} + 16H^+ \rightleftharpoons FeS_2 + 8H_2O$ (29)

$$X_5 \rightleftharpoons Y_5$$

Some of these reactions (oxygen consumption, denitrification, nitrification, sulfato-reduction) are triggered by organisms, while others occur spontaneously. This distinction, however, is irrelevant in an overall budget.

In an ultimate management model of Scheldt water quality, the bacterial heterotrophic activity $H(x, t)$ should be deduced at each point as a function of the organic load and some physico-chemical variables. However, in such a complex system as the Scheldt estuary where the organic matter is so abundant that it does not constitute the limiting factor of bacterial activity and where the fate of organic matter is determined not only by its bacterial degradation, but mainly by flocculation and sedimentation processes, the form of this function is not very easy to assess. In particular, the naive assumption of most "Streeter and Phelps-like" models that bacterial activity is first order with respect to organic matter is surely not applicable. For studying only the effect of bacterial activity on water quality, it is, however, possible to circumvent this problem by directly measuring the bacterial activity all along a longitudinal profile of the estuary and imposing it as a control parameter on the system. Instantaneous bacterial activity has been estimated experimentally using the incorporation of $^{14}C$ bicarbonate in the dark with short incubation time (a few hours) in near *in situ* conditions. When the measured bacterial activity is expressed as an electron flux H(x, t), the problem is reduced to the distribution of this flux between the different chemical and biological processes described by Eqs. (23) to (29) above. As a first approximation, it is hypothesized that thermodynamic equilibrium is realized in every section of the estuary. Thus, if $X_i$ and $Y_i$ represent the concentrations of the oxidized and reduced forms, respectively, and if $E_h$ is the redox potential, one has the corresponding Nernst relations:

$$E_h = a_i + b_i \log \frac{X_i}{Y_i} \qquad (i = 1, 2, \ldots) \qquad (30)$$

where $a_i$ and $b_i$ are known constants. On a logarithmic diagram, Eqs. (30) are represented by linear functions as shown in Figure 4. It must be noted here that some kinetical limitations are already introduced into the model at this stage by choosing the Nernst relation corresponding to the equilibrium of metastable species, instead of the thermodynamically stable ones. This is the case for oxygen (see Sato, 1960 and Breck, 1972) and iron hydroxide.

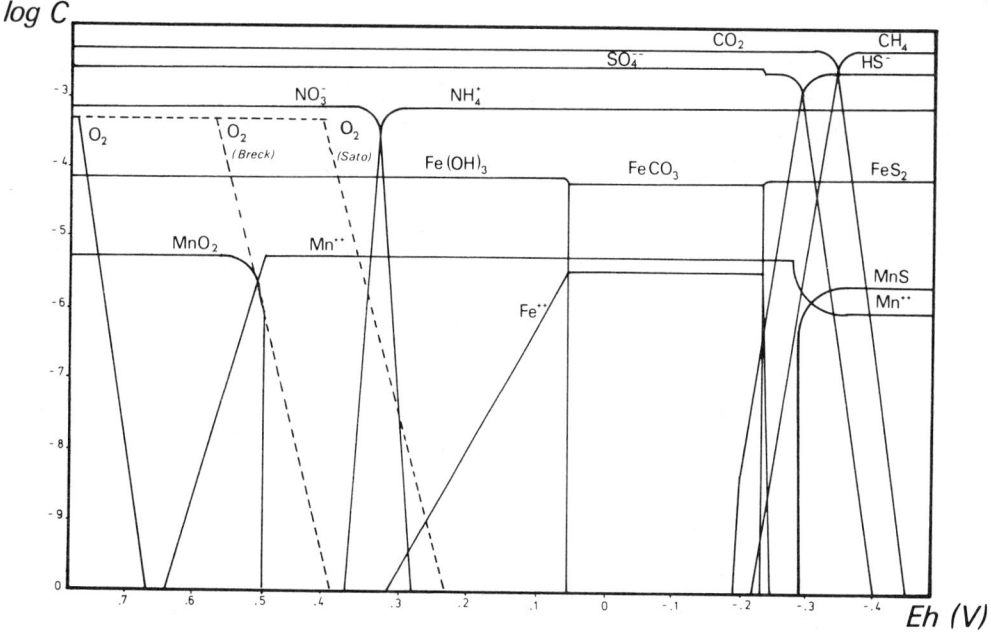

Fig. 4. Log-diagram of redox potential conditions in the hypothesis of local thermodynamic equilibrium.

The longitudinal distribution of any cross-section-averaged concentration can be described by an equation of the form (Nihoul and Ronday, 1976):

$$\mathcal{D}c \equiv \frac{\partial}{\partial t}(\mathcal{A}c) + \frac{\partial}{\partial x_1}(\mathcal{A}uc) - \frac{\partial}{\partial x_1}\left(\Lambda \frac{\partial}{\partial x_1}(\mathcal{A}c)\right) = P - D \tag{31}$$

where $\mathcal{A}$ is the cross-section, u the cross-section-averaged residual velocity, $\Lambda$ the dispersion coefficient and where P and D represent, respectively, the rates of production and destruction of c as a result of biochemical reactions.

If processes such as nutrient uptake by autotrophs, nutrient release by heterotrophs, and sedimentation of insoluble iron and manganese compounds are neglected as a first approximation, one can write for each $X_i$ and $Y_i$

$$\mathcal{D}X_i = P_i - D_i \qquad\qquad (i = 1, 2...) \tag{32}$$

$$\mathcal{D}Y_i = - P_i + D_i \qquad\qquad (i = 1, 2...) \tag{33}$$

and, by addition,

$$\mathcal{D}Z_i = 0 \qquad\qquad (i = 1, 2...) \tag{34}$$

where

$$Z_i = X_i + Y_i \qquad\qquad (i = 1, 2...) \tag{35}$$

Hence, the sum of the concentrations of the oxidized and reduced forms behaves as a passive scalar and can be determined by solving the "passive dispersion" equation (34) where the expressions of $\mathcal{A}$, u, and $\Lambda$ --as functions of $x_1$-- are substituted, u and $\Lambda$ being determined by the hydrodynamic model. Knowing the sum $X_i + Y_i$ and the ratio $X_i/Y_i$ (by Eqs. 30), one determines easily $X_i$ and $Y_i$ (i = 1, 2...). However, the local value of the redox potential $E_h$ in Eqs. (30) is unknown, and an additional relationship between $X_i$'s and $Y_i$'s is needed. This is provided by consideration of the auxiliary function

$$F(x,t) = \sum_i \nu_i X_i \tag{36}$$

where the $\nu_i$'s play the role of stoichiometric coefficients for the electron fluxes. From (32) and (33), one gets

$$\mathcal{D}F = \sum_i \nu_i P_i - \sum_i \nu_i D_i \tag{37}$$

Eq. (37) expresses the global oxydo-reduction budget, and the right-hand side reduces to reaeration (expressed in electron flux) minus H (electron flux due to bacterial activity).

Reaeration can be parameterized by the classical linear relation $K(X_{1,sat} - X_1)$ where K is a known constant.

The system of Eqs. (30), (34), and (37) is complete and allows the determination of the $X_i$'s and the $Y_i$'s as functions of x and t.

Figure 5 shows the solution obtained by this model. Comparison with the experimental profile of Figure 2 shows that the general trends of the longitudinal variations of oxidants concentrations are simulated. The principal discrepancy lies in the fact that our thermodynamic equilibrium model predicts that nitrification must be completed before manganese oxidation and oxygen reappearance. This is far from being the case because of the slow population dynamics of nitrifying bacteria in the Scheldt estuary (Billen, 1975).

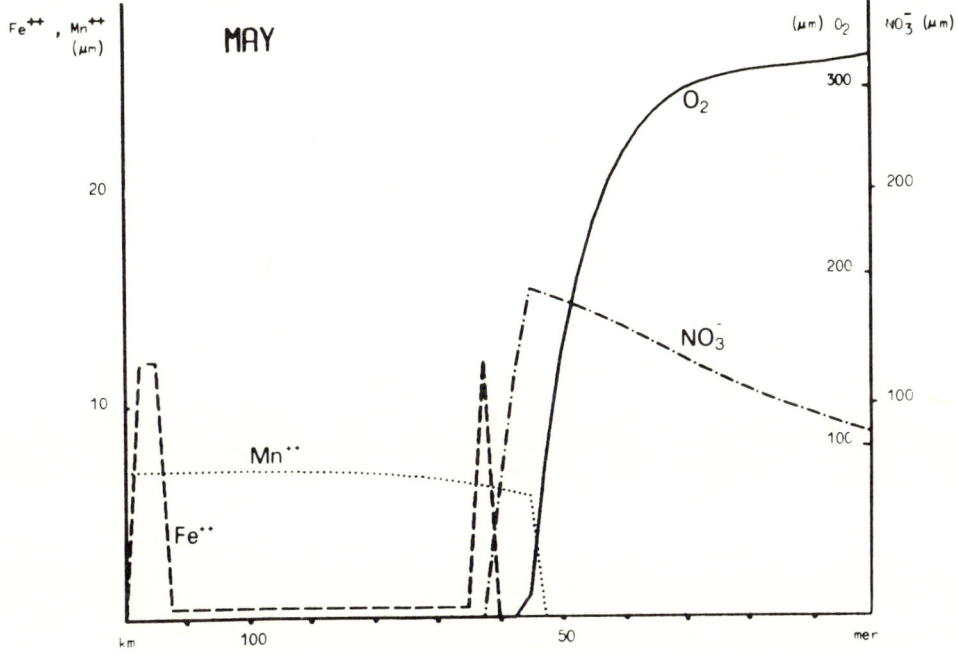

Fig. 5.  Computed cross-section averaged concentrations (thermodynamic equilibrium hypothesis).

However, in a second calculation step, starting from the equilibrium solution, it is possible to introduce a kinetic limitation on the nitrate production term, the $NH_4^+$ / $NO_3^-$ couple being then considered as being outside the thermodynamic equilibrium. As a first attempt following this approach, the nitrate production rate has been merely limited to a maximal value empirically determined. The results of the simulation are given in Figure 6, which shows that the relative position of the nitrate and oxygen profiles is far better simulated than with the equilibrium model. A more sophisticated kinetic model of nitrification in the Scheldt estuary is now in progress.

More details about the numerical model as well as additional simulation results, also including the effect of the thermal discharge of a nuclear power plant at Doel, can be found in Billen and

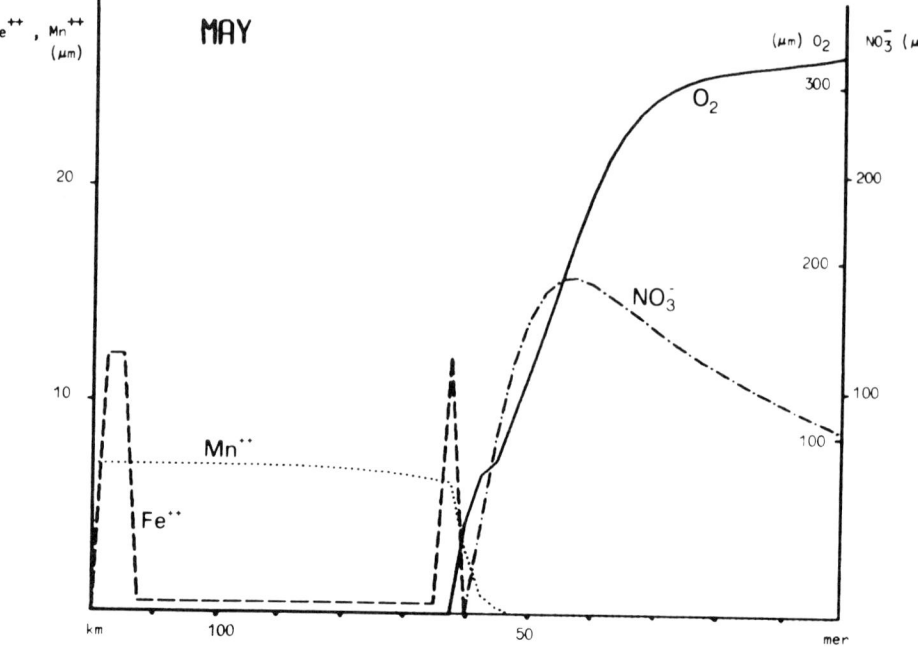

Fig. 6.   Computed cross-section averaged concentrations with kinetic limitation on nitrate production.

Smitz (1976).

## Acknowledgments

The present work was conducted within the scope of the Belgian National Environment Program sponsored by the Science Policy Administration, Office of the Prime Minister.

The authors wish to express their gratitude for the support of their research.

## References

Billen, G. and J. Smitz. 1976. Modéle mathématique de la qualité de l'eau dans un estuaire partiellement stratifié. Hydraul. (Paris) VIb (4).

Breck, W. G. 1972. Redox potentials by equilibration. J. Mar. Res. 30: 121-139.

Nihoul, J. C. J. 1975. *Modelling of Marine Systems*. Elsevier Pub., Amsterdam.

Nihoul, J. C. J. and F. C. Ronday. 1976. Modéle d'un estuaire partiellement stratifié. Application á la circulation résiduelle et á l'étude de l'envasement dans l' Escaut. In: *Recherches et Techniques au service de l'Environnement*, edited by the Consil Scientifique de l'Environnement, U. Leigh Cebedoc Pub. Liége.

Ronday, F. C. 1975. Etude de l'envasement et de la variation longitudinale du coefficient de dispersion dans des estuaires partiellement stratifiés. In: *Annales des Travaux Publics, 4.*

Sato, M. 1960. Oxidation of sulfide ore bodies, 1, Geochemical environment in terms of Eh and pH. Econ. Geol. 55: 928-961.

# Computer Simulation of Urban-Estuarine Interactions in the New Orleans Region

**Richard M. Eckenrod**
**John W. Day, Jr.**
**Leonard M. Bahr, Jr.**
Department of Marine Sciences
Center for Wetland Resources
Louisiana State University
Baton Rouge, Louisiana 70803

The effect of economic development on natural systems is increasingly evident in the New Orleans metropolitan area. To aid in long-range planning for this region, we have developed a computer simulation model which integrates urban-wetland interactions with future energy needs of the region. Three 30-year development plans were simulated: (1) *current trend* in which economic and population growth and related energy demands follow the present pattern for the next 30 years, (2) *high intensity development* in which there are substantial increases in growth rates and per capita energy consumption above the current trend, and (3) a *steady state economy* in which the population and economic output reach an equilibrium condition in the year 2000 and which is accompanied by realistic energy conservation measures. Generally, the high intensity case predicts much greater demands for primary energy and food and significant effects on the natural environment. The steady state economy, with lower population and energy conservation, requires much less energy, fosters the most productive fisheries, and has the least dependence on external economies for food supplies.

## Introduction

The New Orleans region of coastal Louisiana, with its highly productive natural environment, rich fossil fuel reserves, and energy intensive economy, is a focal point of two pressing socioeconomic issues--energy and the environment. Individuals faced with management decisions are challenged to resolve these closely related, and often conflicting, issues in a way that insures the future health of the region's economy and environment.

Several objectives for this study were envisioned, all relating to the resolution of Louisiana's coastal zone management problems. A diverse collection of miscellaneous information on the New Orleans area was already available, and the first objective was to organize these data and incorporate them into a general, highly aggregated mathematical model. This effort required the definition of all key urban-wetland interactions.

A second objective was to demonstrate that dynamic simulation in regional analysis could complement equilibrium approaches such as input/output analysis (Leontief, 1951; Issard, 1960, 1968) and energy cost/benefit analysis as developed by H. T. Odum (1977). Finally, we hoped to bracket the upper and lower limits of New Orlean's energy requirements through the year 2000 by simulating three alternative development plans: (1) the *current trend*, which assumes that economic and population growth and related energy demands will follow the present pattern for the next 30 years, (2) *high intensity development*, which assumes moderate increases in growth rates and per capita energy consumption above the current trend, and (3) a *steady state economy*, in which population and economic output reach an equilibrium condition in the year 2000 and which is accompanied by realistic energy conservation measures.

## DESCRIPTION OF REGION

The study region (Fig. 1) is a 20,000 square kilometer area of southeastern Louisiana bounded on the north by the Pleistocene terrace, on the west by Bayou Lafourche, and on the east by the Chandeleur Islands; the seaward boundary includes that part of the nearshore Gulf of Mexico which contributes to fishery production. For convenience, the computed area of the study region omits open Gulf areas not bounded by land or barrier islands. The study region is fixed at a constant value ($2 \times 10^4$ km$^2$) to which energy flows and storages can be conveniently referenced. Forty-three percent of this total area comprises various wetland types (salt marsh, brackish and intermediate marsh, fresh marsh, and swamp forest) and associated small water bodies (lakes, bays, etc.) also make up 43% of the total, while the remaining 14% is dry land devoted to urban, agricultural, and miscellaneous uses.

## FORMULATION OF MODEL

The generality and breadth of the model call for a high degree of aggregation. Furthermore, our limited data and understanding of some of the processes justify no more than elementary mathematics in the model's formulation. The method used in this study was adapted from the procedure recommended by Kowal (1971) for building dynamic models of ecological systems: (1) specification of variables of interest; (2) construction of energy and material flow diagram; (3) classification of variables; (4) specification of forms of equations; (5) evaluation of coefficients; and (6) modifications for alternative development plans.

The formulation of the model was facilitated by subdividing the overall model into five sub-models: (1) fossil fuel energy flows in

## STUDY AREA

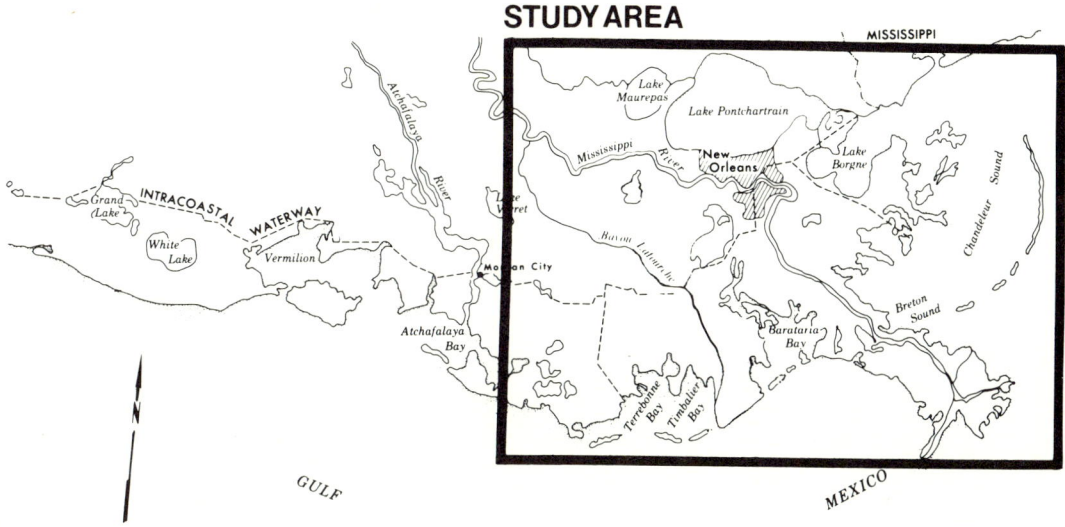

Fig. 1.  The Louisiana coastal zone and study area.

the urban (economic) sector, or the fossil fuel work equivalent of
electrical energy (Odum, 1977); (2) biological energy flows in the
natural sector; (3) food flow through the urban sector; (4) a land
use sub-model; and (5) phosphorus flows through the urban and natural sectors.

Variables used in the model are described in detail under nomen-
clature, but in general they represent rates of energy and/or nutrient flow through the five sub-models.  Forcing functions include
total regional output, fossil fuel production, insolation, and external economics.

### Energy and material flow diagram

An energy and material flow diagram (Fig. 2) was constructed
using Odum's (1971) energy circuit language to show the interrelationships among variables.  Implicit in the diagram are the following assumptions which greatly simplified the formulation of the
model.  (1) Biological energy storages and land use categories were
assumed to be spatially homogeneous and, therefore, representable
as single compartments.  (2) Energy and nutrient fluxes between compartments were likewise assumed to be spatially homogeneous and to
change only with time.

Some potentially significant forcing and control functions were
neglected in order to keep the model to a manageable size.  For example, the possibility of the Mississippi River's changing its
course to the Atchafalaya Basin is not considered, even though preventing this change is becoming increasingly costly and the economic implications of a switch are enormous.

The main driving force of the urban sector sub-model is economic
growth, or increase in Total Regional Output (TRO) of goods and
services (i.e.,final demand).  The energy function shown in the model
is a non-limiting source that expresses the total fossil fuel demand

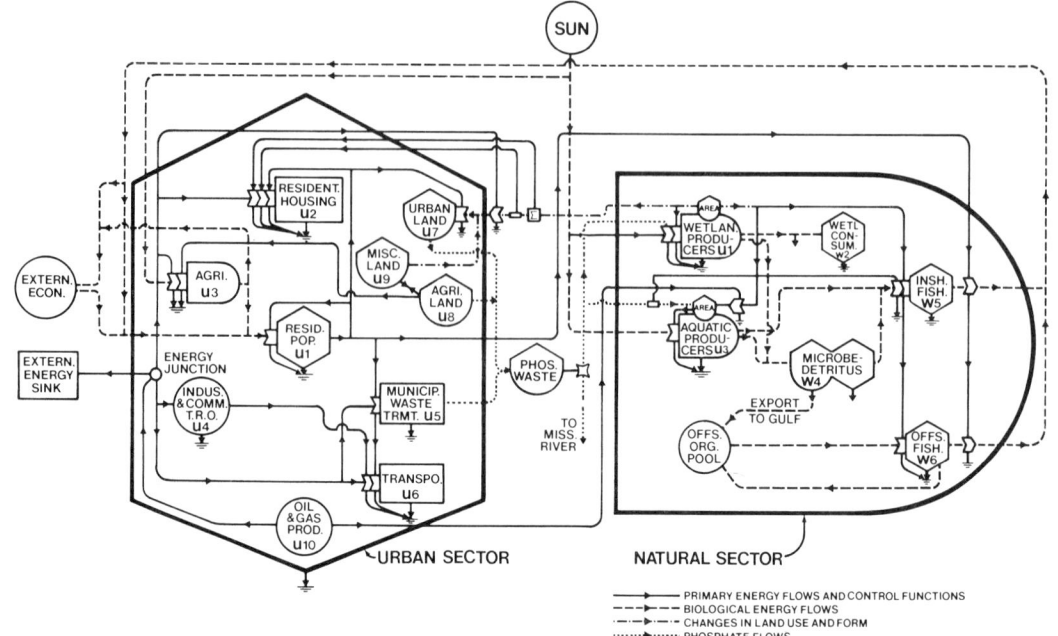

ENERGY AND MATERIAL FLOW DIAGRAM FOR NEW ORLEANS SIMULATION

Fig. 2.   Conceptual model of estuarine urban interactions in the New Orleans region.   See text for discussion.

of the region.   This was included in order to allow the determination of the effect of various growth options on energy demand.   The resident population increases in the model in proportion to the TRO, and these two variables together control primary and food energy consumption of nearly all the remaining urban compartments.   Hence, TRO is the key determinant of regional energy requirements.   Secondary forcing functions in the urban sector include oil and gas production (U1), agricultural land under cultivation (U8), and the external economy that acts both as a donor and receiver of energy from the urban sector.   The natural sector is represented with the minimum number of compartments necessary to model the essential energy flows in that sector.   Solar energy input to the wetland and aquatic producers is represented by gross primary production of those compartments.   The energy source for the offshore fishery population is represented simply as an organic pool in the absence of data on primary production and utilization of detritus by offshore populations.

Flows of energy and materials and the actions of control variables between natural and urban sectors are of particular interest in this model.   Loss of wetlands, through reclamation and natural causes, is an especially important factor that directly affects the yield of nursery-dependent species.   The energy flow diagram also indicates energy cost associated with wetland reclamation, including both flood protection and the increased energy cost of constructing and maintaining buildings on reclaimed wetlands.   Flows of nutrients (as symbolized by phosphorus) through switches controlled by man are shown interacting either positively (increased wetland production) or negatively (reduced fishery yield due to eutrophication) with the natural environment.

Flow rates for energy and food are non-dynamic state variables (i.e., non-energy storing) and are therefore not determined by the state of the system. Rather, they respond only to one or more urban sector forcing functions.

Compartments in the natural sector were assigned dynamic state variable status (energy or material storing) and energy flows between compartments were then defined in terms of these energy storages. Land use categories are the only other dynamic state variables in the model.

### Specification of forms of equations

Fossil fuel and food energy flows in the urban sector were described by simple algebraic equations. Energy storages within urban sector compartments were neglected since they were small compared to the magnitudes of the energy flows. Hence, energy is consumed within each urban compartment at the same rate that it enters.

For the current trend simulation, the Total Regional Output follows a logarithmic function based on projected population growth (Jennings, 1972). Oil and gas production rates follow the optimistic trends predicted by Research Associates (1974). No increase in agricultural land has occurred recently in the study area; therefore, the only factor influencing the energy demand of agriculture was the increasing trend in average crop yields, presumably made possible by fossil fuel subsidies. With minor exceptions, energy consumption rates of the other compartments are assumed to increase in proportion to either the TRO or the resident population. Specific equations are listed in the current trend simulation program (Appendix A).

Two critical phosphorus flows were modeled as indicators of total nutrient flows along urban-to-wetland pathways. The first pathway, influx of phosphorus to Lake Pontchartrain (shown as a dashed line from Phos. Waste to U3) comprises two fractions, natural input from the surrounding wetlands (treated as constant) and the fraction arising from agriculture and urban runoff (treated as a function of population size). The second (alternate) pathway is the natural treatment of waste in wetlands (shown as dashed line to U1).

For the natural sector sub-model, an ordinary differential equation was written for each compartment, equating the time rate of change of biomass to total energy intake minus losses. Gross production is modeled as a function of solar radiation and habitat area. The offshore forcing function is modulated by a nursery control function dependent on wetland area. With one exception, energy flows between compartments were defined by the rules of Hairston *et al.* (1960), i.e., flows from plants (or detritus) to herbivores to predators posers) are recipient-dependent and those from herbivors to predators are both donor- and recipient-dependent. The single exception in this model was to make detritus consumption by aquatic consumers proportional to both state variables.

The role of wetland as nursery ground was incorporated into the model as a work gate on the intake of inshore and offshore fishery

populations. In order for fishery yields to vary linearly with local
wetland area as shown by Turner (1976), it was necessary to make food
intake of aquatic consumers proportional to the square root of wetland area. Hinchee (1977) in a purely linear compartmental model
found that setting food intake by aquatic consumers *directly* proportional to nursery area matched the observed relationships. Some
possible explanations for this difference are discussed under Results
and Discussion.

The effect of eutrophication on the food intake of aquatic consumers in Lake Pontchartrain was modeled by assuming that intake was
inversely proportional to the square root of phosphorus loading beyond
the critical value of 0.48 g P $m^{-2}$ $yr^{-1}$ (Brezonik and Shannon, 1971).
An exact relationship between eutrophication and fishery yield has
yet to be established; however, the inverse square root relationship
produced a reasonable response. This function perhaps represents the
diversion of primary production from food webs culminating in commercial fisheries to food webs culminating in organisms less desirable
as human food. Finally, man's harvest of the inshore and offshore
fisheries was assumed proportional to fishing effort which, in turn,
was made proportional to the resident population. Mathematical expressions of the foregoing relationships are provided in the program
listing (Appendices A and B).

Of the five land use categories built into the model, only three
were functions of model interactions: wetland, water, and urban land
area. Agricultural land and miscellaneous uses were held constant.
On the farm, energy use accounts for only 0.3% of total regional demand, exemplifying the truly urban character of the region as a whole.
The demand for urban land is met exclusively through reclamation of
wetland, which has been the chief source of new urban land in this
region since the turn of the century (Kemp *et al.*, 1976). This process is shown in the model in several ways. Transfer of wetland to
urban land requires fossil energy. This is shown as a work gate from
the energy junction. The rate of construction of residential housing
(U2) is dependent on the rate of wetland reclamation (shown as a sen-
sor in Fig. 2). Additional depreciation of urban structures is shown
by a $\Sigma$ symbol. Additional wetland is converted to aquatic area
through the natural processes of subsidence and erosion and through
canal building for mineral exploration and navigation purposes (Craig
*et al.*, 1977). The time rate of change of wetland area was thus expressed in the model as follows:

$$\frac{d}{dt}(\text{Wetland area}) = - \begin{array}{l}\text{Rate of wetland}\\ \text{reclamation for}\\ \text{urban land}\end{array} - \begin{array}{l}\text{Rate of transforma-}\\ \text{tion of wetlands}\\ \text{to waterbodies}\end{array}$$

Wetland area was determined by numerically integrating the above equation over the simulation interval. Total water area increases at the
rate at which wetland is converted to water, i.e., the last term in
the above equation.

### Evaluation of coefficients
All of the coefficients in the current trend simulation model

and most of those in the high intensity and steady state models are
constants. In the latter two cases, some time-varying coefficients
were used. Typically, coefficients were evaluated by dividing a known
value of the flow by the function's argument(s) measured at some common time, usually the initial conditions. In other cases, the rate
coefficients were determined directly from the available data. For
example, the primary energy consumed directly on the farm for each
acre of harvested land constituted the rate coefficient for that flow.

*Modifications for alternative development plans.*
The models of high intensity and steady state development patterns had the same form as the current trend reference model except
for the modifications listed below.

High Intensity Model:

(1) The total regional output of the economy (and along with it,
population) increases at a rate yielding a 30% greater output relative to the current trend in the year 2000.

(2) The population density decreases by 23% over the term of
the simulation, thereby compounding the need for new urban land.
This density decline is indicative of an increasing trend toward
suburbanization nationwide.

(3) The contribution of mass transit to the total public transportation needs diminishes to 25% of the current level with the
automobile making up the difference.

(4) The per capita energy consumption of residential housing,
industry, and commerce doubles over the next 30 years as it has over
the previous 30 years (Research Associates, 1974).

Steady State Case:

(1) The total regional output reaches a steady state in the
year 2000 at a level ∿30% greater than the 1970 output.

(2) No more wetland is reclaimed for urban development.

(3) The energy subsidy to agriculture levels off at its current
value and so, therefore, does the yield.

(4) Conservation measures are implemented in homes and industry,
which result in a 15% decrease in per capita energy consumption.

(5) The use of mass transit expands to provide 50% of the total
public transportation needs.

(6) In the 10th year of the simulation, 1980, all secondarily
treated sewage, as well as urban runoff, is transferred to wetlands
for tertiary treatment (Meo *et al.*, 1977).

(7) That percentage of the region's total agricultural production which is consumed by the resident population increases from
0.36% to 100%.

These simulations were performed with IBM's Continuous System's
Modeling Program (CSMP), which provided all of the necessary mathematical operations. The execution of the programs placed a nominal
demand on computer resources requiring no more than 150 K bites of
core and running with less than 2 minutes of C.P.U. Details on state
variables, forcing functions, equation forms, and rate constants
are presented in Appendices Tables B-1 through B-7.

## Results and Discussion

The simulation results are most meaningful when the variations among alternative development patterns for important variables are examined individually. Accordingly, graphs have been prepared from the model outputs (Figs. 3 and 4) showing values and trends of key energy and material flows and forcing functions during the simulation term. These results together with the 30-year cumulative energy demands (Table 1) provide useful determinants for development plan evaluation.

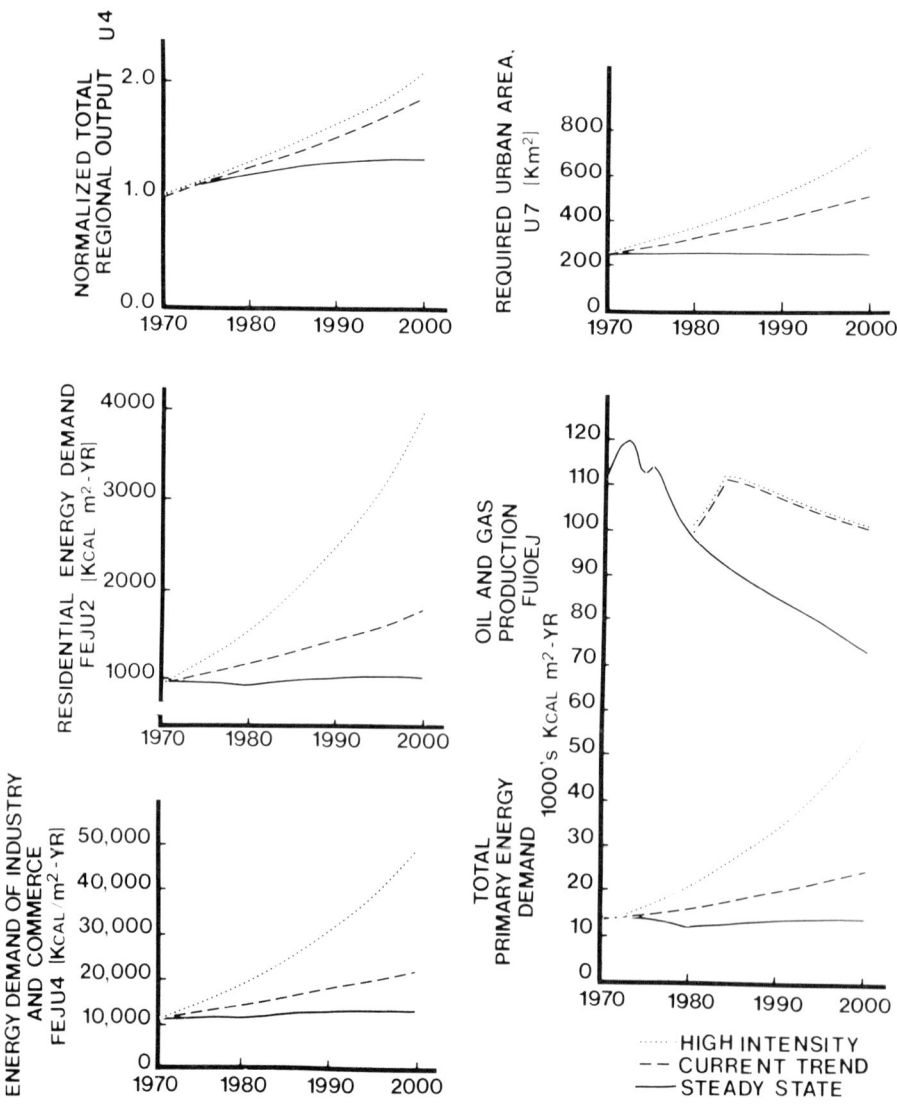

Fig. 3.  Selected results of simulation model.  Refer to text and appendices for discussion and definition of notation used.

PREDICTED PRIMARY ENERGY BUDGET

Graphical representation of the principal urban forcing function, Total Regional Output, is shown in Figure 3. The shape of the
curves is the same as population since we assumed that both would
grow at the same rate.

Significant variation occurs in the final demand for urban

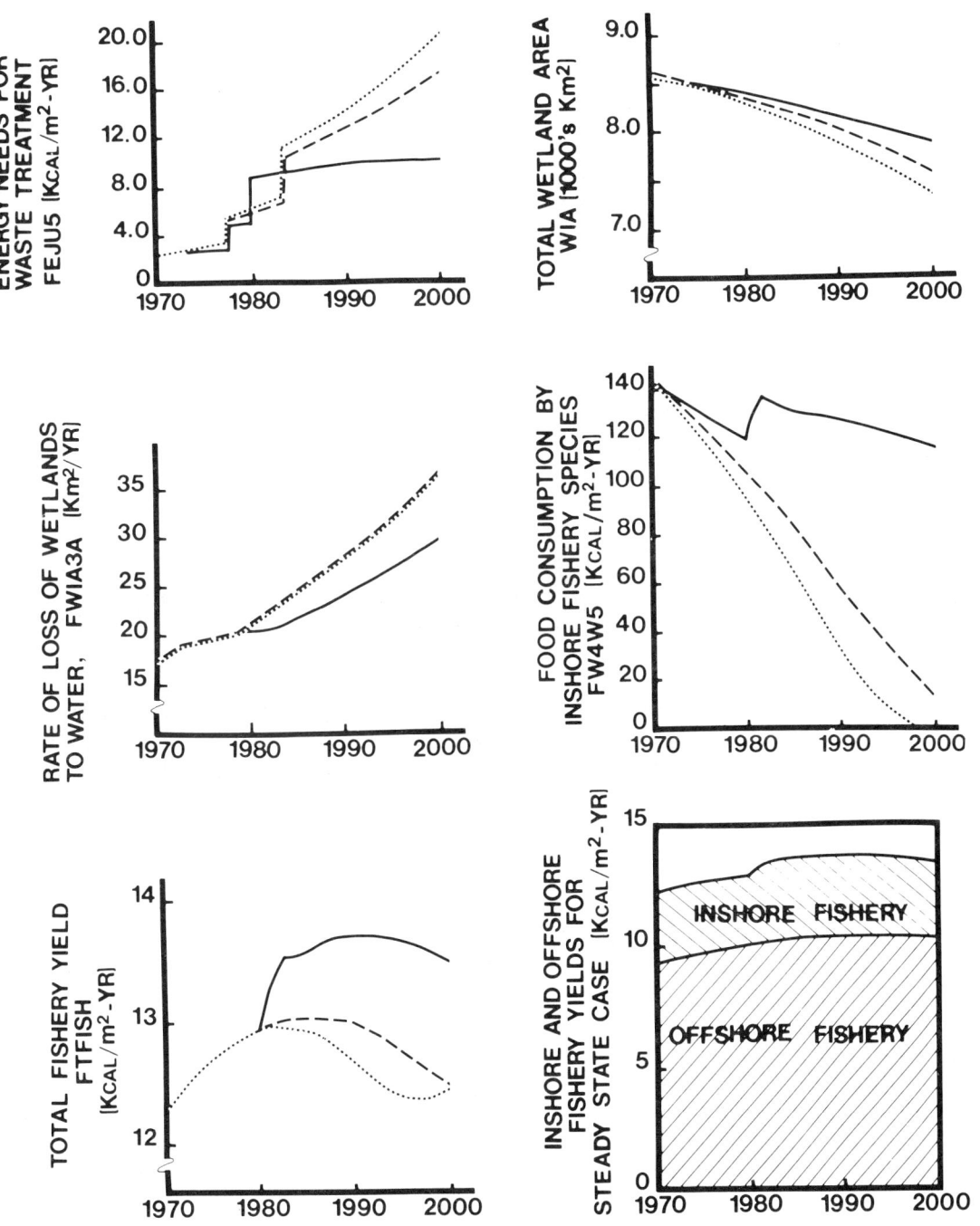

Fig. 4. Selected results of simulation model. See legend
for Figure 3.

Table 1.  Summary of results.[*]

| VARIABLE | 30-YR. SUM UNDER CURRENT TREND | PERCENT CHANGE FROM CURRENT TREND | |
|---|---|---|---|
| | | HIGH INTENSITY | STEADY STATE |
| **URBAN SECTOR:** | | | |
| Resident Population | 2,183,000 | 13.2 | -29.6 |
| Urban Land Area (km$^2$) | 502 | 47.4 | -46.2 |
| Residential Energy Demand | 39,860 | 61.5 | -24.7 |
| Industrial Energy Demand | 488,900 | 61.2 | -24.1 |
| Transportation Energy Demand | 26,200 | 7.8 | -25.3 |
| Total Primary Energy Demand | 555,500 | 58.6 | -24.2 |
| Population Food Requirements | 2,922 | 7.2 | -12.9 |
| Consumption of Regionally Produced Foods | 23.41 | 3.8 | 6660.0 |
| Required Food Imports | 2,899 | 7.2 | -66.8 |
| Food Trade Deficit | 188 | 113.3 | 40.4 |
| | | | |
| **NATURAL SECTOR:** | | | |
| Loss of Wetland Area (km$^2$) | 992. | 23.9 | -31.2 |
| Inshore Fishery Yield | 62.41 | -19.0 | 52.0 |
| Offshore Fishery Yield | 321.8 | 2.5 | -5.6 |
| Total Fishery Yield | 384.2 | -1.0 | 3.8 |

[*]All cumulative energy demands are in units of $Kcal/m_r^2$

land area (Fig. 3).  The rapidly growing population, coupled with a
decrease in the population density in the high intensity case would
require a 274% increase in area of metropolitan New Orleans from
1970 to 2000.  This compares with an 86% increase required under the
current trend.  To meet these urban requirements in the year 2000,
wetland would have to be reclaimed 2.37 times faster under high intensity development than under the current trend.  In the steady
state case there would be no further wetland development, necessi-
tating a 30% increase in population density.

Residential housing is a significant energy consumer in the urban sector, accounting for slightly over 7% of the total demand in
1970.  There is a substantial variation in energy needs of housing
among the three alternatives (Fig. 3).  The current trend simulation
predicts an 87% increase in demand from 1970 to 2000, in keeping
with the 86% increase in population.  The 30-year doubling rate in
per capita consumption assumed in the high intensity case would result in a 424% increase in energy use over the 1970 level.  At the
other extreme, the steady state alternative, reduced population
growth and energy conservation measures would require only an 11%
increase above the 1970 level.  In this last case, energy demand
actually declines over the first ten years of the simulation as the
conservation programs are implemented and then increases and eventually levels out in response to population change.

By far the largest energy-user in the region is the industrial-
commercial sector (which accounted for 88% of the region's total

primary energy demand in 1970). This figure is well above the national average because of the concentration of energy-intensive petrochemical industries in the region (Research Associates, 1974). The predicted trends (Fig. 3) closely follow those for residential housing, and essentially for the same reasons--varying rates of industrial growth, a 30-year doubling rate of per capita energy use in the high intensity case, and energy conservation measures in the steady state case. The extrapolation for the current trend is probably conservative, since per capita energy consumption is expected to continue to increase in Louisiana in spite of the "energy crisis" (Research Associates, 1974). The cumulative industrial demand (i.e. the sum over the 30-year simulation) in the high intensity case is 2.12 times the demand in the steady state case. In addition, difference in the final demand between these two cases is nearly four-fold. Clearly, it would be wise to concentrate energy conservation efforts in the industrial sector where the potential for savings is so great. Recognizing this fact, Louisiana's Department of Conservation has begun to sponsor an annual Industrial Energy Conservation Seminar (*Morning Advocate*, 1976).

The energy demand for waste treatment (Fig. 4) is controlled largely by legal constraints on water quality. The first step increase in energy consumption corresponds to the starting of New Orleans' secondary treatment facility in 1977, as required by the Water Pollution Control Act. The same law calls for a further upgrading of municipal facilities to the "best practical treatment technology" by 1983 (also reflected in Figure 4 by a step increase in energy demand). Both the high intensity and current trends follow this legal timetable, whereas the steady state deviates from it by initiating advanced treatment via the wetlands in 1980. The smaller population growth and lower unit energy cost of wetland treatment combine to yield a substantially reduced energy demand in the steady state case.

Transportation is the region's third largest energy consumer, accounting for 4.7% of the total demand. New Orleans is considered to have a fairly extensive mass transit system (buses and streetcars) even though it currently provides only 2.7% of the region's total transportation needs. In the high intensity and current trend simulations, energy use by this sector is controlled almost exclusively by population. Sharply reducing the current level of service provided by the mass transit system would have very little effect on the total energy demand for transportation and a negligible effect on the total regional energy demand. On the other hand, a dramatic shift to more mass transit would result in a 25% reduction in cumulative transit energy use below the current trend.

The energy cost of keeping New Orleans dry was included only in a token manner, in hopes of addressing it in a more detailed manner later. We included only the energy cost of pumps which drain the city. As a percentage of total energy demand, pumping cost is very small (less than 0.01%), but this is only one, often minor, energy cost associated with flood protection in general (e.g., construction and maintenance of pumping stations, canals, and levees).

The predicted trends in total primary energy demand for each case are dominated by the three largest consumers--industrial-commercial, residential housing, and transportation--which together account for over 99% of the total energy consumption. Therefore, the response pattern of the total demand (Fig. 3) closely resembles the patterns of these three individual demands. Trends in fossil fuel production (Fig. 3) for the high intensity and current trend simulations were assumed to follow the optimistic prediction reported by Research Associates (1974). A more pessimistic, and perhaps more realistic, prediction was made in the steady state case by assuming that the current rate of decline in production and reserves continues. The total demand and production curves illustrate the extent to which the region remains a net producer of energy. They do *not* mean that all of the energy produced is available for local use, even if economic conditions permitted. On the contrary, an increasingly greater fraction of the region's oil and gas is coming under federal control (Research Associates, 1974) which reduces the probability of the fuel's being used locally. Even today, some of the region's low priority users of interstate natural gas (which is under federal control) face imminent curtailments in their gas usage.

## PREDICTED FOOD ENERGY BUDGETS

Primary productivity and provision of nursery grounds are two vital wetland functions which are dependent on total wetland area. Approximately 40% of this loss in the year 2000 would be attributable to reclamation efforts (Fig. 4). The cumulative loss would be 11.6% under the current trend and 14.3% for high intensity. In the steady state case the cessation of wetland reclamation, coupled with decreased fossil fuel production activity in the wetlands, would result in only a 7.9% loss due to natural subsidence and erosion. A more direct illustration of the effect of fossil fuel production on the rate of wetland loss is provided by Figure 4. All of the observed difference between the high intensity and current trends in land loss and steady state trend is attributable to the variation in exploration and production activity.

Primary production of wetland under the high intensity and current trends declines by the same percentage as does wetland area for the respective cases. However, in the steady state case, decline in production was mitigated slightly by increased productivity of those wetland areas used for tertiary treatment service. Thus, while the loss in wetland area from 1970 to 2000 in the steady state case would be 7.9%, the decline in net primary production would be only 7.6%.

In every case, total primary production of inshore waterbodies was directly proportional to the water areas. Aquatic production in the steady state case fell below the high intensity and current trends after ten years, since water area increased at a lower rate as oil and gas production activities declined.

In this model, food intake of the offshore fishery species was

modulated only by the nursery control function and therefore declined
little. Its inshore counterpart, consumption by the inshore fishery
species (Fig. 4), declined dramatically, however, in the current
trend and high intensity simulations. Eutrophication of Lake Pontchartrain and loss of nursery grounds are primarily responsible for
this sensitive behavior. Under the current trend, food intake would
decline to 9% of its present level by 2000 and essentially vanish
under high intensity development. The steady state case, on the
other hand, would show a decline of only 18% by 2000. This relative
stability is due to the following factors: (1) reduced rate of wetland loss, (2) lower rate of increase in fishing effort, (3) rapid
recovery of the Lake Pontchartrain fishery after redirecting urban
runoff to wetlands, and (4) the resulting increase in primary production.

The net result of these differences in food intake by consumers
is reflected in total fishery yields (Fig. 4). The yields in all
three cases are essentially identical through 1980, which suggests
that reduction in fishery biomass for each case is offset by fishing
effort in a way that produces identical results. Beyond 1980, total
yield declines in the high intensity and current trend cases as
the rate of decrease in the inshore yield exceeds the rate of increase in offshore yield. Total yield later increases as offshore
yield again becomes dominant. In the steady state case, the predicted rapid recovery of the Lake Pontchartrain fishery is primarily
responsible for the sharp increase in yield observed in 1980. As
population and fishing effort begin to level off past 1990, the ongoing loss of wetlands due to natural causes leads to a downturn in
total yield. The ratio of offshore to inshore yields for the steady
state case is relatively constant throughout the simulation term
(Fig. 4). Similar results for the current trend and high intensity
cases show sharply increasing ratios.

In a separate simulation experiment, the relationship between
fishery yield and the nursery control function's acting alone were
found to agree with results of simulation of the Lake Pontchartrain
ecosystem by Hinchee (1977). He, too, found that fish biomass and
yield declined in direct proportion to wetland loss.

On a caloric basis the contribution of regional fishery products
to total regional food requirements is small (0.5%) although it represents twice the average American seafood consumption on a per
capita basis. However, the fact that seafood is a high quality,
high cost food increases its relative importance to the local diet
as well as to the economy of the region. In our simulation, total
food needs leveled out in the steady state case, in keeping with
the population trend. By 1980, regionally grown crops were predicted to supply 76% of the total food needs, and 68% of total needs by
the year 2000.

The New Orleans region presently imports 99.2% of its total
food supply from outside of the region. In the steady state economy this dependence would be reduced to 31.5% by the year 2000.
However, to accomplish this, the level of agricultural exports falls

from 99.9% of the total production to zero. Total agricultural production of the region is small compared to total state production.
Therefore, by devoting only a small fraction of Louisiana farmland
to food production for the New Orleans region, food needs of about
30% of the state's population could be met. Under high intensity
and current trends, greatly increased food demands would have to be
met by imports.

The balance between food exports and imports to the region was
also investigated. Under the current development trend it will be
about 1980 before food imports rise to the level of exports, but by
the year 2000, the region's food trade deficit will be 19% of the
total food requirements. In the steady state case the balance point
between imports and exports would be reached in 1974. This rapid
movement to a trade deficit reflects a levelling off of agricultural
yields early in the simulation when the population growth rate is
still high. But as population levels off so does the deficit,
whereas in the high intensity and current trend cases it continues
to decline at an increasing rate. Comparing the 30-year cumulative
food trade deficits reveals that it is greater (by about 40%) in
the steady state case than in the current trend. However, the trends
in the year 2000 suggest that the cumulative deficit under the current trend would soon surpass that of the steady state case.

Significant difference among the three development alternatives
may be summarized by comparing percentage variations in cumulative
energy demands (Table 1). High intensity development results in a
13.2% larger population by 2000 than for the current trend. This
rise is accompanied by a much larger increase in urban land (47.4%)
and in the cumulative energy demand of the two major use categories
--industrial-commercial (61.2%) and residential housing (61.5%).
Increase in energy demand for transportation is up only 7.8%, in
spite of the shift toward greater automobile usage. Food needs increased by 7.18% over the current trend. Since the cumulative consumption of regionally produced food is up only 3.8%, the difference
would be made up by food imports--7.21% greater. Cumulative food
trade deficit increases by a staggering 113.3%. The 1% decline in
cumulative fishery yield explains the net decrease in consumption of
regionally produced food. The 19% decline in inshore fishery yield
is solely responsible for the decrease in total yield, since the
offshore cumulative yield increased by 2.5% over the current trend.

The steady state economy alternative results in a 29.6% smaller
population by the end of the century than the current trend. Conservation measures in the three major energy-consuming sectors could
reduce the cumulative primary energy demand by 24.2% below the current trend demand. Furthermore, the population's cumulative food
needs are down by 12.9%. The 6660% increase in consumption of regionally produced foods helps to cut required food imports 66.8%
below the current trend level. In spite of this substantial reduction in food imports, the cumulative trade deficit is some 40.4%
higher than that of the current trend. However, as pointed out
earlier, the deficit levels off in the steady state case but con-

tinues to increase under the current trend.

In the natural sector, where the loss of wetland area is 31.2% smaller than in the current trend, a 52% increase in the cumulative inshore fishery yield is realized. The net result is a 3.8% increase in the total fishery yield, notwithstanding the considerably smaller fishing effort under this alternative.

## Conclusions

This simulation model has generated some credible predictions of the region's total energy requirements and of the ways that specific urban-wetland interactions affect these needs. Even in this preliminary form, the model has served to bracket the range of energy and food needs as they might develop over the next 30 years. It has also shown probable future states if present growth trends and patterns of energy use continue.

The upper limits of economic and population growth, as modeled in the high intensity case, could only be realized through massive importation of new primary energy supplies (coal, nuclear, etc.), even if oil and gas production meet the most optimistic forecast. Furthermore, the impact on regional wetlands is shown to be substantial. By the year 2000, wetland area is predicted to decline by 15%. This loss of critical nursery area, coupled with further eutrophication of open water and increasing fishing effort, would decimate the inshore fishery and greatly diminish many other values of the wetlands.

At the other extreme, the attainment of a steady state economy by the year 2000, coupled with some realistic energy conservation measures, requires only a modest increase in primary energy and greatly reduces the pressures on the natural environment. Sensible disposal of nutrients from the urban sector can enhance productivity of wetlands and ultimately preserve a high-yielding inshore fishery.

The modeling process has raised new questions and suggested focal points for future research. The second iteration might profitably be concentrated on the following areas:

(1) With the urban sector sub-model, a more realistic representation of the regional economy is in order. To help in this effort researchers at the Urban Studies Institute at the University of New Orleans are developing an input/output model which predicts regional economic growth subject to the constraints of land use, air and water pollution, and lifestyle preference (Mumphrey and Whalan, 1976). This improved sub-model might also include further compartmentalization of the crucial industrial-commercial sector in order to model energy demands, conservation potential, and environmental impacts of specific industries.

(2) Several areas of the natural sector sub-model require further development. More work needs to be done to establish the relationship between eutrophication and fisheries production. Also, a more realistic and consistent formulation of the nursery function of wetlands would improve the credibility of the model.

A final question one might ask is, "How will these simulation results be used?" Ideally, decision-makers, mindful of public well-being and preferences, will factor the results of these and other responsible forecasts into their decisions. To maximize the chances of these results getting due consideration, every effort will be made to have them presented at public forums dealing with general and specific questions of New Orleans regional development.

# References

Allen, R. 1975. Aquatic primary productivity in various marsh environments in Louisiana. M. S. thesis. Louisiana State Univ., Baton Rouge, La.

Brezonik, P. L. and E. E. Shannon. 1971. Trophic states of lakes in North Central Florida. Pub. No. 13, Fla. Water Resources Research Center.

Chabreck, Robert H. 1972. Vegetation, water and soil characteristics of the Louisiana coastal region. Bulletin No. 664, Agr. Exp. Sta., Louisiana State Univ., Baton Rouge, La.

City Planning Commission of New Orleans. 1975. Coastal Zone Management Plan: 1975. Vols. I-III. Printed by the City Planning Commission, New Orleans, La.

Craig, Nancy, R. E. Turner, and J. W. Day, Jr. 1977. Cumulative impact of land-loss in Louisiana's coastal zone. Louisiana State Planning Office, Baton Rouge, La.

Day, John W., Jr., William G. Smith, Paul R. Wagner, and Wilmer C. Stowe. 1973. Community structure and carbon budget of a salt marsh and shallow bay estuarine system in Louisiana. Pub. No. LSU-SG-72-04, Center for Wetland Resources, Louisiana State Univ., Baton Rouge, La.

Day, John W., Jr., T. Butler, and W. Conner. 1977. Productivity and nutrient export studies in a cypress swamp and lake system in Louisiana. In: *Estuarine Processes* (M. Wiley, ed.). Vol. 2, pp. 255-269. Academic Press, New York.

Eckenrod, R. 1977. Computer simulation of urban-estuarine interactions in the New Orleans region. M. S. thesis, Louisiana State Univ., Baton Rouge, La.

Fielder, Lonnie L. 1972. An analysis of trends in yields of major crops in Louisiana. Department of Agricultural Economics Res. Rep. No. 446, Agr. Exp. Sta., Louisiana State Univ., Baton Rouge, La.

Fielder, Lonnie L. and C. O. Parker. 1972. Louisiana crop statistics by parishes through 1970. Dept. Agr. Econ. Res. Rep. No. 446, Agr. Exp. Sta., Louisiana State Univ., Baton Rouge, La.

Gagliano, S. M. 1973. Canals, dredging, and land reclamation in the Louisana coastal zone. Hydrologic and Geologic Studies of Coastal Louisiana, Rep. No. 14, Coastal Resources Unit, Center for Wetland Resources. Louisiana State University, Baton Rouge, La.

Gagliano, S. M., P. Light, and R. Becker. 1973. Controlled diversions in the Mississippi delta system. Hydrologic and Geologic Studies of Coastal Louisiana, Rep. No. 8, Center for Wetland Resources. Louisiana State University, Baton Rouge, La.

Gosselink *et al.* 1976. Ecological Characterization of the Chenier Plain of Southeastern Louisiana and Southwestern Texas. Pilot Study Report: The Calcasieu Drainage Basin. Center for Wetland Resources. Louisiana State University, Baton Rouge, La.

Hairston, N. G., F. E. Smith, and L. B. Slobodkin. 1960. Community structure, population control, and competition. Am. Nat. 94: 421-425.

Happ, G., J. G. Gosselink, and J. W. Day, Jr. 1977. The seasonal distribution of organic carbon in a Louisiana estuary. Estuar. Coast. Mar. Sci. 5: 695-705.

Hinchee, R. 1977. Selected aspects of the biology of Lake Pontchartrain, Louisiana. M. S. thesis, Louisiana State University, Baton Rouge, La.

Hirst, Eric. 1974. Food-related energy requirements. Science 184: 134-184.

Hopkinson, C. S. and J. Day. 1977. A model of the Barataria Bay salt marsh ecosystem. In: *Ecosystem Modelling in Theory and Practice* (C. Hall and J. Day, eds.), pp. 235-265. Wiley Interscience, New York.

Hopkinson, C. S., J. G. Gosselink, and R. T. Parrondo. In press. Above-ground production of seven marsh plant species in coastal Louisiana. Ecology.

Issard, W. 1960. *Methods of Regional Analysis: An Introduction to Regional Science*. MIT Press and Wiley and Sons, Inc., New York.

_____. 1968. Some notes on the linkage of ecologic and economic systems. Regional Sci. Asso. Paper XXII, European Congress, Budapest Conference.

Jennings, William. 1972. Ecological inventory of the city of New Orleans. Tropical Ecology Publication 72-1E, New Orleans Health Dept.

Kemp, Paul, D. Lindstedt, and L. Peeler. 1976. Effects of storm water pumpage on Lake Pontchartrain, La. Paper presented at AIBS meeting, May 30-June 3. New Orleans, La.

Kirby, Conrad J. and James G. Gosselink. 1976. Primary production in a Louisiana Gulf Coast *Spartina alterniflora* marsh. Ecology 57 (5): 1052-1059.

Kowal, Norman E. 1971. Modeling dynamic ecological systems. In: *System Simulation and Analysis in Ecology* (B. C. Patten, ed.), Vol. 1. Academic Press, New York.

Leontief, W. 1951. *Structure of the American Economy, 1919-1939.* Oxford University Press, New York.

Lincoln, G. A. 1973. Energy conservation. Science 180: 155-162.

Lindall, W. N., J. Hall, J. E. Sykes, and E. L. Arnold. 1972. Louisiana Coastal Zone: Analysis of resources and resource development needs in connection with estuarine ecology. Sections 10 and 13 - Fishery Resources and Their Needs. Rep. Comm. Fishery Work Unit, National Marine Fisheries Service Biol. Lab., St. Petersburg Beach, Fla.

Liptak, Bela G. ed. 1974. *Environmental Engineer's Handbook*. Vol. 1. Water Pollution. Chilton Publ., New York.

Louisiana Annual Oil and Gas Report. 1975. Louisiana Department of Conservation, Baton Rouge, La.

Louisiana High School Energy Report. 1975. Louisiana Department of Conservation, Division of Natural Resources and Energy, Baton Rouge, La.

Louisiana Power and Light Co. Annual Report. 1973. New Orleans, La.

Meo, Mark, J. W. Day, and Z. B. Ford. 1977. Land treatment of fish processing waste on dredge spoil sites: Comparative cost evaluation. Coastal Zone Management Journal 3(3): 307-318.

*Morning Advocate*. 1976. Energy savings best in industry. December 8, 1976, p. 1-B. Baton Rouge, La.

Mumphrey, Anthony J. and T. Whalen. 1976. Economic growth, lifestyle preferences, and urban size. Urban Studies Institute, University of New Orleans, New Orleans, La.

New Orleans Water and Sewerage Board Statistics. 1974. New Orleans, La.

Odum, H. T. 1971. *Environment, Power, and Society*. Wiley Interscience, New York.

_____ 1977. Energy value and money. In: *Ecosystems Modelling in Theory and Practice* (C. Hall and John Day, eds.), pp. 173-196, Wiley Interscience, New York.

Paddock, Jerry. 1974. Transit energy consumption in New Orleans. Urban Studies Institute, Louisiana State University, New Orleans, La.

Research Associates. 1974. Louisiana energy study. Prepared for The Ozarks Regional Commission contract No. TA 74-11.

Taylor, Earl G., James N. Morriss, Jr., Jessee Goble, H. Ronald Smalley, and A. Stewart. 1974. Initial facilities for a regional wholesale food distribution center at New Orleans. Marketing Res. Rep. No. 1029, Agr. Res. Service, U. S. Department of Agriculture.

Turner, R. E. 1976. Wetland-water couplings in the coastal zone with an emphasis on commercial fisheries. Louisiana State University Office of Sea Grant publ. (in press).

U. S. Dept. of Agriculture. 1964. Agriculture Statistics. U. S. Government Printing Office, Washington, D. C.

_____ 1971. Agriculture Statistics. U. S. Government Printing Office, Washington, D. C.

U. S. Dept. of Commerce. 1969. Census of Agriculture. Bureau of the Census, Part 35, Section 2, Louisiana, County Data.

Wagner, Frederick W. and Edward J. Durabb. 1975. Environmental costs of urbanization in a wetland area: A case study of New Orleans. Urban Studies Institute, Louisiana State University, New Orlenas. La.

# Appendix A

```
                SIMULATION PROGRAM LISTING*


Program modifications for High Intensity (HI) and Steady State (SS)
simulations are boxed.


TITLE  REGIONAL SIMULATION UNDER CURRENT GROWTH CONDITIONS
         BASE CASE CONDITIONS  - 1970

INITIAL
INCON  ICW1=9772., ICW2=1960., ICW3=15.62, ICW4=976.3,
  ICW5=9.100, ICW6=1.00, ICW1A=85.80E8, ICW3A=87.02E8,
  ICU7=2.70E8, ICU8=3.873E8, ICU9=21.177E8

       Constant for non-biological flow sub-model

CONSTANT U1ABS=1.173085E6, AREA=200.57E8, KAUTO=0.9733,
  KPUB=0.0267, KCOM=157.50, RCEJU2=1.6262E7, RCEJU3=226.2,
  RCSUB1=7.392E-7, RCSUB2=4.435E-8, KEJU7=1.140,
  KSEW=5.9762E14, KPUMP=1.988E-21, KPS=4.9858E-11,
  KPHOS=8.4759E-11, RCEJU4=11757.0,

       Constants for biological energy flow sub-model
          RCW1R=0.36543, RCW3R=66.325, RCW1W2=1.3551,

  RCW1W4=0.077466, RCW3W4=154.07, RCW3W5=1.3725, RCW4R=1.5548,
  RCW4G=0.47476, RCW4W5=0.016168, RCW5R=16.830, RCW5M=0.32418,
  RCW6M=9.30, RCW6R=13.140, RCW6FD=0.64650, RCSRW1=7142.,
  RCSRW3=3455., FCSFW6=63.48, RCW1DP=0.02212, KNONP=0.8214,
  GRATE=0.02070, YTREND=0.01658, RCINU1=1.2017E6,
  RC56U1=480.33, RCU3U1=0.05820E-8, RCU3EE=16.145E-8.
  RC56U4=0.01471....
```

```
SS:
YTREND=0.0
```

```
       Constants for land use sub-model

  RCON=0.04879, CANWID=60.36


*Comment cards are written in lower case.
```

```
Predicted oil production rate in MMB per year

FUNCTION OIL1=(0.0, 1100., 1.0, 1120., 2.0, 1050., 3.0, 900.,
4.0, 878., 5.0, 836., 6.0, 821., 7.0, 816., 8.0, 812., 9.0,
826., 10.0, 838., 15.0, 805., 16.0, 783.0, 30.0, 696.)
```

```
SS:
..., 15.0, 680., 20.0, 633., 30.0, 573.)
```

```
Predicted gas production rate in TCF per year

FUNCTION GAS1=(0.0, 6.8, 1.0, 7.5, 2.0, 7.8, 3.0, 7.9, 4.0...
7.65, 5.0, 7.85, 9.0, 6.8, 12.0, 7.4, 13.0,,7.65, 14.0,...
7.70, 15.0, 7.75, 30.0, 7.12)
```

```
SS:
..., 30.0, 5.00)
```

```
DYNAMIC
```

```
HI:

    Switch on flows and rate constants at time = 10 yrs

  NO SORT
  IF (TIME.GT.10.0) GO TO 2
  KPUB=0.0267 - 0.002002*TIME
  GO TO 4
2 KPUB=0.006675
4 CONTINUE
  SORT
```

```
SS:

    Switch on flows and rate constants at time = 10 yrs

  NO SORT
  IF (TIME.GT.10.0) GO TO 2
  RCEJU2=1.6262E7 - 0.2439E6*TIME
  RCEJU4=11757.0 - 176.3*TIME
  KPUB=0.0267 + 0.04733*TIME
  FPPONT=0.3 + 1.111E4*U1
  FPU5W1=0.0
  FPU7W1=0.0
  RCU3EE=16.145E-8 - 1.6145E-8*TIME
  RCU3U1=0.05820E-8 + 1.6145E-8*TIME
  FEPHOS=0.0
  GO TO 4
2 RCEJU2=1.3823E7
  RCEJU4=0003.0
  KPUB=0.50
  FPPONT=0.3 + 0.6555E4*U1
  FPU5W1=0.03784*QSEW
  FPU7W1=7.0787E12*U1
```

```
    RCU3EE-0.0
    RCU3U1=16.198E-8
    FEPHOS=KPHOS*QSEW
4   CONTINUE
    SORT
```

Non-biological energy flow sub-model

Urban sector forcing function -- gross regional product
normalized to 1970

```
    ARGU1=0.020705*TIME
    U4=EXP(ARGU1)
```

```
HI:
    Urban sector forcing function -- dimensionless

    ARGU1=0.024846*TIME
    U4=EXP(ARGU1)
```

```
SS:

    NO SORT
    IF (TIME.GT.30.0) GO TO 6
    U4=1.3105 - 0.000345*(ABS(TIME-30.0))**2.0
    GO TO 8
6   U4=1.3105
8   CONTINUE
    SORT
```

Human population referred to total regional area

```
    ICU1=U1ABS/AREA
    U1=ICU1*U4
    CU1U2=U1
    SUMWU7=INTGRL(9.0, FW1AU7)
```

```
HI:

    ARGU4=0.023105*TIME
    RCEJU2=ICEJU2*EXP(ARGU4)
```

```
    FEJU2=RCEJU2*CU1U2 + RCSUB1*FW1AU7 + RCSUB2*SUMWU7
    FEJU3=RCEJU3*U8*(1 + YTREND*TIME)/AREA
```

```
HI:
    RCEJU4=ICEJU4*EXP(ARGU4)
```

```
    FEJU4=RCEJU4*U4
    QSEW=KSEW*U1
    FEPUMP=KPUMP*QSEW**2.0
    NO SORT
```

```
    IF (TIME.GT.7.5) GO TO 10
    FEPS=0.0
    GO TO 20
10  FEPS=KPS*QSEW
20  CONTINUE
    IF (TIME.GT.13.5) GO TO 30
    FEPHOS=0.0
    GO TO 40
30  FEPHOS=KPHOS*QSEW
40  CONTINUE
    SORT
    FEJU5=FEPUMP + FEPS + FEPHOS
```

```
HI, SS:
    KAUTO = 1.0 - KPUB
```

```
    FEJU6=U1*(KAUTO*8.1896E6 + KPUB*4.3944E6) + KCOM*U4
    FEJU7=KEJU7*U7/ICU7
    OIL2=AFGEN(OIL1, TIME)
    GAS2=AFGEN(GAS1, TIME)
    FU1OEJ=0.6184E12*OIL2/AREA + 225.3E12*GAS2/AREA
    FEJNET=FU1OEJ-FEJU2-FEJU3-FEJU4-FEJU5-FEJU6-FEJU7
```

Biological energy flow sub-model

Nursery control on fishery intake

```
    CNW5=(W1A/ICW1A)**0.5
    CNW6-CNW5
```

Lake Pontchartrain eutrophication

```
    FPPONT=0.3 + 1.111E4*U1
    KPONT=0.17408/FPPONT**0.5
    CPW5=KPONT + KNONP
```

Human fishing effort

```
    CU1W5M=U4
    CU1W6M=U4
```

```
SS:
    Wetland waste treatment

    FPUW=FPU5W1 + FPU7W1
    FRACF=FPUW/(44.83*W1A)
    FRACNF=1.0-FRACF
```

Forcing functions
FSRW1=RCSRW1*(W1A/ICW1A)

```
SS:
    FSRW1=RCSRW1*(W1A/ICW1A)*(FRACNF + 1.50*FRACF)
```

FSRW3=RCSRW3*(W3A/ICW3A)
FSFW6=FCSFW6*CNW6

Non-dynamic state variables

FW1R=RCW1R*W1
FW3R=RCW3R*W3
FW1W2=RCW1W2*W2
FW1DEP=RCW1DP*FSRW1
FW1W4=RCW1W4*W1
FW3W4=RCW3W4*W3
FW3W5=RCW3W5*W5*CNW5*CPW5
FW2R=FW1W2
FW4R=RCW4R*W4
CW4IEG=FW1W4 + FW3W4
FW4G=RCW4G*CW4IEG
FW4W5=RCW4W5*W4*W5*CNW5*CPW5
FW5R=RCW5R*W5
FW5M=RCW5M*W5*CU1W5M
FW6M=RCW6M*W6*CU1W6M
FTFISH=FW5M + FW6M
FW6R=RCW6R*W6
FW6FD=RCW6FD*FSFW6

Dynamic state variables

W1=INTGRL(ICW1, FSRW1-FW1R-FW1W2-FW1W4-FW1DEP)
W2=INTGRL(ICW2, FW1W2-FW2R)
W3=INTGRL(ICW3, FSRW3-FW3R-FW3W4-FW3W5)
W4=INTGRL(ICW4, FW1W4+FW3W4-FW4R-FW4G-FW4W5)
W5=INTGRL(ICW5, FW3W5+FW4W5-FW5R-FW5M)
W6=INTGRL (ICW6, FSFW6-FW6R-FW6FD-FW6M)

Biological energy flow in urban sector

F1NU1=RCINU1*U1
F56U1=RC56U1*U1*FTFISH
F56U4=RC56U4*U4*FTFISH
FU3U1=RCU3U1*U8*(1+YTREND*TIME)
FEEU1=FINU1-F56U1-FU3U1
F56EE=FW5M+FW6M-F56U1
FU3EE=RCU3EE*U8*(1+YTREND*TIME)
FSRU3=FU3EE+FU3U1
FTOTEE=F56EE+FU3EE
FFDELT=FTOTEE-FEEU1

Land use sub-model

Harvested acreage - constant

U8=ICU8

Demand for urban land in the New Orleans S.M.S.A.

CU1U7=U1/ICU1
U7=ICU7*CU1U7

```
HI:
    ARGU3=0.03359*TIME
```

```
SS:
    U7=ICU7
```

FW1AU7=DERIV(0.05643E8, U7)
FU8U9=0.0
ARGU2=RCON*TIME
FW1A3A=0.10413E8 + CANWID*FU10EJ*EXP(ARGU2)
SW1A3A=INTGRL(0.0, FW1A3A)
FU9U7=0.0
W1A=INTGRL(ICW1A, -FW1AU7-FW1A3A)
W3A=INTGRL(ICW3A, FW1A3A)
U9=INTGRL(ICU9, FU8U9-FU9U7)

```
SS:
    U9=ICU9
```

AREACK=W1A+W3A+U7+U8+U9

Summation of energy flows

Non-biological flows

SEJU2=INTGRL(0.0, FEJU2)
SEJU3=INTGRL(0.0, FEJU3)
SEJU4=INTGRL(0.0, FEJU4)
SEJU5=INTGRL(0.0, FEJU5)
SEJU6=INTGRL(0.0, FEJU6)
SEJU7=INTGRL(0.0, FEJU7)
S10EJ=INTGRL(0.0, FU10EJ)

Biological flows

S56U1=INTGRL(0.0, F56U1)
SU3U1=INTGRL(0.0, FU3U1)
STOTEE=INTGRL(0.0, FTOTEE)
SFDELT=INTGRL(0.0, FFDELT)
SW5M=INTGRL(0.0, FW5M)
SW6M=INTGRL(0.0, FW6M)
STFISH=INTGRL(0.0, FTFISH)
SINU1=INTGRL(0.0, FINU1)
SEEU1=INTGRL(0.0, FEEU1)
SPPONT=INTGRL(0.0, FPPONT)

Execution and output control

```
METHOD RKSFX
PRINT FW1AU7, FEJU4, FEJU2, FEJU3, FEJU6, FEJU7,...
    FU10EJ, CNW5, FSRW1, FSRW3, FSFW6, FW1R, FW3R,...
    FW1DEP, FW1W4, FW3W4, FW3W5, FW4R, FW4G, FW4W5, FW5R,...
    FW5M, AREACK, SUMWU7, FW6M, FW6R, FW6FD, FINU1, F56U1,...
    F56U4, FU3U1, FEEU1, F56EE, FSRU3, FTOTEE, FFDELT,...
    QSEW, FEPUMP, FEPS, FEPHOS, FEJU5, FPPONT, KPONT, CPW5,...
    RCEJU2, RCEJU4, KPUB
PRTPLOT W1, W3, W4, W5, W6, W1A, W3A, U7, U4
TIMER FINTIM=30.001, DELT=0.01, PRDEL=0.25, OUTDEL=0.16667
CONTINUE
RESET
PRINT SEJU2, SEJU3, SEJU4, SEJU5, SEJU6, SEJU7, S10EJ,...
    S56U1, SU3U1, STOTEE, SFDELT, SW5M, SW6M
END
STOP
```

# Appendix B

TABLE B-1

Energy and Material Storages
(Dynamic State Variables)

| VARIABLE NAME | DESCRIPTION | INITIAL CONDITION[*] | SOURCE |
|---|---|---|---|
| W1 | Biomass of primary producers in the wetlands proper. Includes salt marsh, brackish marsh, intermediate marsh, fresh marsh, and swamp forest habitats | 9772.0 | Hopkinson et al. 1977a; Kirby & Gosselink 1976 Table B-7 |
| W2 | Combined biomass of all wetland consumers. | 1960.0 | Day et al. 1976 |
| W3 | Biomass of all primary producers in the associated waterbodies of each marsh habitat. | 15.62 | Hinchee 1977 |
| W4 | Combined biomass of the microbe-detritus complexes in the aquatic zone of all habitats (includes zooplankton). | 976.3 | Day et al. 1973 |
| W5 | Combined biomass of all consumers in waterbodies associated with the wetlands. Does not include zooplankton but does include those microbes feeding on dead consumers. | 9.10 | Day et al. 1973; Hinchee 1977 |
| W6 | Combined biomass of the major components of the offshore fishery including menhaden, shrimp, and blue crabs. | 1.0 | Eckenrod 1977; Day et al. 1973 |
| | Land Use Sub-model | | |
| W1A | The surface area of all wetland habitats in the region including closely associated waterbodies ($<$ 10 acres). | 8580 ($km^2$) | |
| W3A | The area of all large waterbodies in the region. | 8702 ($km^2$) | Table B-7 |
| U7 | Urban land area in the tri-parish metropolitan area of New Orleans. | 270 ($km^2$) | |
| U8 | Total harvested agricultural land. | 387.3 | |
| U9 | Area of upland under all other uses. | 2117.7 | |

[*]All biomasses in units of $Kcal/m_F^2$ and land areas in $km^2$

TABLE B-2

Forcing Functions

| VARIABLE NAME | DESCRIPTION | FUNCTIONAL FORM | INITIAL VALUE | SOURCE |
|---|---|---|---|---|
| U4 | The Total Regional Output(T.R.O.) normalized to a 1970 value of 1.0. | EXP(0.02070*TIME) | 1.0 | Jennings 1972 |
| FU1OEJ | The predicted oil and gas production rate for the region. | 0.6184E12*OIL2/AREA + 225.3E12*GAS2/AREA | 1.103E5 $Kcal/m_F^2$-yr | La. Oil & Gas 1975; Res. Assoc. 1974 |
| FSRW1 | Average gross primary production of all wetland producers. | 7142.*(W1A/ICW1A) | 7142. $Kcal/m_F^2$-yr | Hopkinson et al. 1977a; Day et al. 1973, 1977; Chabreck 1972 |
| FSRW3 | Average gross primary production of all aquatic producers. | 3455.*(W3A/ICW3A) | 3455. $Kcal/m_F^2$-yr | Allen 1975; Day et al. 1973, 1977 |
| FSFW6 | Biological energy input to the offshore fishery from the offshore organic pool. | 63.48*CW6 | 63.48 $Kcal/m_F^2$-yr | Eckenrod 1977 |

TABLE B-3

Energy and Material Flows
(Non-dynamic State Variables)

| VARIABLE NAME | DESCRIPTION | FUNCTIONAL FORM | INITIAL VALUE* | RATE CONSTANT(S) | SOURCE |
|---|---|---|---|---|---|
| | Primary Energy Flow Sub-model | | | | |
| FEJU2 | Primary energy demand of residential housing for operation of utilities, the additional cost of wetland vs. upland development, and the incremental maintenance cost due to subsidence. | RCEJU2*U1+RCSUB1*FW1AU7 + RCSUB2*SLMWU7 | 955.1 | RCEJU2= 1.67627 RCSUB1= 2.3928 RCSUB2= 4.4358-8 | La. High School Energy Rept. 1975; Res. Assoc. 1974; Wagner & Durabb 1975 |
| FEJU3 | Direct, on-the-farm energy demand for agricultural production. | RCEJU3*UN* (1 + YTREND*TIME) | 5.37 | 226.2 | Hirst 1974; USDA Ag Stat. 1964; Fielder 1972 |
| FEJU4 | Energy demands of the region's industry and commerce. Includes electric utilities and all other users not specifically accounted for in this model. | RCEJU4*U5 | 11,757.0 | 11757.0 | Eckenrod 1977 |
| FEJU5 | Energy cost of municipal waste treatment including collection, secondary, and tertiary costs. | FEPUMP + FEPS + FEPHOS | 2.429 | See App. C | Jennings 1972; SOWERS 1974; Liptak 1974 |
| FEJU6 | Energy requirements for all transportation within the region. Includes private automobile, public mass transit, and commercial. | U1*(KAUTO*8.1896E6 + KPUB*4.3944E6) + KCOM*U4 | 610.6 | KAUTO= 0.9733 KPUB= 0.0267 KCOM= 157.50 | Lincoln 1973; Paddock 1974 |
| FEJU7 | Energy cost of flood protection. Includes only pump operation, not levee construction. | KEJU7*U7/1CU7 | 1.14 | 1.140 | Wagner & Durabb 1975; CPC, New Orleans, 1975; La. Power & Light 1973 |
| | Biological Energy Flows in Natural Sector | | | | |
| FW1R | Respiration losses of wetland primary producers. | RCW1R*W1 | 3571. | 0.36543 | Unpub. data; Day et al. 1973; Hopkinson and Day 1977 |
| FW3R | Respiration losses of aquatic producers | RCW3R*W3 | 1036. | 66.125 | Ibid. |
| FW1W2 | Intake of primary production by all wetland consumers. | RCW1W2*W2 | 2656. | 1.3551 | Computed by difference |
| FW1DEP | Organic deposition in marsh sediments. | RCW1DP*FSR.1 | 158.0 | 0.02212 | Chabreck 1972; Gagliano 1973; Day et al 1973 |
| FW1W4 | Export of detritus from wetlands to inshore waterbodies. | RCW1W4*W1 | 757.0 | 0.077466 | Gosselink et al. 1976; Day et al. 1973, 1977 |
| FW3W4 | Primary aquatic production passing to the aquatic detritus compartment. | RCW3W4*W3 | 2406.5 | 154.07 | Computed by compartment balance |
| FW3W5 | Intake of primary aquatic production directly by aquatic consumers. | RCW3W5*W5*CNW5*CPW5 | 12.49 | 1.3725 | Hinchee 1977 |
| FW2R | Respiration losses of wetland consumers. | FW1W2 | 2814. | -- | Computed |
| FW4R | Respiration losses of the microbe-detritus complex. | RCW4R*W4 | 1518. | 1.5548 | Day et al. 1973 |
| FW4G | Export of detritus from inshore waterbodies to the Gulf of Mexico. | RCW4G*(FW1W4 + FW3W4) | 1501.9 | 0.47476 | Happ et al. 1977 |
| FW4W5 | Intake of detritus by aquatic, inshore consumers. | RCW4W5*W4*W5*CNW5*CPW5 | 143.64 | 0.016168 | Hinchee 1977 |
| FW5R | Respiration losses of aquatic, inshore consumers. | RCW5R*W5 | 153.15 | 16.830 | Day et al. 1973; Hinchee 1977 |
| FW5M | Harvest of inshore fishery by man. Includes menhaden, shrimp, blue crabs, and oysters. | RCW5M*W5*CU1W5M | 2.95 | 0.32418 | Lindall et al. 1972; Day et al. 1973 |
| FW6M | Harvest of offshore fishery including menhaden, shrimp, and crabs. | RCW6M*W6*CU1W6M | 9.30 | 9.30 | Lindall et al. 1972 |
| FW6R | Respiration losses of offshore fishery species. | RCW6R*W6 | 13.14 | 13.14 | Day et al. 1973 |
| FW6FD | Energy flows through feces and death from the offshore fishery. | RCW6FD*FSFW6 | 41.04 | 0.64650 | Ibid. |
| FTFISH | Total fishery yield from the region's waters. | FW5M + FW6M | 12.25 | -- | -- |
| | Food Energy Flows in Urban Sector | | | | |
| FINU1 | Total food requirements of resident population of the region. | RCINU1*U1 | 70.284 | 1.2017E6 | USDA 1971 |
| FCP1 | Consumption rate of regional fishery products. | RC56U1*U1*FTFISH | 0.3441 | 480.33 | USDA 1971; Taylor et al. 1974; Eckenrod 1977 |
| FCT | Consumption of regional seafood products by tourists. | RC56U4*U4*FTFISH | 0.1802 | 0.01471 | Ibid. |
| FCU1 | Residents' consumption of regionally grown agricultural products. | RCU3U1*U8 *(1+YTREND*TIME) | 0.2254 | 0.05820E-8 | Fielder 1972; US Dept. Comm. 1969 |
| FUBU1 | Food imported from external economy and consumed locally. | FINU1-F56U1-FUBU1 | 69.715 | -- | -- |
| F56U1 | Export of regional fishery products to the external economy. | FTFISH - F56U1 | 11.906 | -- | -- |
| FU3EE | Export of regional agricultural products to external economies. | RCU3EE*U8 *(1+YTREND*TIME) | 62.530 | 16.145E-8 | Fielder and Parker 1972 |
| FTOTEE | Total food exports (seafood plus agricultural products) from the region. | F56EE + FU3EE | 74.435 | -- | -- |
| FFDEIT | The region's balance of food trade—exports minus imports. | FTOTEE - FEEU1 | 4.7206 | -- | -- |

# Table B-3 (cont.)

Rates of Land Use Change and Phosphate Fluxes

| | | | | | |
|---|---|---|---|---|---|
| FW1AU7 | Rate of wetland reclamation for urban land uses in New Orleans metropolitan area. | $\dfrac{d(U7)}{dt}$ | 5.59 km$^2$/yr | -- | Hinchee 1977 |
| FW1A3A | Rate of wetland transformation to water area due to natural and man-induced causes. | 0.10413E8 + CANWID*FU1OEJ*EXP(ARGU2) ARGU2 = RCON*TIME | 17.071 km$^2$/yr | CANWID= 60.36 RCON= 0.04879 | Gagliano et al. 1973 |
| FPPONT | Total phosphate flux through Lake Pontchartrain. | 0.3 + 1.111E4*U1 | 0.9498 g/m$^2$-yr | -- | Craig et al. 1977 |
| QSEW | Flow rate of sewage collected in New Orleans sanitary sewer system. | KSEW*U1 | 3.495E10 gal/yr | 5.9762E14 | Jennings 19~ |

## TABLE B-4

### Control Functions

| VARIABLE NAME | DESCRIPTION | FUNCTIONAL FORM | INITIAL VALUE |
|---|---|---|---|
| CNW5 CNW6 | Effect of wetland nursery on energy flow to inshore and offshore fisheries. | (W1A/ICW1A)**0.5 | 1.0 |
| CPW5 | The effect of nutrient-induced eutrophication on Lake Pontchartrain's fishery production, where KPONT equals the fractional contribution of the Lake's yield to the total inshore yield. | CPW5=KPONT + KNONP KPONT=0.17408/FPPONT**0.5 KNONP=0.8214 (Appendix C(38)) | 1.0 |
| CU1W5M CU1W6M | The relationship between the region's industrial-commercial growth and the inshore and offshore fishing efforts. | U4 | 1.0 |
| CU1U7 | The effect of population growth on the demand for urban land. | U1/ICU1 | 1.0 |

## TABLE B-5

### Variable Additions and Modifications for High Intensity Development Simulation

| VARIABLE NAME | DESCRIPTION | FUNCTIONAL FORM | RATE CONSTANT(S) |
|---|---|---|---|
| U4 | Total Regional Output--primary urban sector forcing function. | U4=EXP(ARGU1) ARGU1=0.024846*TIME | -- |
| FEJU2 | Primary energy demand of resiential housing. | Same as current trend reference case | RCEJU2=ICEJU2* EXP(0.023105*TIME) ICEJU2=1.6262E7 |
| FEJU4 | Energy demands of industry and commerce. | Same | RCEJU4=ICEJU4* EXP(0.023105*TIME) ICEJU4=11757.0 |
| FEJU6 | Energy demands for transportation. | Same | For 0<TIME<10, KPUB=0.0267 - 0.002002*TIME For TIME>10, KPUB=0.006675 At all times, KAUTO=1.0 -KPUB |
| U7 | Demand for urban land in New Orleans metro area. | ICU7*EXP(ARGU3) ARGU3=0.03359*TIME | -- |

## TABLE B-6

### Variable Additions and Modifications for Steady State Economy Simulation

| VARIABLE NAME | DESCRIPTION | FUNCTIONAL FORM | RATE CONSTANT(S) |
|---|---|---|---|
| U4 | Total Regional Output | 1.3105 - 0.000345* (ABS(TIME-30.0))**2.0 | -- |
| FEJU2 | Energy demand for residential housing. | Same | TIME<10: RCEJU2= 1.6262E7-0.2439E6*TIME TIME>10: RCEJU2=1.3823E7 |
| FEJU4 | Energy demand of industry and commerce. | Same | TIME<10: RCEJU4= 11757.0-176.3*TIME TIME>10: RCEJU4=9993.0 |
| FEJU5 | Energy requirements for waste treatment. | Same | See Appendix C, Item 15 |
| FEJU6 | Energy demands of transportation. | Same | TIME<10: KPUB= 0.0267+0.04733*TIME TIME>10: KPUB=0.50 At all times: KAUTO=1-KPUB |
| FU1OEJ | Oil and gas production rates. | Compare listings of NOASIS and LOWCAL in Appendix D | -- |
| FSRW1 | Gross primary production of wetlands. | RCSRW1*(W1A/ICW1A)* (FRACNF+1.50*FRACF) FRACF: Appendix C, Item 39 FRACNF=1.0 - FRACF | Same |

TABLE B-6 (Continued)

| VARIABLE NAME | DESCRIPTION | FUNCTIONAL FORM | RATE CONSTANT(S) |
|---|---|---|---|
| YTREND | Trend in agricultural yield (average of major crops). | YTREND=0.0 | -- |
| FU3EE | Export of regional agricultural products. | Same | TIME<10: RCU3EE=0.0<br>TIME>10: RCU3EE=<br>16.145E-8-1.6145E-8*TIME |
| FU3U1 | Consumption rate of local agricultural products. | Same | TIME<10: RCU3U1=<br>0.0582E-8+1.6145E-8*TIME<br>TIME >0: RCU3U1=16.198E-8 |
| U7 | Urban land area of New Orleans metro area. | U7=ICU7 | -- |
| FPPONT | Phosphorous flux through Lake Pontchartrain. | TIME<10 yrs:<br>FPPONT=0.3+1.111E4*U1<br>TIME>10 yrs:<br>FPPONT=0.3+0.6555E4*U1 | -- |

TABLE B-7

Areal Distribution of Wetland
and Aquatic Habitats

| HABITAT | HYDROLOGIC UNIT AREAS (km$^2$) | | | | SUBTOTALS |
|---|---|---|---|---|---|
| | 1 | 2 | 3 | 4 | |
| **WETLANDS (W1A):** | | | | | |
| Saline Marsh | 301 | 279 | 0 | 629 | 1209 |
| Brackish Plus Intermediate | 503 | 475 | 216 | 771 | 1965 |
| Fresh Marsh | 18 | 1 | 115 | 695 | 829 |
| Swamp Forest | 1389 | 12 | 0 | 980 | 2381 |
| Plus Waterbodies <10 acres | -- | -- | -- | -- | 2196 |
| SUBTOTALS | -- | -- | -- | -- | 8580 |
| **WATERBODIES (W3A):** | | | | | |
| Saline | 3412 | 1409 | 46 | 815 | 5682 |
| Brackish Plus Intermediate | 1918 | 231 | 560 | 564 | 3273 |
| Fresh (Marsh) | 139 | 1 | 395 | 564 | 1099 |
| Fresh (Swamp Forest) | 822 | 0 | 0 | 22 | 844 |
| Minus Waterbodies <10 acres | -- | -- | -- | -- | 2196 |
| SUBTOTALS | -- | -- | -- | -- | 8702 |
| Dry Land (U7 + U8 + U9) | 1233 | 186 | 110 | 1246 | 2775 |
| TOTALS | -- | -- | -- | -- | 20057 |

Data Sources:  Gane (1976), Chabreck (1972), Gagliano et al. (1974).

# Development of a "Living" Salt Marsh Ecosystem Model: A Microecosystem Approach

**Wiley M. Kitchens**
**Richard T. Edwards**
**William V. Johnson**
Belle W. Baruch Institute for Marine Biology and Coastal Research
Marine Field Station
University of South Carolina
Georgetown, South Carolina 29440

This study was designed to ultimately test replicability of salt marsh microecosystem units and the closeness with which the structural and functional processes within the microecosystems "tracked" that of a natural marsh site. Four replicate microecosystems were constructed by utilizing "sodded" *Spartina alterniflora* salt marsh plots (with attendant fauna and flora) contained within plastic combination holding and metabolism chambers constructed outdoors and completely open to the atmosphere. This paper reports an approach to the design of the units and rationale for selected community structure and functional process measurements within the units. Each unit was provided with a flow-through seawater system replete with timer and valves in such a fashion as to provide a semidiurnal tide to the marshes which simulated the following characteristics of the tides at the natural site from which the marshes were extracted: (1) sheet flow, (2) real time synchrony with the natural tides, (3) depth of inundation, and (4) duration of inundation. In addition, each unit was complete with flow measurement devices and water sampling ports which were utilized to budget the waters entering and exiting the units. Samples were taken to determine the water-borne fluxes of the various forms of nitrogen and phosphorous for a selected tide. In addition, each unit is complete with flow-through air sampling devices, filon fiber glass lids, sequence timers, pumps, an infrared analyzer, and digital data recording system.

## Introduction

The intrinsic value of the coastal salt marshes as natural resources essential to the economy of the coastal fisheries has been well documented (Odum and Odum, 1972; Gosselink *et al.*, 1973). Likewise, the vulnerability of these systems to the ever-increasing encroachment of man and his attendant influences has also been established (Odum, 1973; Vernberg, 1976). In response to an obvious need for predictive information for functional coastal zone management strategies (particularly in regard to salt marshes), several mathematical simulation salt marsh ecosystem models have been developed (Day *et al.*, 1973; Nixon and Oviatt, 1973; Weigert *et al.*, 1975; Vernberg *et al.*, 1977). These models are attempts to synthesize existing knowledge into a systematic and integrative scheme which will enable the modeler to make assessments at an holistic level regarding ecosystem responses to selected perturbations. Although these models are valuable in interpreting the interactive components of the ecosystem as well as in pinpointing "sensitive" areas or parameters which are control mechanisms (Wiegert *et al.*, 1975), they simply do not have the resolution required for simulation and prediction of very specific environmental perturbations. Mann (1978) has drawn attention to specifics regarding the limitations of these and other models in application to these problems. He has suggested that one problem area is the lack of realistic validation procedures for ecosystem models.

Another approach utilizing an holistic strategy to assess ecosystem functional and structural properties, as well as responses to perturbations, is the "living" model or microecosystem approach. Basically, this technique involves "capturing" a viable part of the ecosystem in question and subjecting it to environmental regimes that simulate as closely as possible the natural system, while at the same time maintaining some boundary control over these regimes for manipulative purposes. Ideally, these systems should be large enough to incorporate as many of the ecosystem components as possible without sacrificing ease of replication, manipulation, and response measurements (Cooke, 1971). In the past, this approach has been successfully employed to investigate aspects of such fundamental ecosystem processes as nutrient cycling in aquatic systems (Whittaker, 1961; Eley *et al.*, 1975), community metabolism (Beyers, 1963, 1965), patterns of ecological succession under various environmental regimes and stress (Cooke, 1967; Wilhm and Long, 1969; Kitchens, unpublished data), response to low freshwater flow regimes in estuaries (Cooper, 1970), and assorted community responses to environmental alterations in flowing streams (Odum and Hoskin, 1957; Lauff and Cummings, 1964; Kevern and Ball, 1965; McIntire *et al.*, 1964; and McIntire and Phinney, 1965). In addition, these "living" models can be excellent tools for the verification of mathematical models.

In the following sections, we will present our design for a short *Spartina alterniflora* salt marsh microecosystem. The discussions will also include rationale and data for measurements detailing community structure and selected functional processes within the microecosystems. (See Figure 1 for conceptual model.) These units were developed in conjunction with the Environmental Protection Agency and the following criteria were incorporated within the design: (1) the units would have to be replicable for use as a bioassay tool; (2) the units would have to be practical enough to be installed at various laboratory locations along the coast to

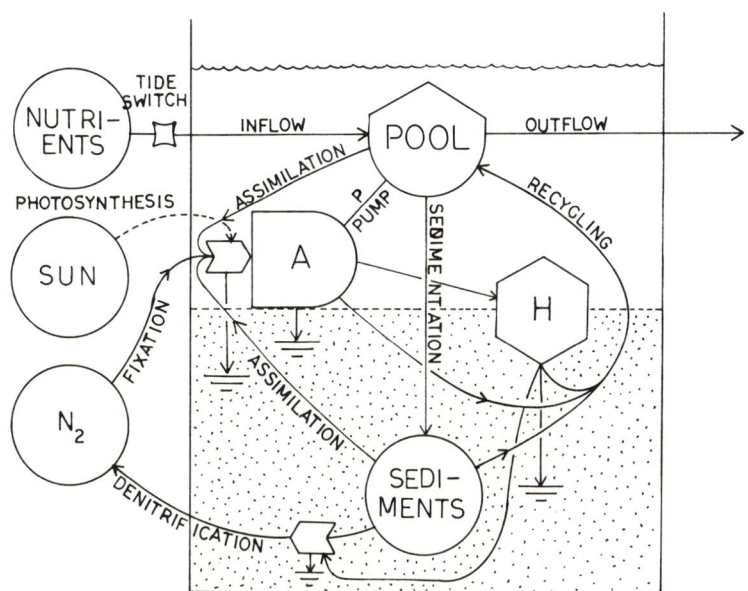

Fig. 1. Generalized conceptual model depicting nutrient flow pathways.

assess local pollution problems; (3) the holistic response measurements would have to incorporate as much automation as feasible; (4) the construction and maintenance costs would have to be kept at a minimum; and (5) the results of tests within the units would have to be extrapolated to the natural ecosystem (community verification would be required).

## Methodology

### Rationale

We designed the "living" microecosystems to study nutrient flux. Figure 1 represents a very simple conceptual model of the community structure and energy flow pathways within the microecosystems. Inputs to the microecosystems are indicated on the left while exports are indicated on the right. Since these systems are semi-enclosed by tank walls, the culmination of all the pathways is the difference between inputs and exports. Since Copeland (1967) defined an environmental stress as any factor that alters these normal pathways, we have speculated that for these "living" models the best index of any stress (whether applied or natural) is reflected as a discrepancy between the input-output characteristics before and after any perturbations. Hence, we designed the seawater systems in such a manner as to be able to budget the fluxes of water and any of its constituents in and out of the systems at any time. We initiated some preliminary tidal budget studies for the fluxes of selected nutrients.

In addition to the nutrient flux studies, we also designed the systems for the determination of community metabolism for the following reasons: (1) community metabolism is a sensitive indicator of community imbalances in various aquatic systems (Copeland and Dorris, 1964; Copeland, 1965; Copeland, 1967); (2) replicated laboratory aquatic systems do not differ significantly with respect to levels of community metabolism (Abbot, 1966); and (3) automation facilitates the measurement of responses.

In addition to these holistic response determinations, we selected to monitor the macro- and meiobenthic faunal communities for the following reasons: (1) since these organisms are sessile, they are directly subject to any environmental stress applied to the systems; (2) the communities have sufficient regeneration times which would facilitate monitoring subtle stress responses or recovery of the systems; and (3) these surveys allow one to compare the microecosystems directly with the natural systems.

The results presented in the following sections are not intended as comprehensive studies but rather as examples of the type of sampling used to assess ecosystem dynamics as well as responses to perturbations.

### DESIGN AND PRELIMINARY RESULTS

The microecosystems consist of four square tanks, each containing approximately six square meters of short *Spartina alterniflora* marsh. The units function as both holding tanks and metabolism chambers for productivity measurements. All support and monitoring systems are automated to operate with a minimum of supervision.

The size of the marsh plots minimizes edge effects found in smaller units without sacrificing sampling convenience and experimental control. Marsh was transplanted from the high marsh zone in the form of sod blocks (approximately 30 cm on a side) with substrate intact to a depth of 20 cm. Observation during transplanting showed that this was deep enough to include the majority of the root mass and virtually all of the attendant faunal community. Sod blocks were reassembled in the tanks using a random numbering arrangement. Initial attempts at establishing a viable marsh with the sod resting directly on the tank bottoms failed due to water pooling which damaged the root stock. A layer of pea gravel was provided to improve drainage and prevent pooling. This type of drainage design simulates interstitial drainage as observed at the control site. A cross section of a tank unit is illustrated in Figure 2.

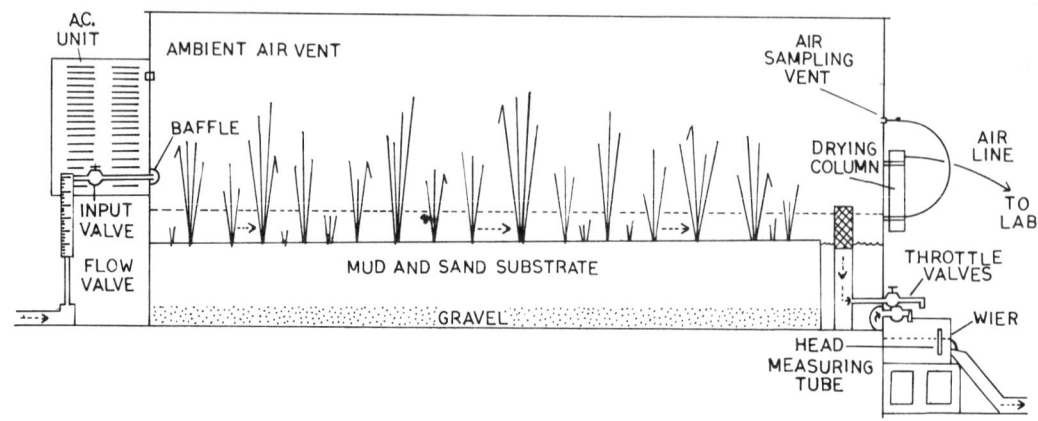

Fig. 2. Cross section of a tank microecosystem unit.

The microecosystems are maintained with a flow-through seawater system designed to simulate natural tidal regimes. A semi-diurnal tide, coincident with the natural tide, was provided by flooding the tanks at 12-hour and 20-minute intervals. Duration and depth of inundation represent average annual conditions at the control site. With minor variations, daily tides 10 cm in depth inundate the substrate for two and one-half hours per cycle. Water movement across the substrate simulated sheet flow. This flow-through design maintains high water quality. The presence of viable meroplankton insures natural recruitment of benthic organisms.

A diagram of the seawater system is shown in Figure 3. Water is pumped continuously from the source creek for the control marsh to a 3000 liter head tank. Water flows into the bottom of the tank and overflows through an outlet at the top into a holding pond. Residence time within the head tank is approximately 15 minutes. To create a flood tide, water is diverted from the head tank through a solenoid valve and into a branching supply network to the tank inputs. After passing over the marsh surface, water flows out through baffled drains into the holding pond. In-line flow meters on the input side and calibrated wires on the output side provide accurate measurements of water flow.

The air sampling system for the productivity measurements is also a flow-through design. During productivity measurements, air

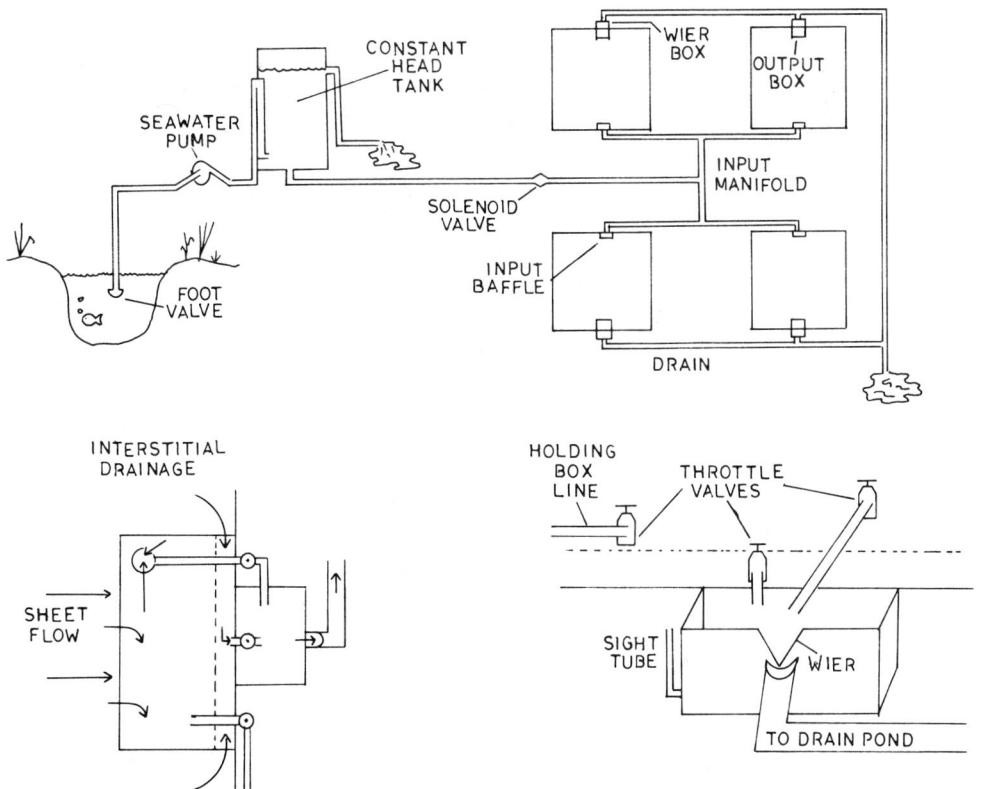

Fig. 3. Schematic of seawater system.

trapped in the chambers is continuously pumped through polyethylene tubing into the lab, where the $CO_2$ concentration is recorded at hourly intervals. Air removed for analysis is replaced by ambient air through vents in the tank walls. Ambient $CO_2$ levels are also measured hourly so that a correction can be made for atmospheric carbon drawn in through the wall vents. The constant addition of ambient $CO_2$ prevents the depletion of $CO_2$ in the chamber due to high photosynthetic rates. The air sampling system is diagramed in Figure 4.

Diaphragm air pumps situated in the lab draw air out of the sampling vents, through drying columns, and into the lab at a rate of $0.3m^3$ per hour. Air is then pumped into a switching manifold where it can be routed either to a Beckman 865 infrared analyzer or an exhaust vent. Each of the five sample streams is diverted through the analyzer for a period of twelve minutes. Halfway through each sampling period, the $CO_2$ concentration of the sample is recorded along with the date, time, and air temperature within the tank.

To trap air during productivity runs, translucent tank lids are bolted to the chambers. The lids are constructed of Filon fiberglass greenhouse panels with a transmittance of 95% in the visible light

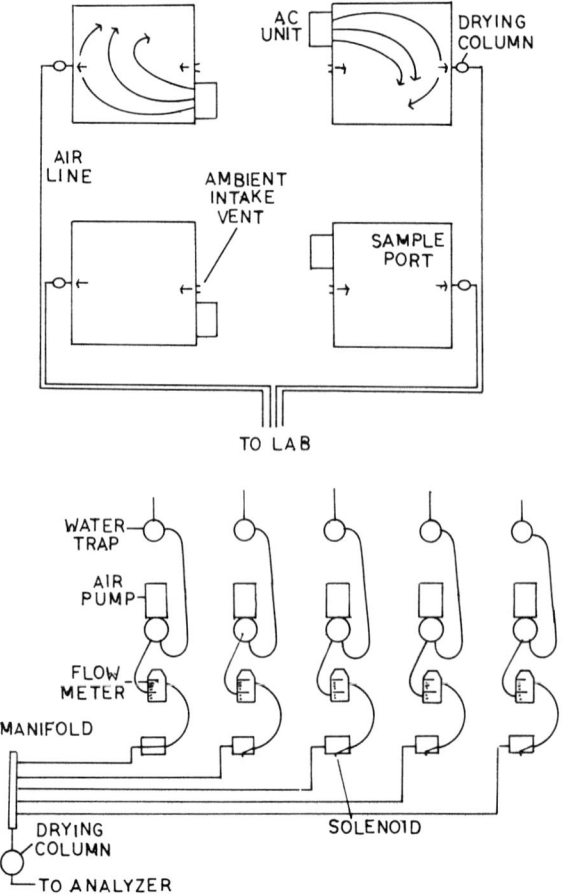

Fig. 4. Schematic of air sampling system.

range.  Since the lids are in place only during productivity measurements, the microecosystems are subjected to normal ambient weather
conditions at all other times.

Window-type air conditioners installed in the tank walls compensate for heat buildup due to the greenhouse effect.  Cooling is
controlled by a differential temperature controller situated on one
of the tank walls.  Measured variations from tank to tank do not
exceed 1°C and from tank to ambient are within ± 2°C.

Net fluxes of dissolved nutrients were determined by comparing
areas under rate change curves plotted for flood and ebb components
of the tide.  Water samples from the input line and from each outflow
drain were collected every 15 minutes for the flux determinations
depicted in Figure 5.  The flow rate at each location was recorded,
along with the date and time.  Using standard analytical techniques

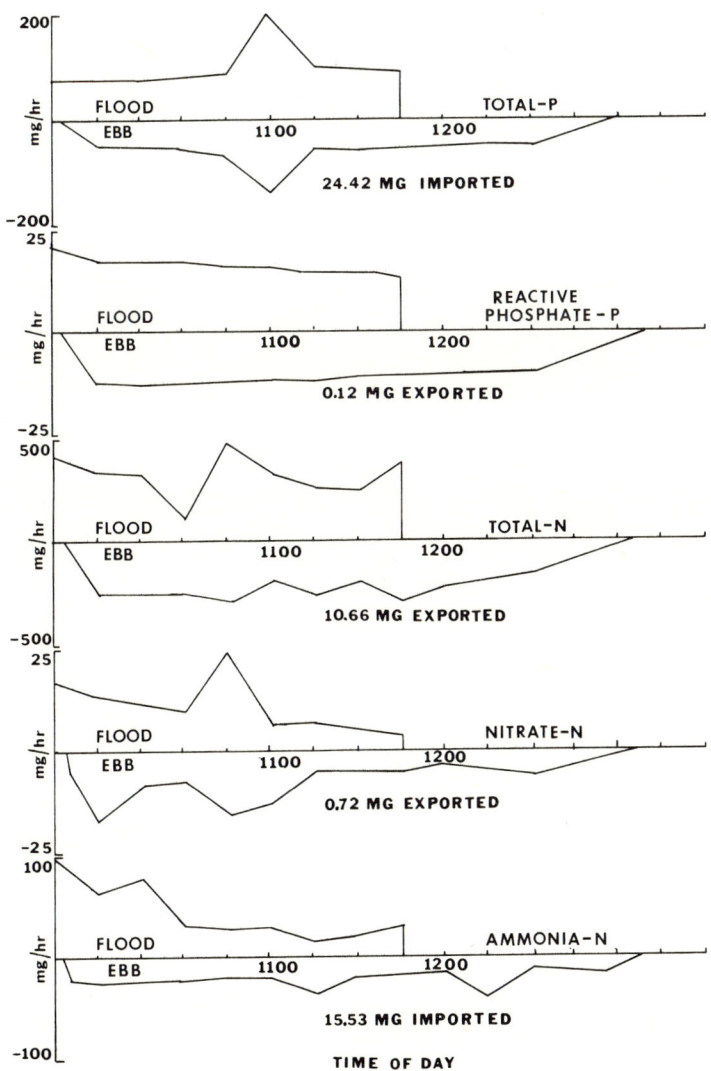

Fig. 5.  Exemplary nutrient flux curves.

(Strickland and Parsons, 1968), concentrations of the following nutrients were determined: nitrates, nitrites, ammonia, total phosphorous, and reactive phosphate phosphorous. Concentration and flow values were then used to calculate the nutrient flow rate at each sampling interval. Rate change curves were then plotted for the inflow and outflow of each tank over the sampling period. These curves were then integrated, with a comparison of the areas under the curves yielding net values for import or export of each nutrient in each tank. In this particular example, 24 mg of total phosphorous were imported into the microecosystems; 0.1 mg of reactive phosphorous were exported; 11 mg of total nitrogen were exported; 0.7 mg of nitrate nitrogen were exported; and 15 mg of ammonia nitrogen were imported. We do not have enough data to make a comprehensive statement regarding these fluxes at this time.

A similar but more complicated method was used to determine community metabolism values. The average rate of change in the mass of carbon in the enclosed air was determined in hourly increments as shown in Figure 6. These values were then corrected for changes

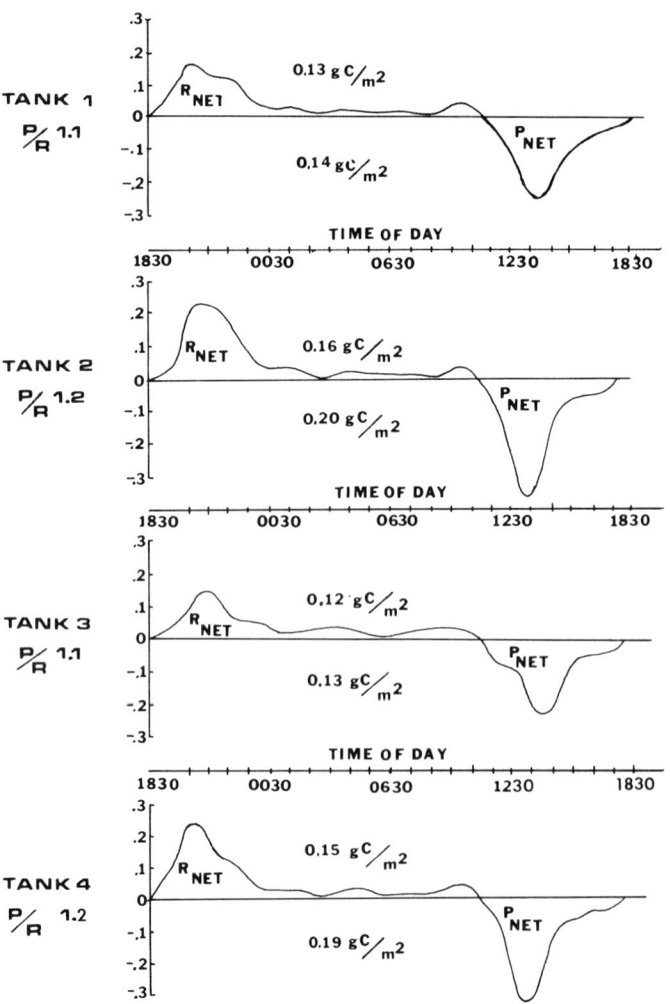

Fig. 6.  Exemplary community metabolism curves.

caused by net fluxes of carbon due to the flow-through air design. Negative rates indicated net production, and positive values showed net respiration. The corrected values were then used to plot rate change curves for the sampling period. Comparisons of the integrated curves were then made to find photosynthesis/respiration ratios for each tank.

In this particular example, net photosynthesis ranged between 0.14 g $C/m^2$/day to 0.20 g $C/m^2$/day for the four replicate micro-ecosystems. Respiration values ranged between 0.12 g $C/m^2$/day. One would expect some variations in these absolute values due to slight discrepancies in the total marsh areas within each unit. The P/R ratio normalizes these values, and the values recorded for the four replicates varied only between 1.1 and 1.2.

### PRELIMINARY RESULTS - VALIDATION STUDIES

Tables 1 and 2 represent a species list for the control site and the microecosystem units. The fiddler crab, *Uca pugilator*, and the polychaetes, *Heteromastus filiformis* and *Neris succinea* were clearly the dominant species in terms of numbers of individuals per $m^2$ in both the microecosystem samples and the control site.

The three species were encountered in all the samples for both sites for all the time periods. The other species were only sporadically encountered although the relative abundances between the two sites remained essentially the same. Inspection of the ANOVA tables (Table 1) reveals that the communities within the microecosystems and the control site are statistically inseparable at the 0.05 level in terms of species densities, with overall mean values of 114 individuals/$m^2$ for the control site and 174 individuals/$m^2$ within the microecosystems.

The results of the meiobenthic survey contrasting the micro-ecosystems with control site are summarized in Table 2. Again in terms of total species densities the two sites are statistically inseparable at the 0.05 level, with a mean value of 525 individuals/ 10 $cm^2$ for the control site and 352 individuals/10 $cm^2$ for the micro-ecosystems.

Table 1. Macrobenthic faunal abundance and diversity in control site and microecosystem for four samples in each site.[*]

| | Date | | | | |
|---|---|---|---|---|---|
| | 6/29/76 | 10/28/76 | | 2/3/77 | |
| Species | Initial | Control | Microecosystem | Control | Microecosystem |
| *Corphium* sp. | – | – | 31.75 | – | – |
| *Geukensia dimissa* | – | 31.75 | – | – | – |
| *Heteromastus filiformis* | 158.75 | – | 63.50 | 22.25 | 317.50 |
| Insects | – | – | – | – | 63.50 |
| *Melampus bidentatus* | – | – | 603.25 | 31.75 | – |
| *Nereis succinea* | 31.75 | 95.25 | 95.25 | 31.75 | 127.00 |
| *Orchestia grillus* | – | – | – | 31.75 | 317.50 |
| *Uca pugilator* | 349.25 | 127.00 | 31.75 | 158.75 | 95.25 |
| *Uca pugnax* | 158.75 | – | – | 31.75 | – |
| Mean #/$m^2$ | 174.63 | 84.67 | 165.10 | 84.67 | 184.15 |
| Diversity ($\overline{H}$) | 0.70 | 0.16 | 0.37 | 0.71 | 0.73 |

[*]Macrobenthic faunal abundance $F_{0.05}(1.22) = 4.30$.

Table 2.  Meiofaunal abundance in control site and microecosystem ($F_{0.05}$ (1.69) = 3.98).

| Site | Time (No. of Samples) | | | | | | | |
|------|---------|---------|---------|----------|----------|--------|--------|--------|
|      | 7/76(6) | 8/76(6) | 9/76(3) | 10/76(3) | 12/76(3) | 1/77(3) | 2/77(3) | 3/77(3) |
| Control | 387.3 | 554.5 | 612.0 | 615.6 | 279.5 | 599.0 | 438.6 | 315.0 |
| Microecosystem | 447.0 | 491.0 | 581.0 | 568.9 | 761.0 | 424.0 | 330.8 | 340.7 |

## Concluding Remarks

It is felt that the design and initial testing of a viable and functional microecosystem, or "living" model of a selected salt marsh community, has proved this system feasible.  This model seems to "track" the natural marsh in terms of its structural components, and testing certain functional aspects such as nutrient assimilation and community metabolism between the model and natural marsh should further validate the model.  This design can be a potentially valuable tool for the assessment and prediction of certain environmental perturbations within the salt marsh ecosystem.

## Acknowledgment

The authors express gratitude to the Environmental Protection Agency and Belle W. Baruch Institute for Marine Biology and Coastal Research for supporting this project.  We also wish to thank F. J. Vernberg, S. S. Bell, B. C. Coull, L. R. Gardner, and J. Matheson for their contributions to this study.  Contribution No. 221 of the Belle W. Baruch Institute of Marine Biology and Coastal Research.

## References

Abbot, W.  1966.  Microcosm studies on estuarine water.  I. The replicability of microcosms.  J.W. P.C.F. 38:258-270.

Beyers, R. J.  1963.  A characteristic diurnal metabolic pattern in balanced microcosms.  Publ. Inst. Mar. Sci. Univ. Tex. 9:19-27.

_____.  1965.  The pattern of photosynthesis and respiration in laboratory microecosystems.  In: R. R. Goldman (ed.) *Primary Productivity in Aquatic Environments*, pp. 61-74.  Univ. Calif. Press, Berkeley.

Cooke, G. D.  1967.  The pattern of autotrophic succession in laboratory microcosms.  BioScience 17: 717-721.

_____.  1971.  Aquatic laboratory microecosystems and communities.  In: J. Cairns (ed.) *The Structure and Function of Freshwater Microbial Communities*.  Am. Soc. Symp., Virginia Polytech. Inst., Blacksburg, Va.

Cooper, D. C.  1970.  Responses of continuous-series estuarine microecosystems to point-source input variations.  Ph.D. dissertation, Univ. of Texas.

Copeland, B. J.  1965.  Evidence for regulation of community metabolism in a marine ecosystem.  Ecology 46(4): 563-564.

_____.  1967.  Biological and physiological basis of indicator communities.  In: T. A. Olson and F. J. Burgess (eds.) *Pollution and marine ecology*.  Interscience, New York.

_____ and T. C. Dorris.  1964.  Community metabolism in ecosystems receiving oil refinery effluents.  Limnol. Oceanogr. 9(3): 431-447.

Day, J. W., W. G. Smith, P. R. Wagner, and W. C. Stowe.  1973.  Community structure and carbon budget of a salt marsh and shallow bay estuarine system in Louisiana.  Center for Wetland Resources. L.S.U., Baton Rouge, La.

Eley, R. L., J. W. Falco, and C. J. Kirby.  1975.  The role of physical modelling in marsh-estuarine mineral-cycling research.  In: F. G. Howell, J. B. Gentry, and M. H. Smith (eds.) *Mineral Cycling in Southeastern Ecosystems*.  USAEC Symposium Series, CONF.-740513.

Gosselink, J. G., E. P. Odum, and R. M. Pope.  1973.  The value of the tidal marsh.  Pre-publication draft.  Urban and Regional Development Center.  Gainesville, Florida.

Kevern, N. R. and R. C. Ball.  1965.  Primary productivity and energy relationships in artificial streams.  Limnol. Oceanogr. 10: 74-87.

Lauff, S. H. and K. W. Cummings. 1964. A model stream for studies in lotic ecology. Ecology 45: 188-190.

Mann. K. H. 1978. Qualitative aspects of estuarine modelling. In: R. F. Dame (ed.) *Marsh-Estuarine Systems Simulation*. University of South Carolina Press, Columbia.

McIntire, C. D., R. L. Garrison, H. K. Phinney, and C. E. Warren. 1964. Primary production in laboratory streams. Limnol. Oceanogr. 9:92-102.

_____ and H. K. Phinney. 1965. Laboratory studies of periphyton production and community metabolism in lotic environments. Ecol. Mongr. 35: 237-258.

Nixon, S. W. and C. A. Oviatt. 1973. Ecology of a New England salt marsh. Ecol. Mongr. 43: 463-498.

Odum, E. P. and H. T. Odum. 1972. Natural areas as necessary components of man's total environment. Trans. North Amer. Wildlife and Nat. Res. Conf. 37: 178-189.

Odum, H. T. and C. M. Hoskin. 1958. Comparative studies on the metabolism of marine waters. Publ. Univ. Tex. Mar. Sci. Inst. 5: 16-46.

Odum, W. E. 1973. Insidious alteration of the estuarine environment. Trans. Amer. Fish. Soc. 99: 836-847.

Strickland, J. D. H. and T. R. Parsons. 1968. A Manual of Sea Water Analysis. Bull. Fish. Res. Bd. Canada 125.

Teal, J. M. 1962. Energy flow in the salt marsh ecosystem of Georgia. Ecology 43: 614-624.

Vernberg, F. J. 1976. Characterization of the natural estuary in terms of energy flow and pollution impact. Chapter 3. In: *Report to Congress of Estuaries*. Environmental Protection Agency.

_____, R. Bonnell, B. Coull, R. Dame, Jr., P. DeCoursey, W. Kitchens, Jr., B. Kjerfve, H. Stevenson, W. Vernberg, R. Zingmark. 1977. The dynamics of an estuary as a natural ecosystem. Final Report. EPA. Gulf Breeze, Florida.

Weigert, R. G., R. R. Christian, J. L. Gallager, J. R. Hall, R. D. H. Jones, and R. L. Wetzel. 1975. A preliminary ecosystem model of coastal Georgia *Spartina* marsh. In: L. Eugene Cronin (ed.) *Estuarine Research* 1: 583-601.

Whittaker, R. H. 1961. Experiments with radiophosphorous tracer in aquarium microcosms. Ecol. Mongr. 31: 157-188.

Wilhm, J. L. and J. Long. 1969. Succession in algal mat communities at three different nutrient levels. Ecology 50: 645-652.

# Effects of Perturbations on Estuarine Microcosms

**Donald R. Heinle, David A. Flemer, Rogers T. Huff,
Shelly T. Sulkin, Robert E. Ulanowicz**
University of Maryland
Center for Environmental and Estuarine Studies
Chesapeake Biological Laboratory
Solomons, Maryland 20688

Microcosms containing planktonic communities from Chesapeake Bay responded to enrichment with sewage by developing larger standing crops of phytoplankton and zooplankton. Data suggest that increased productivity would be reflected up the food chain but might increase existing problems with dissolved oxygen and might lead to qualitative changes in the composition of the zooplankton.

Either phosphorus or nitrogen was removed more rapidly from solution depending on where and when the experimental water was obtained. Increases in standing crop of algae were associated with loss of nitrogen from solution in two experiments and losses of both nitrogen and phosphorus from solution in one experiment.

## Introduction

It is now generally conceded that in most unpolluted fresh waters phosphorus is the major nutrient that most often limits algal growth (e.g., Deevey, 1972; Fuhs $et$ $al.$, 1972). Similarly, nitrogen is usually considered to be the limiting nutrient in coastal marine waters (e.g., Ryther and Dunstan, 1971; Thayer, 1971; Flemer, 1972). An estuary with its gradient of fresh to salt water thus probably contains a spatial gradient in limiting nutrients. In addition, there are temporal changes in the concentration of nutrients within the saline portion of an estuary (Taft $et$ $al.$, 1976; Taft and Taylor, 1976). Recent evidence indicates that some tributaries and portions of Chesapeake Bay have already been considerably altered by cultural eutrophication (Jaworski $et$ $al.$, 1972; Flemer and Heinle, 1974). The first readily observable effects of eutrophication have been increased standing stocks of algae as measured by chlorophyll a (Flemer, 1972; Flemer and Heinle, 1974). In Back River, a small tributary of Chesapeake Bay heavily loaded by treated sewage, excess production of algae has led to periods when nighttime respiration by algae has led to anoxic conditions. The tidal freshwater portions of some tributaries have developed noxious blooms of blue-green algae (Brehmer, 1967; Jaworski *et al.*, 1972).

The undesirability of blooms of blue-green algae and anoxic waters is fairly obvious, as they represent or cause major changes in the trophic structure of estuaries that directly affect human use. Less dramatic changes, such as increases in standing stocks and production of algae without apparent qualitative changes, are not as clearly perceived. In this paper we describe some experiments designed to determine the short-term effects of nutrient additions to planktonic ecosystems in Chesapeake Bay. At the time of the experiments, approximately 2.8% of the fresh water entering Chesapeake Bay was sewage (Brush, 1974). Our experiments were designed to determine how the standing crops of phytoplankton and zooplankton might respond to the further addition of sewage and which major nutrients might be limiting standing stocks and productivity, and also to obtain time-series data for use in developing and testing a modeling technique. Methods and results of the modeling have been described by Mobley (1973), Ulanowicz *et al.* (1975), and Ulanowicz *et al.* (1978). This paper describes the data in greater detail, including some of the interrelationships between nutrients, phytoplankton, and zooplankton that were not fully described in the papers on modeling.

## Methods and Materials

The three experiments reported in this paper were done from September 16 to November 3, 1972 (experiment 1), July 19 to August 3, 1973 (experiment 2), and September 25 to October 9, 1973 (experiment 3). Experimental procedures were modified slightly each time as a result of experience gained during preceding experiments in an effort to improve replication and reduce day-to-day variations in data. Microcosms were chosen over field studies as they provided a means of repeatedly sampling the same mass of water, albeit small.

We used six 750-liter polyethylene tanks (cylinder 0.94m in diameter by 1.22 m high) for microcosms (Plate 1). Early experiments indicated that mixing was important for sampling purposes and necessary for the retention of realistic planktonic populations in the microcosms. The tanks were stirred four times daily by large polyethylene propellers mounted on flexible fiberglass shafts driven at 24 rpm by gear motors. During the first two experiments each stirring was for one hour, and each was for 20 minutes during the third experiment. Each tank was illuminated by a 500 w quartz-iodide floodlight mounted above the surface of the water on a cycle of 12 hrs. light and 12 hrs. dark. One-eighth inch plexiglass covers attenuated the illumination to about 0.15 g cal cm$^{-2}$·min$^{-1}$ at the surface of the water. Dawn and dusk were simulated by

Plate 1.   CBL estuarine microcosms.

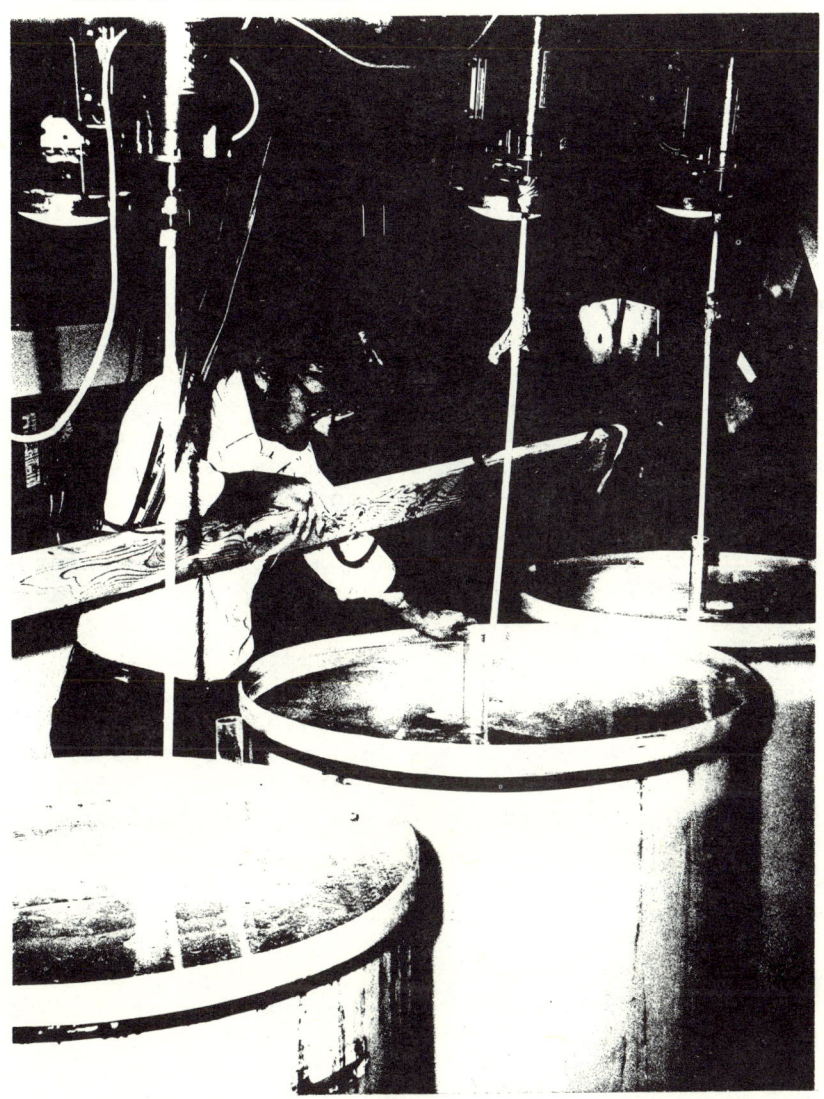

a single 20 w plant-gro flourescent light that burned for 30 minutes
before and after the quartz-iodide lights.  Control of temperatures
was achieved by cold water circulated through 0.5cm inside diameter
polyethylene tubing coiled near the perimeter of the tanks at the
surface.  The walls and bottoms of the tanks were scrubbed daily with
a string mop of synthetic sponge during the third experiment and
less frequently during the first two experiments.  Samples were taken
daily between 0900 and 1000 hours  during a mixing period by
vigorously agitating the contents further with a bucket and then
dipping the sample from the surface with the bucket.

Chlorophyll $\underline{a}$ was measured by a modification of the flourometric
techniques of Yentsch and Menzel (1963) and Holm-Hansen $et$ $al.$ (1965)
as described by Heinle and Flemer (1975).  Samples collected on
Whatman GF/C filters with magnesium carbonate were frozen for analysis.

Particulate organic carbon was determined by the method of Menzel and Vaccaro (1964) using a Beckman Model IR215 infra-red analyzer.

The Dumas method of high temperature oxidation was used to determine particulate nitrogen. Analyses were carried out on a Coleman Model 29A Nitrogen Analyzer equipped with a Model 29 combustion tube and syringe. Particulate material was concentrated on Whatman GF/C filters. These filters were then desiccated and frozen until analyzed. The filters were rolled and put in the combustion tube for oxidation at 850°C.

Total phosphorus was determined with the oxidation method of Menzel and Corwin (1965). The same method was used to oxidize dissolved phosphorus after the sample had been passed through a GF/C filter. Dissolved inorganic reactive phosphorus was determined with the composite reagent method (Strickland and Parsons, 1968).

All samples of dissolved nutrients were quick-frozen and stored for later analysis. Ammonia plus amino acids were measured by the methods of Solorzano (1969). Nitrite and nitrate were analyzed by the methods given in Strickland and Parsons (1968). We employed a modification of the ultra-violet light oxidation method for the determination of soluble organic nitrogen (Strickland and Parsons, 1968). A half-strength seawater solution was used for the solvent for the blanks, ammonium sulfate, and pyridine standards. The seawater solution was made up, according to Strickland and Parsons' (1968) nitrate method, and then diluted by one-half with double distilled, deionized water. This solution was used to dilute river water samples and to add salts to facilitate the uv oxidation. We diluted 20 ml of Bay water sample to 100 ml with half-strength seawater. Two drops of 30% hydrogen peroxide were added to the sample in the quartz tube; the sample was capped and irradiated 7 cm from a 1,200 watt Hanevia-Englehardt 189A lamp for three hours. Strickland and Parsons' (1968) procedure was followed for the remainder of the analysis.

Relative primary productivity was measured by suspending two light and one dark bottles to which $^{14}$C had been added just below the surface of the water in each tank. Phytoplankton were exposed to the label for 4 hr. Between 20 and 40 ml, depending on the density of phytoplankton, were filtered through 0.45-μm Millipore filters. Samples were counted on a Nuclear Chicago planchet counter with a model 8707 decode scaler, a model 1042 planchet changer, and a model 470 gas flow detector.

One-liter samples of zooplankton were concentrated in a net cup fitted with a No.20 (70-74 μm aperture) Monel screen. The concentrated samples were preserved in 4% formaldehyde. The entire content of each sample was later counted by species and major age group (eggs, nauplii, copepodids and adults of copepods; rotifers and eggs of rotifers). All the individuals in sparse samples and representative numbers in dense samples were measured so that herbivore biomass could be calculated from known length-weight relationships of copepods (Heinle, 1969; Heinle and Flemer, 1975) and from the spherical volume of rotifers and an assumed density of 1.0 and dry weight of 20% of mass.

Temperature and concentrations of dissolved oxygen were measured daily with a Yellow Springs Instrument Model 54 oxygen analyzer. These data were taken primarily to assure that the environmental control was adequate, and as little variation occurred, they will not be reported in detail.

The first two experiments were done with water collected from a depth of 1m at the end of the pier at the Chesapeake Biological Laboratory at Solomons, Maryland, near the confluence of the Patuxent River estuary with Chesapeake Bay. The third experiment was done with water collected from a depth of 4m above the outfall of the Broad Neck Sewage Treatment plant at the confluence of the Magothy River estuary with Chesapeake Bay, hopefully to predict effects of that plant on the immediate area (Flemer and Heinle, 1974). For all three experiments water was pumped into a mixing tank and thence into the experimental tank. During the first two experiments, we attempted to regulate the initial population of herbivores by filtering the water through a series of successively finer mesh nets down to 28 um. Copepods were then added from mass cultures. As we were unable to completely regulate the resulting populations, water was filtered through a 0.5mm net only to remove predatory zooplankton (ctenophores and jellyfish) during the third experiment. In all three experiments one to three small carnivorous fish (*Menidia menidia*) were placed in each tank. We attempted to achieve a 10 to 1 ratio of herbivore biomass to carnivore biomass and so used fish 20 to 40 mm long. Fish that died during the second experiment were replaced. No fish died during the first and third experiments.

During the first experiment, two tanks were perturbed by the addition of 10 µg-atoms of nitrogen as $NH_4Cl$ and 1 µg-atom of phosphorus as $KH_2PO_4$ x $1^{-1}$, respectively. There was no replication of the first experiment. A single tank received no additions. We thus attempted to determine if nutrients added singly would affect the ecosystems. During the second experiment, two tanks each received 1% and 10% by volume of treated sewage from the Broad Neck treatment plant. Two tanks received no additions and served as controls. The 10% dose caused changes that appeared to be related to chlorine rather than nutrients, so during the third experiment, two tanks each received 1% and 0.1% by volume of treated sewage from the Broad Neck treatment plant. Two tanks again were unperturbed and served as controls.

Prior to the third experiment, the inner walls of the tanks were scrubbed thoroughly with 70% ethanol to reduce populations of microorganisms.

Table 1 is a summary of the analytical precision we achieved for fractions of nitrogen and phosphorus, using several replicates of individual representative samples. Analytical precision as percent of means was somewhat better for nitrogen compounds than for phosphorus.

## Results

During the first experiment, the concentrations of chlorophyll a increased in all three tanks during the first three days (Fig. 1),

Table 1. Analytical precision estimated from replicates of representative samples.

| Compound | Mean (µg at /1) | Standard deviation | No. of replicates | 95% confidence limits(±) |
|---|---|---|---|---|
| NO$_2$ | 1.12 | 0.006 | 10 | 0.395 |
| NO$_3$ | 13.89 | 0.105 | 9 | 0.207 |
| NH$_3$ | 9.49 | 0.202 | 12 | 0.395 |
| Total dissolved nitrogen | 8.03* 80.3(x10) | 0.633 6.33 | 10 10 | 1.24 12.4 |
| Dissolved inorganic phosphorus | 2.01 | 0.074 | 10 | 0.146 |
| Total dissolved phosphorus | 2.79 | 0.503 | 13 | 0.986 |
| Total phosphorus | 2.79 | 0.365 | 15 | 0.715 |

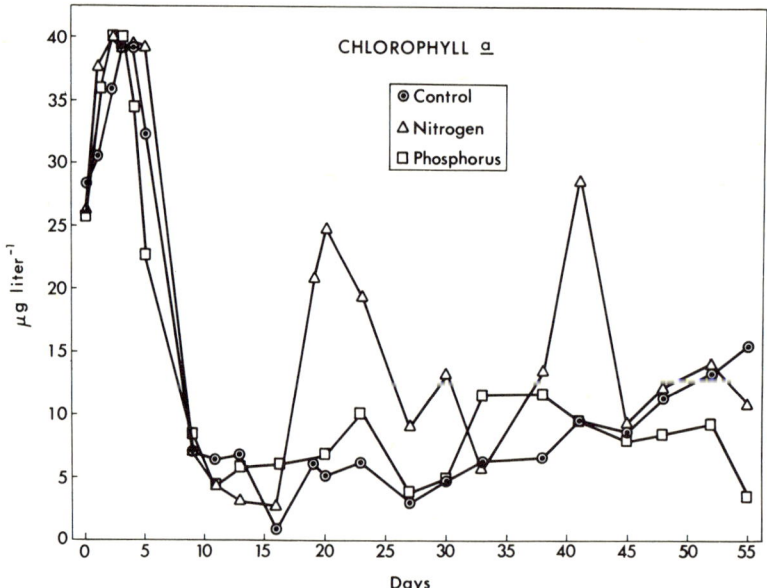

Fig. 1. Concentrations of chlorophyll a (µg l$^{-1}$) versus time in tanks enriched with nutrients and a control tank, September 16 to November 3, 1972. Experiment 1.

presumably in response to a release from grazing. The rise was most rapid in the tank to which NH$_4$Cl was added, less rapid in the tank to which KH$_2$PO$_4$ was added, and least rapid in the control tank. The concentrations of chlorophyll a were between 37 and 40 µg l$^{-1}$ in the control and phosphorus tanks for only 2 days but remained at those levels for 5 days in the nitrogen tank. Concentrations fell to below 10 µg l$^{-1}$ in all tanks by the ninth day of the experiment. The concentrations in the control and phosphorus tanks then rose gradually with fluctuations until the end of the experiment while the nitrogen tank experienced two more "blooms" on approximately the 20th and 40th days.

The nitrogen tank consistently had the lowest biomass of copepods through the first 40 days (Fig. 2). The species copepod that was added to the tanks on day 5 was *Eurytemora affinis*, but as

*Acartia tonsa* always was much more abundant, the herbivore populations developed primarily from eggs that were accidentally introduced into the tanks. The increase in biomass of copepods during the gap in sampling between days 4 and 9 was due mainly to growth and reproduction of *A. tonsa*. Most of the decline in concentrations of chlorophyll a between days 4 and 9 was presumably the consequence of grazing by the copepods.

The concentrations of ammonia are reflected on day 0 in Figure 3. The tank to which $NH_4Cl$ was added had an ammonia concentration of 5.35 µg at $1^{-1}$ prior to the addition, nearly identical to the control tank. Concentrations of ammonia remained reasonably high, but variable, in the nitrogen tank through day 13 and then fell to less than 1 µg at $1^{-1}$ by day 20. After an initial drop, the concentrations of ammonia rose between days 3 and 13 in both the control and phosphorus tanks, presumably as a result of excretion by the large population of copepods. Concentrations of ammonia also fell in the control and phosphorus tanks by day 20. With the exception of one sample from the control tank, less than 2.5 µg at $1^{-1}$ of ammonia were observed in all three tanks for the duration of the experiment.

The concentration of nitrate varied slightly around 3 µg at $1^{-1}$ in the nitrogen tank until day 13 while it fell very rapidly in the phosphorus tank and equally rapidly but from a higher initial value in the control tank (Fig. 4). Preferential use of ammonia added to the nitrogen tank probably accounts for the relatively constant concentration of nitrate in that tank. The timing of the removal of nitrate in the phosphorus and control tanks may reflect, in part, the one day difference between them in algal growth (Fig. 1).

The phosphorus added was used rapidly, and concentrations were similar in all tanks by the 4th day (Fig. 5). The concentrations of phosphorus rose again in the phosphorus and nitrogen tanks on days 11 and 13, but were low again by day 20. The source of the large pulse of phosphorus in the control tank on day 38 is unknown, but apparently real. By the 41st day most had been converted to dis-

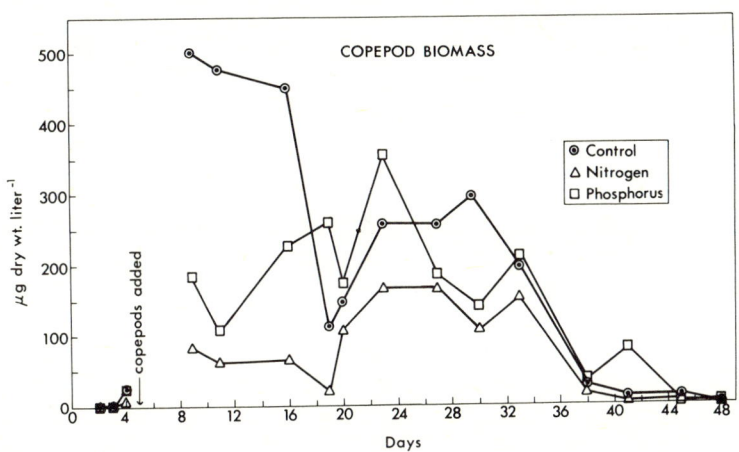

Fig. 2. Biomass of herbivores (µg dry weight $1^{-1}$) versus time. Experiment 1.

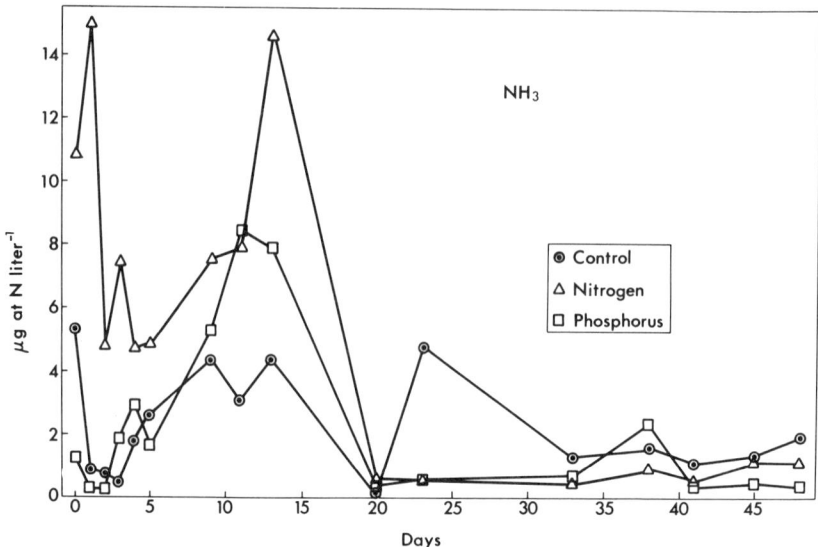

Fig. 3. Concentrations of ammonia (μg at $1^{-1}$) versus time. Experiment 1.

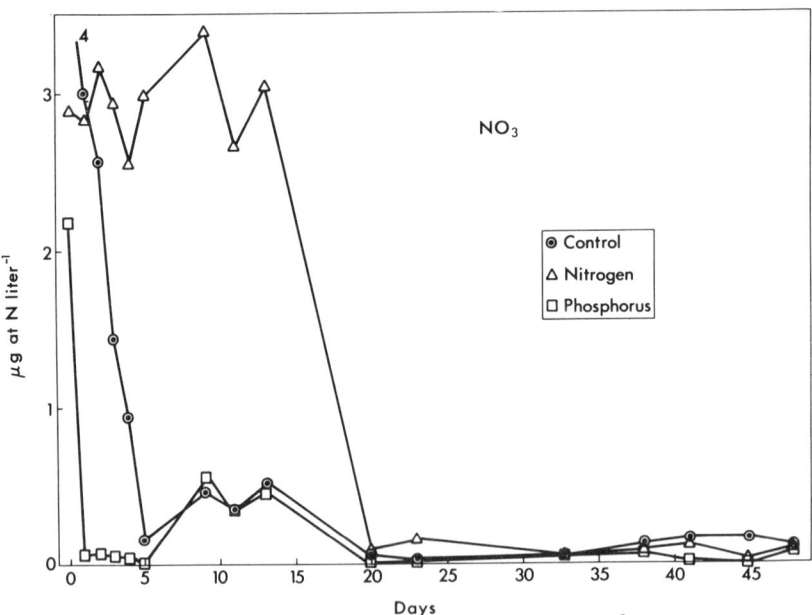

Fig. 4. Concentrations of nitrate (μg at $1^{-1}$) versus time. Experiment 1.

solved organic phosphorus (the difference between total dissolved phosphorus and dissolved inorganic phosphorus), and concentrations were similar in all three tanks after day 45. Phosphate concentrations that we measured were higher than those for total dissolved phosphorus, probably a result of arsenate interference (Goulden and Brooksbank, 1974) and therefore not reported here.

Three fish were added to each of the tanks on the 28th day of the experiment. The fish were between 20 and 42 mm long. Comparison of length-weight relationships at the end of the experiment with those of the wild fish placed in the tank indicated that the fish

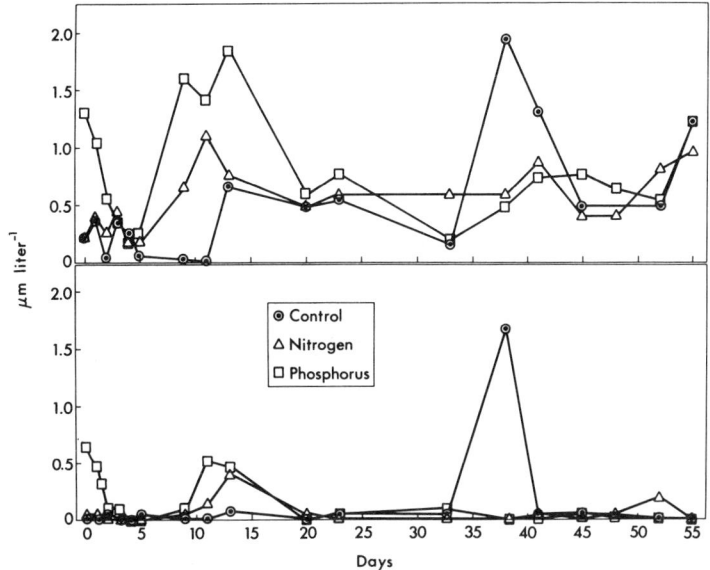

Fig. 5. Concentrations of dissolved inorganic phosphorus
(lower panel) and total dissolved phosphorus (upper panel) ($\mu$g at $1^{-1}$)
versus time. Experiment 1.

were slightly fatter than wild fish in all three tanks, but fattest
in the control tanks (Table 2). The final condition of the fish
thus indicated feeding during the experiments.

From the first experiment we learned that rapid changes in
populations of primary producers, grazers, and nutrients occurred
during the first two weeks. A limiting nutrient could not be
clearly detected in this experiment although phosphorus was reduced
to low levels more rapidly than nitrogen. Large effects related
to the addition of nutrients were not apparent in either the concentration of chlorophyll a (Fig. 1) or the biomass of herbivores
(Fig. 2).

In the second experiment, the concentrations of chlorophyll a
fell from about 30 $\mu$g $1^{-1}$ to about 10 $\mu$g $1^{-1}$ in the tank that received 10% sewage (Fig. 6). The treated sewage had a chlorine concentration of "higher than 3 parts per million," according to the
operator of the sewage treatment plant. We could smell chlorine
over the tanks that received the 10% by volume of sewage for a few
hours after the addition. The concentrations of chlorophyll a remained lower in those tanks until day 7. After day 7 the concentrations in one tank (Tank B) were similar to those in the control

Table 2. Length-weight relationship of *Menidia menidia*
placed in the tanks during experiment 1 (wild fish) and after 16
days in the tanks. Dry weight (mg) = b length (mm)-a, r = linear
correlation coefficient, n = number of fish.

|  | b | a | r | n |
|---|---|---|---|---|
| Wild fish | 3.48 | − 59.4 | 0.995 | 5 |
| Control tank | 5.57 | −109.4 | 0.998 | 3 |
| Nitrogen tank | 4.25 | − 89.2 | 0.996 | 3 |
| Phosphorus tank | 4.51 | − 87.0 | 0.996 | 3 |

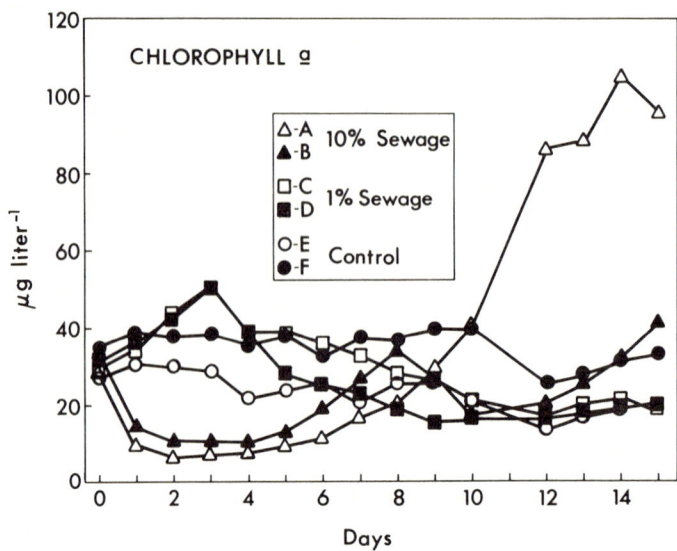

Fig. 6. Concentrations of chlorophyll a (µg l$^{-1}$) versus time in tanks perturbed by 10% and 1% sewage and in controls July 19 through August 3, 1973. Experiment 2.

tanks and the tanks that received 1% sewage while those in the other tank (Tank A) rose to over 90 µg l$^{-1}$. The concentrations of chlorophyll a rose to 50 µg l$^{-1}$ by day 3 in the tanks that received 1% sewage and then declined gradually for the remainder of the experiment. The concentrations of chlorophyll a in the control tanks remained between 20 and 40 µg l$^{-1}$ for most of the experiment. Primary productivity followed the trends of chlorophyll a.

Eurytemora affinis were added from mass cultures on the first day of the experiment. After day 7, Tanks A and B (10% sewage) still contained primarily E. affinis, Tanks C and D (1% sewage) were mixed E. affinis and Acartia tonsa, and the control tanks (E and F) had sparse populations of nearly pure A. tonsa with some Oithona colcarva.

The biomass of copepods remained low in all tanks until day 7 (Fig. 7). The biomass of the copepods increased to over 600 µg dry weight per l in Tank B (10% sewage) and Tanks C and D (1% sewage) during the last seven days of the experiment. Biomass of copepods remained low in the other tank that received 10% sewage (Tank A) and in the two controls for the duration of the experiment.

Two fish were added to each tank on day 0. By day 3 all of the fish had died in the tanks that received 10% sewage, one died in one of the tanks that received 1% sewage, and one died in each of the controls. All were replaced on day 3. The fish were not observed on day 5 and presumed dead, but not replaced. A live fish was later observed in Tank A (10% sewage) and removed on day 11.

In this experiment the tanks that received sewage developed eithe greater concentrations of chlorophyll a (Tank A), greater biomass of copepods (Tank B), or both (Tanks C and D). The failure of the copepods (E. affinis) to outgrow the predation by fish in Tank A apparently caused that tank to develop quite differently from its repli-

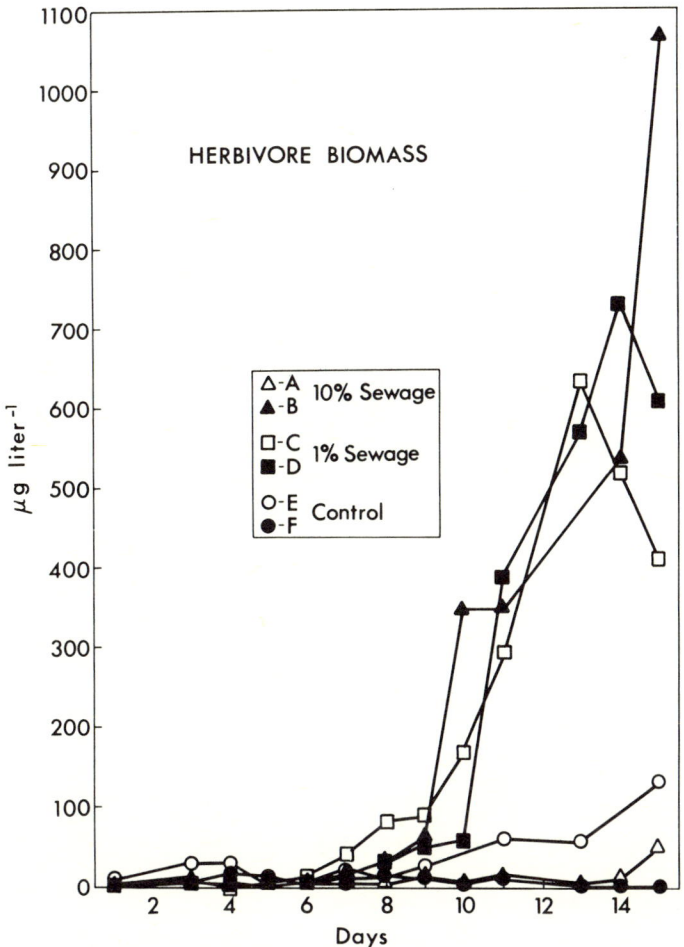

HERBIVORE BIOMASS

△-A
▲-B  10% Sewage

□-C
■-D  1% Sewage

O-E
●-F  Control

$\mu$g liter$^{-1}$

Days

Fig. 7.  Biomass of herbivores ($\mu$g dry weight l$^{-1}$) versus time.
Experiment 2.

cate (Tank B).  The concentrations of total dissolved phosphorus
in all six tanks reflected the addition of sewage between days 0 and
1 and then remained nearly constant for the duration of the experiment (Fig. 8).  Clearly, there was no net removal of phosphorus related to the increases that occurred in algal or copepod biomass in
the tanks that received sewage.  The concentrations of dissolved inorganic phosphorus and total phosphorus were similar to those for
total dissolved phosphorus shown in Figure 8 and therefore are not
presented.

The concentrations of ammonia did not reflect the addition of
sewage and varied widely--between 1 and 20 $\mu$g at l$^{-1}$ in all six
tanks with no apparent patterns.  The concentrations of nitrate did
reflect the addition of sewage, however, and rose to 180 - 193
$\mu$g at l$^{-1}$ in the tanks that received 10% sewage and to 15.5 - 16.5
$\mu$g at l$^{-1}$ in the tanks that received 1% sewage (Fig. 9).  Concentrations of nitrate in the controls varied between 0.02 and 0.24 $\mu$g at
l$^{-1}$ throughout the experiment with no apparent trends with time.
Concentrations in the tanks that received 10% sewage remained over

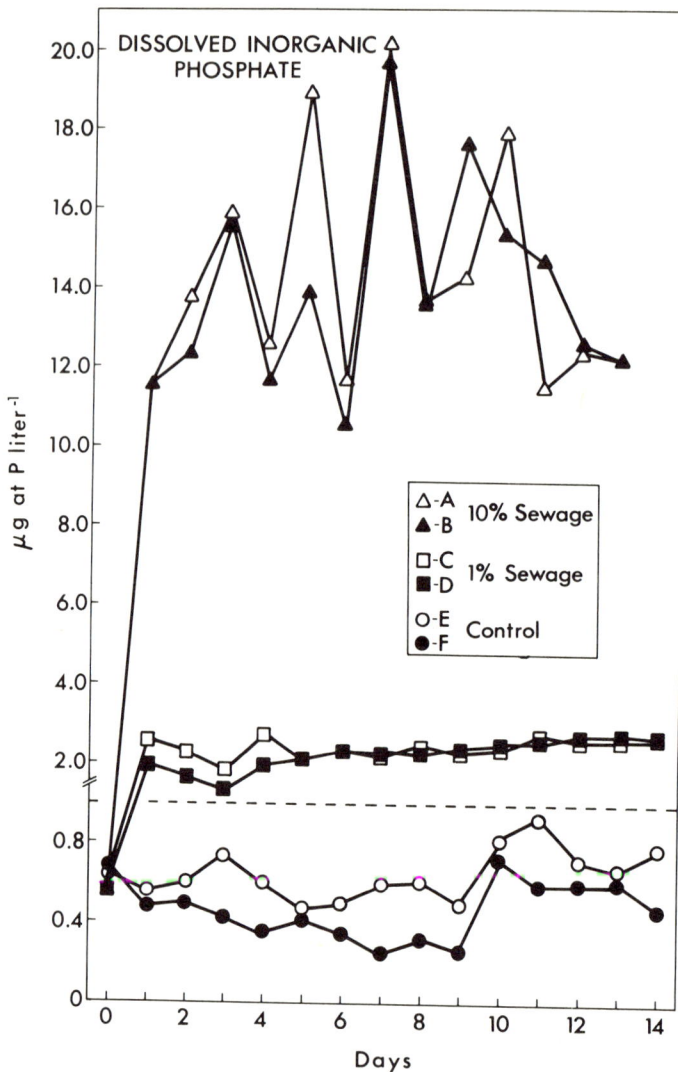

Fig. 8. Concentrations of total dissolved phosphorus (μg at P
$1^{-1}$) versus time. Experiment 2.

130 μg at $1^{-1}$ for the first six days, fell briefly to about 60 μg
at $1^{-1}$ during days 7 and 8, and then varied between 116 and 152 μg
at $1^{-1}$ for the remainder of the experiment. The events in the tanks
that received 1% sewage were more interesting. Concentrations of
nitrate fell to between 1 and 4 μg at $1^{-1}$ by day 4 (coinciding with
the increase in concentration of chlorophyll a in those tanks) and
then remained relatively constant (Fig. 9) as the algal biomass
(Fig. 6) was converted to copepods (Fig. 7) during the remainder
of the experiment. Concentrations of nitrite were below (usually
well below) 1 μg at $1^{-1}$ during the entire experiment with no apparent trends. Concentrations of total dissolved nitrogen reflected
the differences caused by the initial addition of sewage, but the
controls and the tank that received 1% sewage tended to converge
somewhat by about day 7. Concentrations of total dissolved nitro-

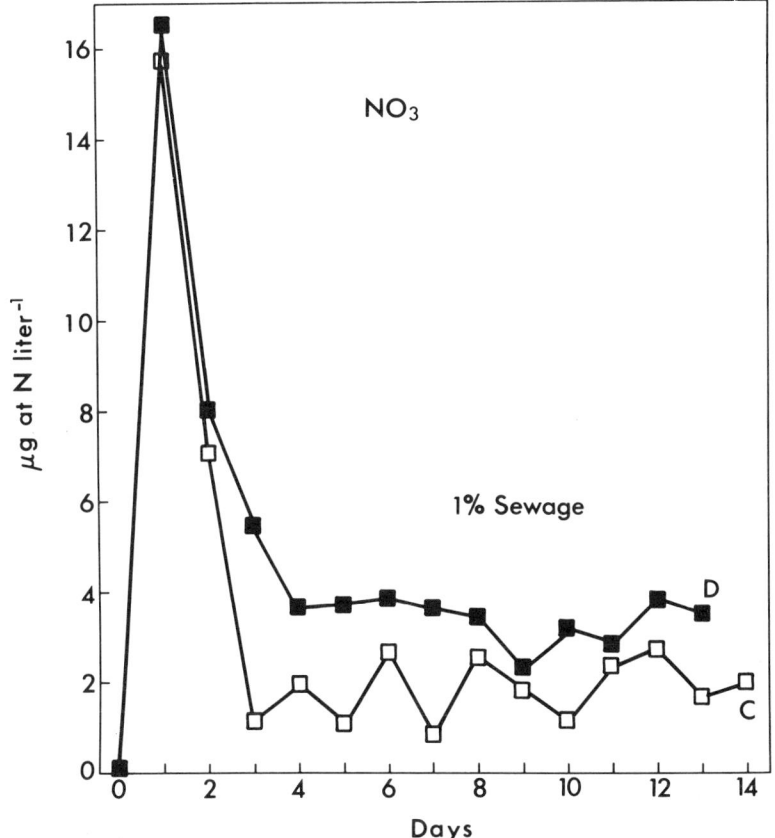

Fig. 9. Concentrations of nitrate (μg at N $1^{-1}$) versus time in tanks perturbed by 0.1% sewage. Experiment 2.

gen were rarely below 20 μg at $1^{-1}$ in the tanks that received 1% sewage, reflecting the high concentrations of nitrate in those tanks. Patterns of concentrations of particulate nitrogen generally follow-ed those of chlorophyll a (Fig. 6) and are not reported here in de-tail.

The constancy of the concentrations of phosphorus (Fig. 8) and the decline in concentration of nitrate (Fig. 9) which accompanied the increases in concentrations of chlorophyll a in the tanks per-turbed by 1% sewage (Fig. 6) suggest that nitrate-nitrogen was a major limiting nutrient in the water used. Phosphorus was apparent-ly present in surplus as there was no net removal of phosphorus from solution in any of the tanks.

The third experiment was done between September 25 and October 9, 1973, at temperatures of 25 to 30°C. These data have been re-ported in part by Flemer and Heinle (1974) and Ulanowicz *et al.* (1977). The remaining results are all from the third experiment. Concentrations of chlorophyll a rose from 20 to 100 μg $1^{-1}$ within 3 days in the tanks that received 1% sewage and from 35 to 40 μg $1^{-1}$ within 3 days in the tanks that received 0.1% sewage. Concentrations then fell gradually in both pairs of perturbed tanks (Fig. 10). The concentrations of chlorophyll a rose from 20 to 30 ug $1^{-1}$ in the control tanks on the second day and then declined to

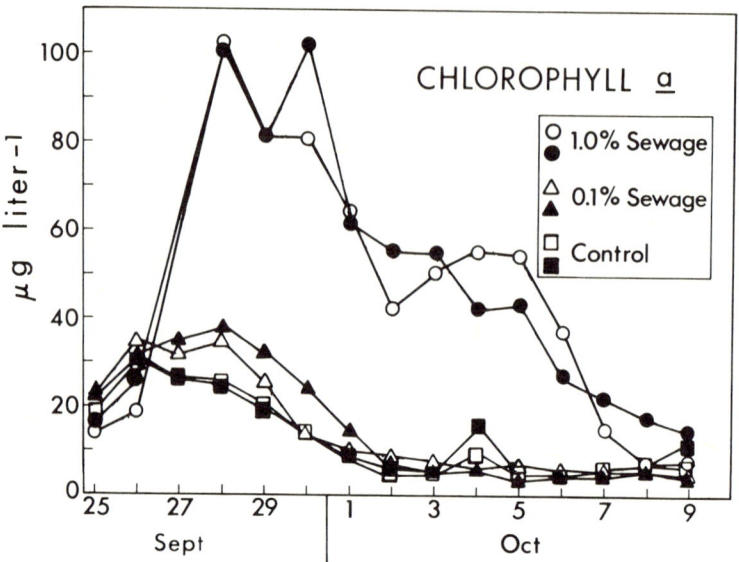

Fig. 10. Concentrations of chlorophyll a (µg 1⁻¹) versus time September 25 through October 9, 1973. Experiment 3.

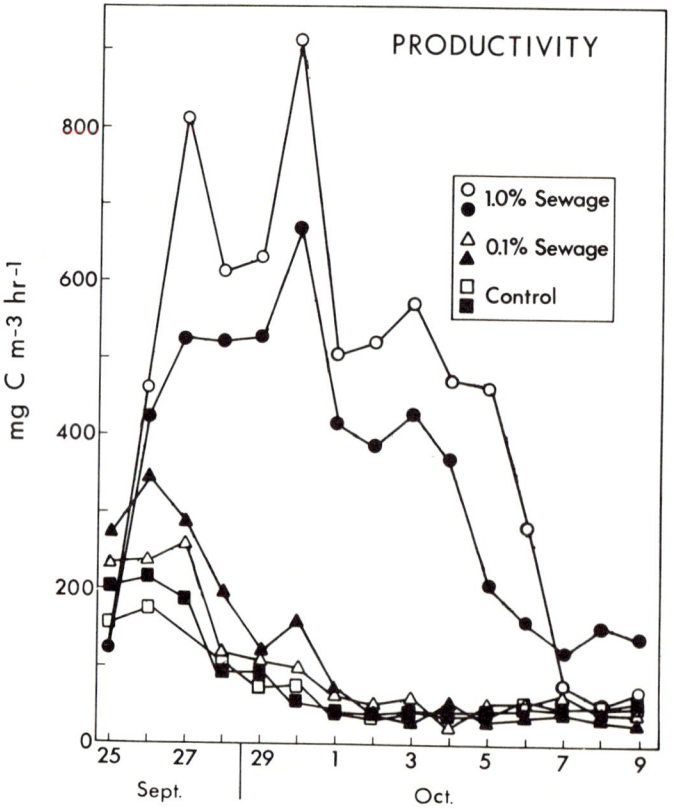

Fig. 11. Primary productivity (mg C m⁻³1⁻¹) versus time. Experiment 3.

133

about 5 µg $l^{-1}$ by day 7 (October 2) and remained low thereafter. Concentrations of chlorophyll a were similar in the control and 0.1% sewage tanks after day 7, but the tanks that received 1% sewage continued to have higher concentrations until the end of the experiment.

Primary productivity followed chlorophyll a in pattern (Fig. 11) although values differed more in replicate tanks. Maximum productivities of 650 to 900 mg C $m^{-3}hr^{-1}$, 250 to 350 mg C $m^{-3}hr^{-1}$, and 150 to 210 mg C $m^{-3}hr^{-1}$ were achieved in the 1% sewage, 0.1% sewage, and control tanks, respectively.

The biomass of zooplankton (A. tonsa and rotifers) rose in all tanks until day 8 with considerable overlap and variation. By day 8 (October 3), the biomass in the perturbed tanks was up to twice as great as in the controls (Fig. 12). After day 8 the biomass of zooplankton declined in the control tanks. In the tanks that received 0.1% sewage, the biomass of zooplankton was highest on days 9 and 10. The one very high value of 625 µg carbon $l^{-1}$ in Tank A (1% sewage) is probably the result of sampling error. Ignoring that point, the biomass of zooplankton in the tanks that received 1% sew-

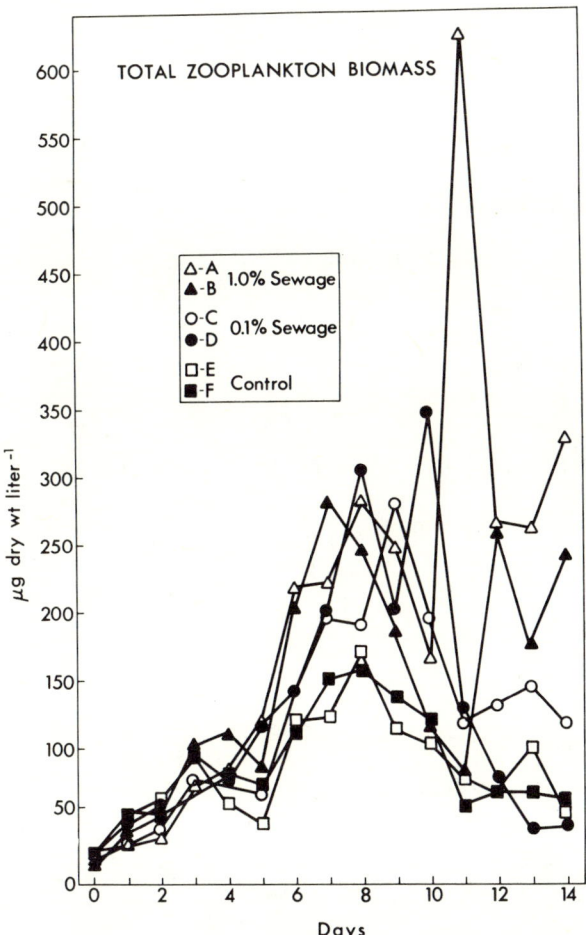

Fig. 12. Biomass of herbivores (µg dry weight $l^{-1}$ of copepods plus rotifers) versus time. Experiment 3.

age declined from days 7 and 8 to days 9 and 10 and then rose again
until the end of the experiment. We believe that the final rise in
biomass of zooplankton (essentially pure *A. tonsa*) during days 10
through 14 was an artifact of calculating biomass from the length-
weight relationship of wild copepods. Production of eggs by the
copepods ceased in all tanks by day 10, suggesting inadequate food
(Dagg, 1977), and adult copepods collected at the end of the experiment weighed less than half as much as calculated from the length-
weight relationship. The biomass of rotifers peaked at 30 µg C $1^{-1}$
in the control tanks on day 2 (September 27), at 44 to 61 µg C $1^{-1}$
in the 0.1% sewage tanks on day 3, and at 140 to 184 µg C $1^{-1}$ on
days 7 and 8 in the 1% sewage tanks (Fig. 13). Rotifers declined
to 1 to 5 µg C$^{-1}$ in the control tanks and one of the 0.1% sewage
tanks by day 6 and rose gradually after day 8. The biomass of rotifers never fell below 9 µg C $1^{-1}$ in the other tank that received
0.1% sewage. The biomass of rotifers diverged after day 7 in the
tanks that received 1% sewage, falling rapidly in Tank B and remaining much higher in Tank A. The total biomass of zooplankton on day
7 was thus over 50% rotifers in the tanks that received 1% sewage
and predominantly copepods in the other tanks.

Concentrations of particulate carbon rose from 1.0-1.5 mg $1^{-1}$ to
2.0 mg $1^{-1}$ in all tanks between days 0 and 2 and then fell gradually
to 1.5 mg $1^{-1}$ in the control and 0.1% sewage tanks (Fig. 14).
Particulate carbon continued to rise over 4 mg $1^{-1}$ on day 6 in the

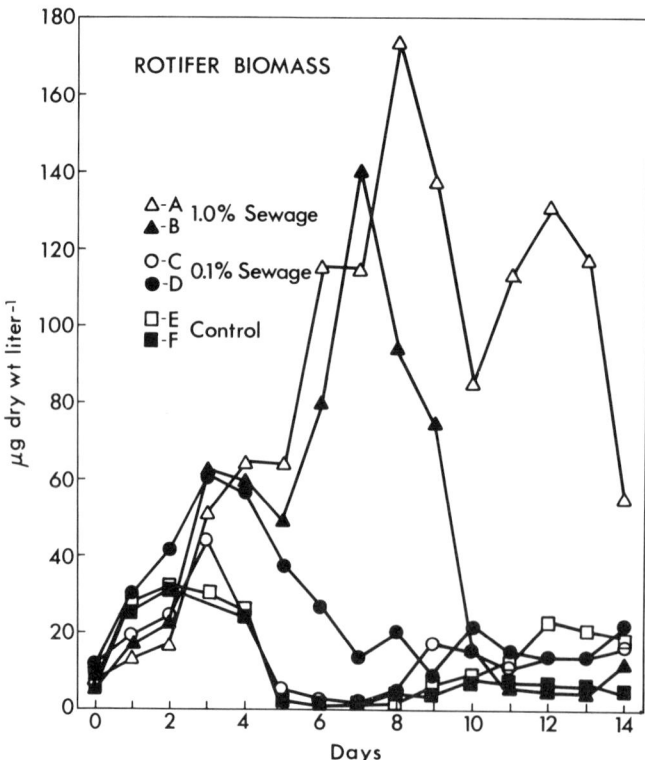

Fig. 13. Biomass of rotifers (µg dry weight $1^{-1}$) versus time.
Experiment 3).

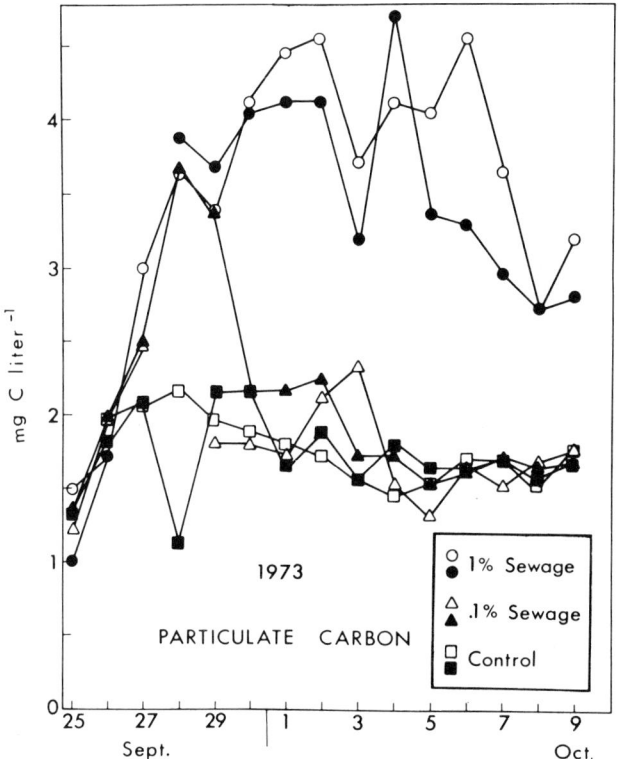

Fig. 14. Concentrations of particulate carbon (mg C $l^{-1}$) versus time. Experiment 3.

tanks that received 1% sewage and remained above 3 mg $l^{-1}$ throughout the remainder of the experiment.

Concentrations of particulate nitrogen (Fig. 15) were relatively constant (about 0.28 mg N $l^{-1}$) in the control tanks, less constant and higher in the tanks that were perturbed by 0.1% sewage, and highest in the tanks that received 1% sewage. The concentrations of particulate carbon and nitrogen reflect primarily the combined algal and zooplankton biomass. Estimating algal carbon by 50x chlorophyll a (Strickland, 1965) and summing with zooplankton carbon suggest that about 75% of the particulate carbon was in those fractions on day 7.

One large fish (about 60mm) was added to each tank on day 0. The fish lived for the duration of the experiment and appeared to be healthy throughout.

The concentrations of dissolved inorganic phosphorus reflected the addition of sewage on day 0 (Fig. 16). Concentrations in the control tanks were 0 to 0.05 µg at $l^{-1}$ on days 0 through 3 and remained so low they were not graphed. Concentrations in all tanks fell to 0 by day 6 and remained 0 through day 9. Concentrations rose in all tanks after day 10 but only reached 0.1 µg at $l^{-1}$ in the control tanks while rising to 0.20 to 0.36 µg at $l^{-1}$ in the 0.1% sewage tanks and 0.78 to 0.88 µg at $l^{-1}$ in the 1% sewage tanks.

Nitrates similarly reflected the addition of sewage (Fig. 17) but declined to lower values in the 0.1% sewage than in the con-

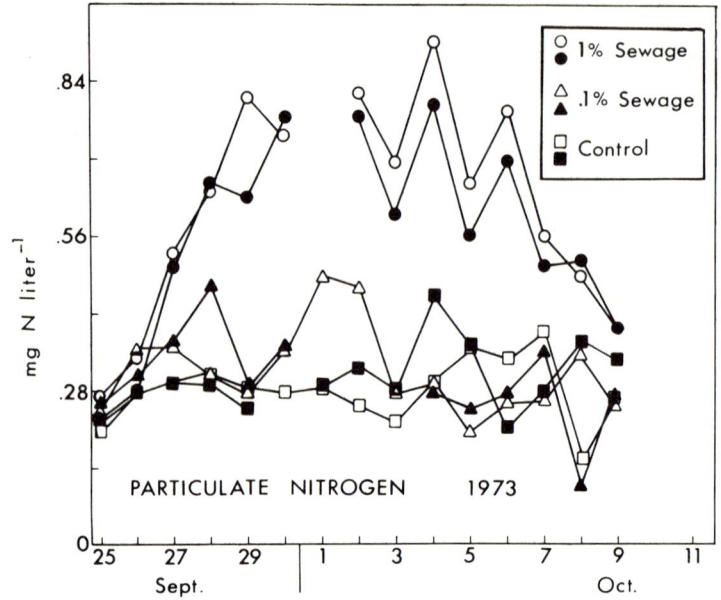

Fig. 15. Concentrations of particulate nitrogen (mg N $1^{-1}$) versus time. Experiment 3.

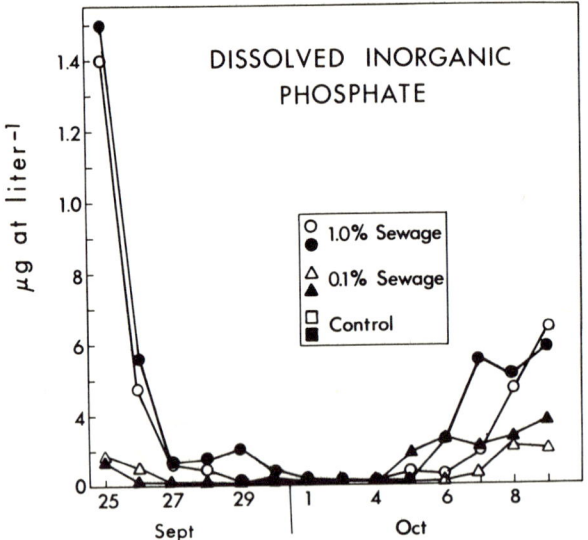

Fig. 16. Concentrations of dissolved inorganic phosphorus (µg at P $1^{-1}$) versus time. Experiment 3.

trols and to lowest values in the 1% sewage tanks. Concentrations of nitrates in the controls remained between 10 and 15 µg at $1^{-1}$ as N throughout the experiment, with the exception of one point on day 8.

Nitrite concentrations were not increased by the addition of sewage (Fig. 18). Like the concentrations of nitrate, those of nitrite decreased in the tanks that were perturbed by sewage. Lowest concentrations were observed in the tanks that received 1% sewage and highest concentrations in the controls.

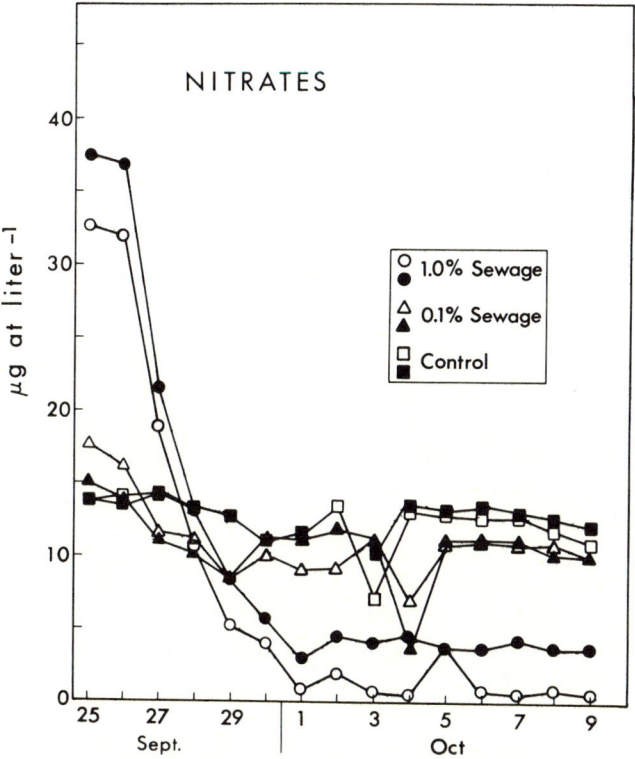

Fig. 17.   Concentrations of nitrates (μg at N l$^{-1}$) versus time.
Experiment 3.

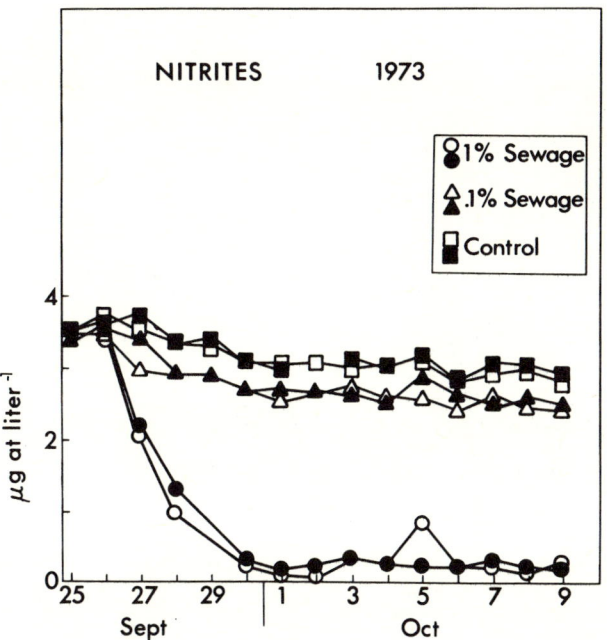

Fig. 18.   Concentrations of nitrites (μg at N l$^{-1}$) versus time.
Experiment 3.

The concentrations of ammonia also failed to reflect the addition of sewage and, while somewhat variable, dropped from 4.5 to 8.0 µg at $1^{-1}$ on day 0 to 1.0 to 2.3 µg at $1^{-1}$ on day 3 in all tanks except one control with no clear distinction between treatments except that concentrations in the 1% sewage tanks were toward the lower part of the range (Fig. 19).  Thereafter, the concentrations of ammonia varied considerably with low (4.0 µg at $1^{-1}$) and high (15 µg at $1^{-1}$) concentrations being observed in all tanks at various times.  After day 7 (October 2) the concentrations of ammonia trended upward from 0 to 2 µg at $1^{-1}$ in all tanks.

It appears that the addition of phosphorus with the sewage stimulated the growth of algae that caused nitrogen concentrations to decrease below those in the controls.  While all forms of nitrogen may have been used, nitrate and nitrite nitrogen, both of which were present in moderately high concentrations in the Bay water, were decreased more rapidly than ammonia in the enriched tanks (Figs. 17 and 18).

Increases in algal standing crops were roughly proportional to the amounts of sewage added (Fig. 11).  Some of the algal increase was reflected at the herbivore level (Fig. 12), but stimulation of the ecosystem by sewage considerably influenced the composition of the zooplankton (Fig. 13).

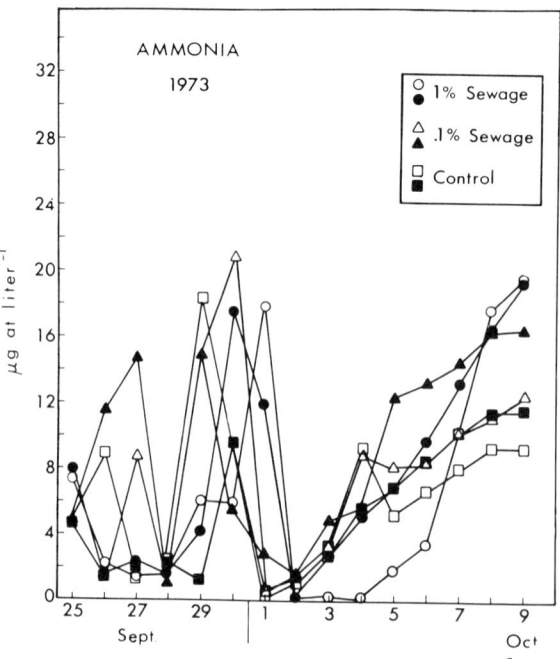

Fig. 19.  Concentrations of ammonia (µg at N $1^{-1}$) versus time.  Experiment 3.

# Discussion

Our results suggest that continued cultural eutrophication of Chesapeake Bay will lead to additional increases in primary production and standing stocks of algae. Secondary producers might also be increased. As we were unable to identify the species of algae in our experiments, we do not know if the qualitative changes we observed in the zooplankton (more rotifers in enriched microcosms) were caused by some unknown qualitative change in the phytoplankton or by the increased quantity of phytoplankton.

Our microcosms were not large enough to allow detailed study of the carnivores we added. It seems likely that increased secondary production without changes in species composition must result in increased biomass of zooplankton as most species in Chesapeake Bay are often growing at physiologically maximum rates (Heinle, 1966, 1974; Heinle and Flemer, 1975).

Qualitative changes in zooplankton, as observed in experiment 3 (Fig. 13), could, however, lead to increased secondary production without increases in the biomass of herbivores, as rotifers are generally more active metabolically than copepods (Allan, 1976). As prey density and composition may directly affect the efficiency of planktivorous fish (Ivlev, 1955), increased productivity should be reflected up the food chain.

We have not attempted to evaluate quantitatively the effects of increased productivity of surface waters on the duration or extent of the mass of deep water in Chesapeake Bay that is nearly anaerobic during the summer months. It seems likely, however, that increased productivity would enhance the depletion of oxygen in deep waters as organic material sinks through the halocline. Increases in productivity of plankton might thus have serious negative effects on existing benthic communities in Chesapeake Bay.

Ulanowicz *et al*. (1978) concluded from the modeling of data from the third experiment that increases in chlorophyll _a_ were most closely related to the rates of change in concentration of nitrates and nitrites. Their model, derived from the tanks that received 1% sewage, did not predict the changes in the concentrations of nitrate and nitrite in the other four tanks, however. While nitrite and nitrate nitrogen disappeared from solution more rapidly in the enriched tanks during the third experiment, all forms including ammonia reached minimum concentrations on day 6 or 7 (October 1 or 2) (Figs. 17, 18, 19) coincident with a decrease in primary productivity in the tanks enriched with 1% sewage (Fig. 11). It appears that phosphorus was the "limiting nutrient" in an absolute sense at the beginning of that experiment, but it was removed from solution so rapidly during the first 3 days of the experiment that later changes in chlorophyll _a_ did not correlate well with concentrations of phosphorus. Nitrogen may have limited primary productivity on days 6 and 7. Increasing concentrations of ammonia (Fig. 19) and biomass of herbivores (Fig. 12) toward the end of experiment three suggest that grazing rather than nutrients limited primary production during that period.

By contrast, forms of nitrogen appear to be the limiting nutrients in the second experiment, and possibly the first. In some cases, the "preferred" forms of nitrogen appeared to be nitrate and nitrite rather than ammonia, as these forms were removed from solution more rapidly than ammonia (second and third experiments). The biomass of herbivores was sufficiently low during the early part of all experiments so that regeneration of ammonia was not a major source of nitrogen. Past day 7, however, ammonia excreted by zooplankton may have been the major source of nitrogen in some tanks. Other studies on Chesapeake Bay have shown that ammonia is generally the preferred form of nitrogen, based on rates of uptake (McCarthy *et al.*, 1975, 1977). Our results are not inconsistent with those of McCarthy *et al.* (1975, 1977). Our experiments were done during August and October when nitrogen preferences were less distinct. We believe that all forms of nitrogen may have been used by algae in our experiments, but ammonia was excreted by zooplankton, resulting in greater variation in the concentrations of that form.

Our experimental design did not allow us to evaluate the role of benthic community processes in nutrient regeneration. That role is known to be important in weakly stratified, aerobic estuaries (Nixon, Oviatt, and Hale, 1976). Future studies of microcosms should include a benthic component (see for example papers by Nixon *et al.* and Oviatt *et al.* in this volume).

Our results suggest that Chesapeake Bay, while already impacted by cultural eutrophication (Flemer and Heinle, 1974), is capable of responding to the addition of increased amounts of sewage with increased primary production. Production at higher trophic levels will probably increase also, but qualitative changes may occur whose consequences are difficult to predict.

We are convinced that the keys to the excellent replication obtained in the third experiment were the careful cleaning of the tanks prior to the experiment, rigorous control of environmental variables, and minimum manipulation of the composition of the planktonic community.

## Acknowledgments

We would like to thank M. Linton Beaven and R. A. Murtagh for their technical assistance. This work was supported by NSF - RANN through a grant to the Chesapeake Research Consortium. Contribution No. 803, University of Maryland, Center for Environmentaland Estuarine Studies.

## References

Allan, J. D. 1976. Life history patterns in zooplankton. Amer. Natur. 110(971): 165–180.

Brehmer, M. L. 1967. Nutrient assimilation in a Virginia tidal system. In: *Proc. Natl. Symp. Estuarine Poll.* (P. L. McCarty and R. Kennedy, eds.), pp. 23–25. Stanford Univ., California.

Brush, L. M. 1974. Inventory of sewage treatment plants for Chesapeake Bay. Chesapeake Res. Consortium Publ. 28, The Johns Hopkins University, Baltimore, Md.

Curl, H., Jr. and E. W. Davey. 1967. Improved calibration and sample-injection systems for non-destructive analysis of permanent gases, total $CO_2$, and dissolved organic carbon in water. Limnol. Oceanogr. 12: 545–548.

Dagg, M. 1977. Some effects of patchy food environments on copepods. Limnol. Oceanogr. 22: 99–107.

Deevey, E. S. 1972. Biogeochemistry of lakes: Major substances. In: *Nutrients and Eutrophication* (G. E. Likens, ed.), pp. 14–20. Limnol. Oceanogr. Spec. Symp., No. 1.

Flemer, D. A. 1972. Current status of knowledge concerning the cause and biological effects of eutrophication in Chesapeake Bay. Chesapeake Sci. 13(Suppl.): 144-149.

Flemer, D. A. and D. R. Heinle. 1974. Effects of waste water on estuarine ecosystems. Chesapeake Res. Consortium Publ. 33, Univ. Md. Nat. Resour. Inst. Ref. 74-79.

Fuhs, G. W., S. D. Demmerle, E. Canelli, and M. Chen. 1972. Characterization of phosphorus-limiting-nutrient concept). In: *Nutrients and Eutrophication* (G. E. Likens, ed.), Limnol. Oceanogr. Spec. Symp., No. 1.

Goulden, P. D. and P. Brooksbank. 1974. Automated phosphate analysis in the presence of arsenic. Limnol. Oceanogr. 19: 705-706.

Heinle, D. R. 1966. Production of a calanoid copepod, *Acartia tonsa*, in the Patuxent River estuary. Chesapeake Sci. 7: 59-74.

Heinle, D. R. 1969. Culture of calanoid copepods in synthetic sea water. J. Fish. Res. Bd. Can. 26: 150-153.

Heinle, D. R. 1974. An alternate grazing hypothesis for the Patuxent estuary. Chesapeake Sci. 15: 146-150.

Heinle, D. R. and D. A. Flemer. 1975. Carbon requirements of a population of the estuarine copepod, *Eurytemora affinis*. Mar. Biol. 31: 235-247.

Holm-Hansen, O., C. J. Lornezen, R. W. Holmes, and J. D. H. Strickland. 1965. Fluorometric determination of chlorophyll. J. cons. perm. Expl. Mer 30: 3-15.

Ivlev, V. S. 1955. *Experimental Ecology and Nutrition of Fishes.* Pishchemizdat Moscow (Transl., D. Scott, Yale Univ. Press, New Haven, 1961.)

Jaworski, N. A., D. W. Lear, and O. Villa, Jr. 1972. Nutrient management in the Potomac estuary. In: *Nutrients and Eutrophication* (G. E. Likens, ed.), Vol. I. Limnol. Oceanogr. Spec. Symp., No. 1.

McCarthy, James J., W. R. Taylor, and J. L. Taft. 1975. The dynamics of nitrogen and phosphorus cycling in the open waters of the Chesapeake Bay. In: *Marine Chemistry in the Coastal Environment* (Thomas M. Church, ed.). ACS Symp. Ser., No. 18, pp. 664-681.

McCarthy, J. J., W. R. Taylor, and J. L. Taft. 1977. Nitrogenous nutrition of the plankton in the Chesapeake Bay. 1. Nutrient availability and plankton preference. Limnol. Oceanogr (in press).

Menzel, D. W. and N. Corwin. 1965. The measurement of total phosphorus in seawater based on the liberation of organically bound fractions by persulfate oxidation. Limnol. Oceanogr. 10: 280-282.

Menzel, D. W. and F. F. Vaccaro. 1964. The measurement of dissolved organic and particulate carbon in seawater. Limnol. Oceanogr. 9: 138-142.

Mobley, D. D. 1973. A systematic approach to ecosystems analysis. J. Theo. Biol. 41: 119-136.

Nixon, S. W., C. A. Oviatt, and S. S. Sale. 1976. Nitrogen regeneration and the metabolism of coastal marine bottom communities. In: *The Role of Terrestial and Aquatic Organisms in Decomposition Processes* (J. M. Anderson and A. Macfadyen, eds.),pp. 269-283. Blackwell Scientific Publications, Oxford.

Ryther, J. H. and W. M. Dunstan. 1971. Nitrogen, phosphorus, and eutrophication in the coastal marine environment. Science 171: 1008-1013.

Solorzano, L. 1969. Determination of ammonia in natural waters by the phenolhypochlorite method. Limnol. Oceanogr. 14: 799-801.

Strickland, J.D.H. 1965. Production of organic matter in the primary stages of the marine food chain. In: *Chemical Oceanography* (J. P. Riley and G. Skirrow, ed.), Vol. 1, pp. 477-610. Academic Press, New York.

Strickland, J.D.H. and T. R. Parsons. 1968. A practical handbook of seawater analysis. Fish. Res. Bd. Can. Bull. 167.

Stross, R. G. and J. R. Stottlemyer. 1965. Primary production in the Patuxent River. Chesapeake Sci. 6: 125-140.

Taft, J. L. and W. R. Taylor. 1976. Phosphorus distribution in the Chesapeake Bay. Chesapeake Sci. 17: 67-73.

Taft, J. L., W. R. Taylor, and J. J. McCarthy. 1976. Uptake and release of phosphorus by phytoplankton in the Chesapeake Bay Estuary, U.S.A. Mar. Biol. 33: 21-32.

Thayer, G. W. 1971. Phytoplankton production and the distribution of nutrients in a shallow unstratified estuarine system near Beaufort, N.C. Chesapeake Sci. 12: 240-253.

Ulanowicz, R. E., D. A. Flemer, D. R. Heinle, and R. T. Huff. 1978. The empirical modeling of an ecosystem. Ecol. Model. 4: 29-40.

Ulanowicz, R. E., D. A. Flemer, D. R. Heinle, and C. D. Mobley. 1975. The *a posteriori* aspects of estuarine modeling. In: *Estuarine Research* (L. E. Cronin, ed.), Vol. I, pp. 602-616. Academic Press, New York.

Yentsch, C. S. and D. W. Menzel. 1963. A method for the determination of phytoplankton, chlorophyll and phaeophytin by fluorescence. Deep-Sea Res. 10: 221-231.

# On the Season and Nature of Perturbations in Microcosm Experiments

**Candace A. Oviatt**
**Scott W. Nixon**

Graduate School of Oceanography
University of Rhode Island
Kingston, Rhode Island 02881

**Kenneth T. Perez**

Environmental Protection Agency
Narragansett, Rhode Island 02882

**Betty Buckley**

Graduate School of Oceanography
University of Rhode Island
Kingston, Rhode Island 02881

Microcosms have been designed to mimic and miniaturize an estuarine ecosystem. In these microcosms the effects of perturbations operating singly or in combination and at different times during the year have been investigated with a view toward extrapolating the results to real ecosystems.

While sewage additions produced prolonged phytoplankton blooms in a spring-summer experiment, it produced no bloom in the fall. Ancillary experiments with fall microcosm water suggested that a phytoplankton growth-controlling substance was present in the water.

A series of three experiments explored microcosm response to artificial predation on the plankton during spring, summer, and fall. Predation was size-selective in the first experiment, of variable intensity in the second experiment, and imposed on a eutrophication gradient (developed from different levels of sewage addition) in the third experiment. Although the artificial predation (which was applied 2 or 3 times per week) was more severe than natural levels, in each of the experiments there was little to no measurable microcosm response. Rather, zooplankton which grazed continuously appeared to be controlling phytoplankton concentrations in the first two experiments. These results suggest that plankton have a high stability with regard to artificial predation stress during each season.

Seasonality in stability and resilience was apparent in ecosystem behavior when a multivariate analysis was used. As a consequence, microcosm experiments must be repeated seasonally or carried out continuously before results can be extrapolated to the ecosystem.

## Introduction

Numerous approaches have been used to study Narragansett Bay, Rhode Island, including intensive field surveys and numerical simulation models (Kremer and Nixon, 1978). These approaches are primarily designed to reveal information about the dynamics of the ecosystem in its present state and are very limited in terms of allowing us to make predictive statements about the behavior of the system if it is perturbed or altered in some fundamental way. The natural system is too unwieldy for experimentation, and mathematical formulations which permit self-design characteristics in simulation models are unknown. We have now tried a third approach to overcome these drawbacks, the establishment of living models or microcosms of the real system. Microcosms are easily manipulated under controlled conditions yet retain much of the complexity of the real system which is not yet incorporated into computer simulation models. As such, microcosms become a powerful tool for investigating the effects of perturbations operating singly and in combination on the ecosystem the microcosms miniaturize.

The most common approach in microcosm experiments, and one that is efficient in terms of exploring a large number of questions, has been to do short-term experiments of the order of several weeks (Davies *et al.*, 1975; McAllister *et al.*, 1961; Takahashi *et al.*, 1975; Ulanowicz *et al.*, 1975, and others). Yet the state and response of the system can be expected to be quite different at different times of the year in areas with strong seasonal programming, as field studies by Miller (1967) indicate. However, the literature is silent on the fundamental supposition of repeatability of microcosm experiments at different seasons of the year. In our early microcosm work we established a eutrophication gradient in microcosms by imposing three levels of sewage on replicated tanks, and we chose to do a long-term experiment from early spring to early fall (Oviatt *et al.*, 1977). In our latest series of experiments we have chosen to do short-term studies and to repeat the experiments occasionally.

A eutrophication experiment which was carried out in the fall for comparison with the earlier spring experiment is described in the first section of this paper. The three different levels of sewage (1%, 5%, 10% by volume) were chosen to be comparable to initial levels in the first experiment. The behavior of Narragansett Bay during the fall period has long intrigued local ecologists. The expected pattern of a fall diatom bloom which occurs in most temperate estuaries (Cushing, 1975) usually fails to occur, yet the numerical model of Narragansett Bay predicts on the basis of light, grazing pressure, and nutrients that it should, in fact, be observed (Kremer and Nixon, 1978). When a bloom again failed to occur this past fall, several experiments were performed with microcosm water in an attempt to discover why. Thus, the living models became useful in discovering new information about seasonal programming in the real system. This first section reports on the results of these microcosm experiments with the eutrophication gradient established in the fall and on the results of a number of ancillary experiments with microcosm water.

The second section of this paper deals with short-term artificial predation experiments carried out in the spring, summer, and fall.  In these experiments we sought the effects of a generalized stress which might result from predation or general mortality associated with a toxic spill.  In the spring experiment the level of predation was constant and size-selective on planktonic organisms. In the summer experiment the level of predation was variable and size-selective for all planktonic organisms.  In the fall experiment predation was imposed along a sewage gradient at a constant level and made size-selective for larger phytoplankton.

The final section of this paper deals with the importance of seasonal changes on microcosm experiments  investigating the impacts of stress.  The use of multivariate techniques is suggested as a possibility for evaluating these seasonal interactions.

## Section I.  Fall Eutrophication Experiment

The primary objective of the fall experiment was to repeat the earlier spring sewage experiment in order to assess the effects of season on repeatability.  A secondary objective was to investigate the impact of artificial predation on the different levels of sewage impacted microcosms.  Results of this objective will be discussed in Section II.

DESIGN AND METHODS

The sewage levels for the fall (8/17-10/27/76) experiment of 0%, 1%, 5%, and 10% sewage by volume were similar to initial levels in the first sewage experiment of 0%, 3%, 7%, and 13% sewage.  All the sewage was collected at one time from the Cranston, RI, sewage plant on July 23, 1976, and frozen as aliquots for sequential additions.  Composition of the treated urban sewage which was added to the microcosms is given in Table 1.

Characteristics of the 20 microcosms are summarized in Table 2. Microcosms were filled with hand-bucketed water (to prevent damage to plankton) collected from the surface of the lower West Passage of the bay.  Three times a week (Mondays, Wednesdays and Fridays) 10 1 of new water were added to each microcosm along with the appropriate aliquots of sewage.  The microcosms were thoroughly intermixed and all microcosms were run as replicates for the first fourteen days to ensure a uniform distribution among tanks prior to sewage treatment.  A heterotrophic benthic community was included in each microcosm in an opaque box.  This community was collected in an area with unpolluted silt-clay sediments dominated numerically by the bivalve, *Nucula annulata*.  An undisturbed community was obtained by using the benthic box as a diver-implemented box corer and then by attaching a bottom to the box.  A vacuum pumping system forced microcosm water through the benthic boxes.  Turbulence in the microcosms was created by reversing plastic mesh paddles.

The parameters measured, frequency of measurement, and methods used during microcosm experiments are summarized in Table 3.  The sewage treatment was carried out in triplicate on 12 of 20 microcosms.  In the other 8 microcosms, a 60% predation per week was

Plate 1.  A 150 l microcosm showing: a) air driven pump which caused water to flow through, b) the benthic box, and c) the plastic mesh paddles in the center of the microcosms.  The microcosms sit in an ambient bay temperature water bath.

Table 1.  Composition of secondary treated sewage added to Narragansett Bay microcosms.  The sewage was collected from the Cranston, RI, sewage plant July 23, 1976.

| Nutrients | μg at/l | |
|---|---|---|
| $NH_4$-N | 1226 | |
| $NO_2$-N | 31.2 | |
| $NO_3$-N | 18.5 | |
| $PO_4$-P (DIP) | 11.8 | |
| Organic-P (DOP) | 5.8 | |
| $SiO_3$-Si | 137.7 | |
| Paticulate Dry Weight | 28.0 mg/l | |
| C/N | 6.7 | |
| Trace Metals* | Dissolved | Particulate (>0.4μ) |
| | μg/l | |
| Ag | --- | 3 |
| Cd | <1 | <1 |
| Cr | <1 | 4 |
| Cu | 22 | 18 |
| Fe | 4 | 327 |
| Mn | 67 | 11 |
| Ni | 194 | 17 |
| Pb | 1 | 8 |
| Zn | 23 | 19 |

*Mean of duplicate samples analyzed by the U.S. Environmental Protection Agency, Narragansett, R.I.

imposed on phytoplankton-sized particles in the 4 levels of sewage (0%, 1%, 5%, and 10%). During the experiment, temperatures dropped from 21 to 15$^o$C and light was adjusted from 19 to 7 ly/day.

Table 2. Characteristics of the microcosms.

| | |
|---|---|
| Volume | 150 1 |
| Depth | 0.7 m |
| Turnover Time | 37 days |
| Source of Input Water | Surface-dipped samples from the lower West Passage of Narragansett Bay. |
| Turbulence | The microcosms were stirred with 0.14 m$^2$, 1.8 cm plastic mesh paddles at 32 rpm. The paddles were rotated for 30 sec., then stopped for 6 sec., then reversed for 30 sec. in a continuous cycle. |
| Sediment Area | 167 cm$^2$ - scaled to the sediment/volume ratios for Narragansett Bay. (Maintained in opaque box cores immersed in each microcosm. A vacuum pump system flowed microcosm water over the benthic community at 0.7 1/min.). |
| Light Input | Intensity matched empirically to have the same effect in the microcosms as in Narragansett Bay water column. |
| Temperature | 1.5$^o$C. above ambient bay water. |
| Salinity | $\pm 2^o$/oo of ambient bay water. |

Table 3. Summary of the analytical methods and sampling procedures.

| Parameter | Frequency | Method |
|---|---|---|
| Nutrients: | | |
| ammonia | 1/week | Solorzano, 1969 |
| Plankton: | | |
| chlorophyll a | 2/week | Extracted fluorescence calibrated with spectro-photometric determinations (Strickland and Parsons, 1969). |
| Phytoplankton | 1/week | Cell counts and species identifications on samples preserved in Lugol's solution. |
| zooplankton | 1/week | Identification of major groups and counts of relative abundance in 1 liter samples. |
| productivity $\simeq$ | 1/week | $^{14}$C., Strickland and Parsons, 1968. |
| Benthos: | | |
| oxygen $\simeq$ | 1/week | Winkler titration (Strickland and Parsons, 1968). |
| ammonia flux $\simeq$ | 1/week | Solorzano, 1969. |

RESULTS AND DISCUSSION

During the replication period from August 17 to September 2, a marked phytoplankton bloom occurred concurrently in the microcosms and in the bay, apparently stimulated by hurricane Belle, which passed through the area the previous week. Chlorophyll concentrations peaked at about 100 µg/l just prior to treatment (Fig. 1). As a consequence, ammonia concentrations were very low throughout this period (Fig. 2). Zooplankton, taking advantage of the bloom, increased throughout the replication period and peaked at 25-30/1 the day before treatments began (Fig. 3). All tanks replicated each other and the bay extremely well (Figs. 1,2,3).

As in previous years, phytoplankton did not grow during the fall treatment period. Consequently, just as treatment was begun, chlorophyll concentrations dropped simultaneously in all microcosms and in the bay to 1 µg/l or less by early September and remained low throughout the experiment (Fig. 1). Ammonia levels rose according to sewage treatment in the individual microcosms (Fig. 2). These peak levels dropped by early September, and measurements of nitrate indicated that ammonia was being oxidized to nitrate in the water column as we have also observed in the bay water column. Those microcosms which received only filtration stress had slightly rising levels of ammonia, higher than the controls and the bay, which also had ammonia levels which rose slightly throughout the experimental period. Zooplankton dropped from their peak levels prior to treatment to uniformly low levels of 5-10/1 in all microcosms and were generally similar to levels in the bay (Fig. 3).

Despite very low levels of chlorophyll, canonical variates analysis indicated some treatment effect (Fig. 4). Although not detectable in time-series plots (Fig. 1), a greater growth of phytoplankton did occur in more eutrophic microcosms. The sensitive multivariate analysis separates this effect along the first axis. Structural coefficients (the elements of the canonical variate scaled as correlation coefficients between the variables and the canonical variate) indicated that chlorophyll concentrations, primary production, and phytoplankton counts were primarily responsible for variation along the first axis. Low sewage treatment replicates are ordered to the lower right but overlap control replicates on the first axis as in the first experiment (Fig. 5). All microcosms which overlap on the second axis are separated from input-dock water or bay values along the second axis, perhaps due to a greater variability in bay values.

The earlier spring sewage-addition experiment had resulted in eutrophic microcosms with large, prolonged blooms of phytoplankton. Very small increases, really not detectable in time-series plots (Fig. 1), occurred in this second experiment. Generally there is less separation and more overlap in the canonical variates analysis of the second experiment when compared to that of the first experiment where initial sewage concentrations were similar (0%, 2.7%, 6.7%, 13%, Fig. 5). Multivariate analyses of several data sets in the earlier sewage experiments consistently showed large treatment

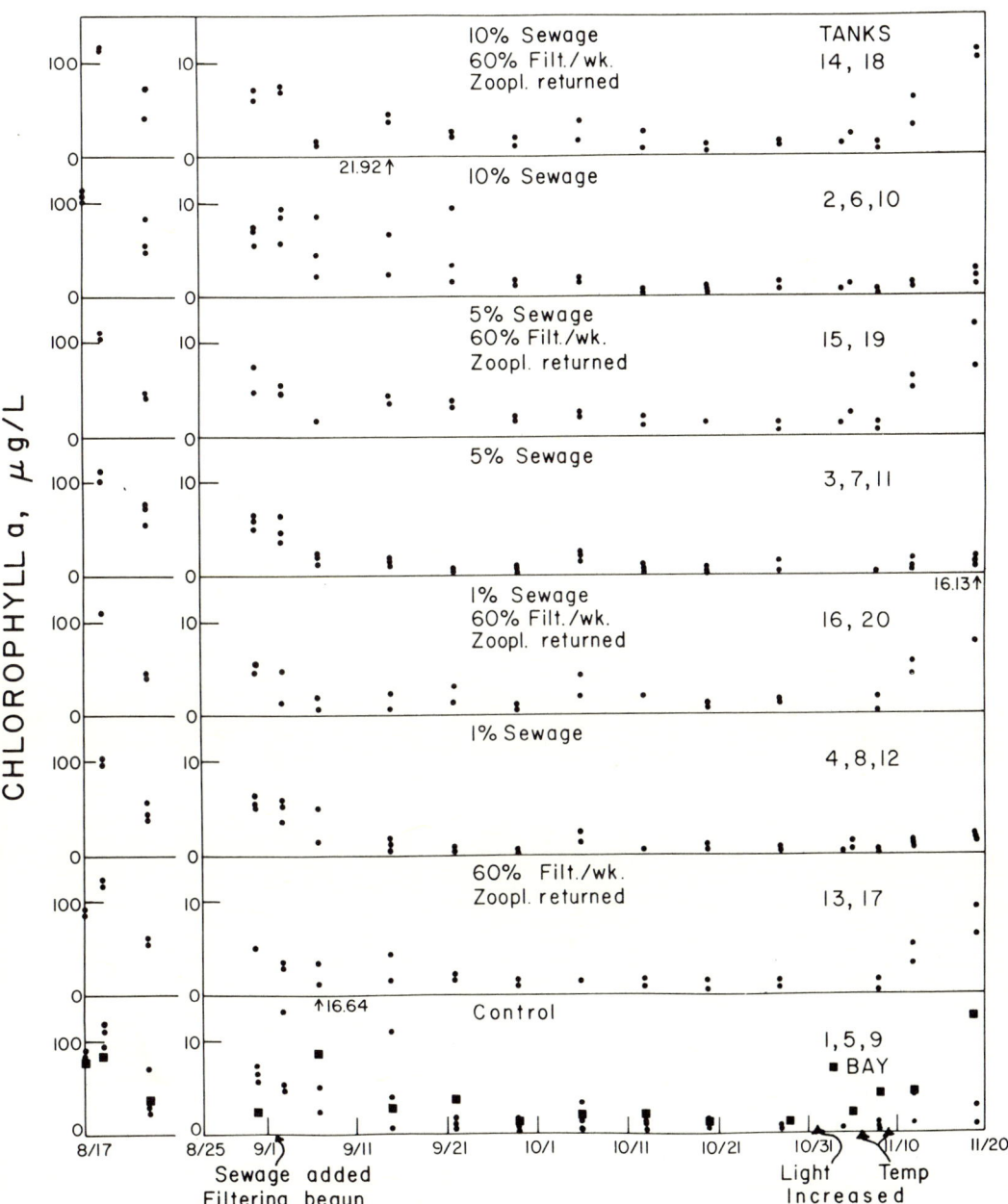

Fig. 1.   Chlorophyll concentrations in microcosms receiving different levels of secondary treated sewage and a filtration level of 60% per week.   In many cases, replicate tanks overlap and only one position is plotted.

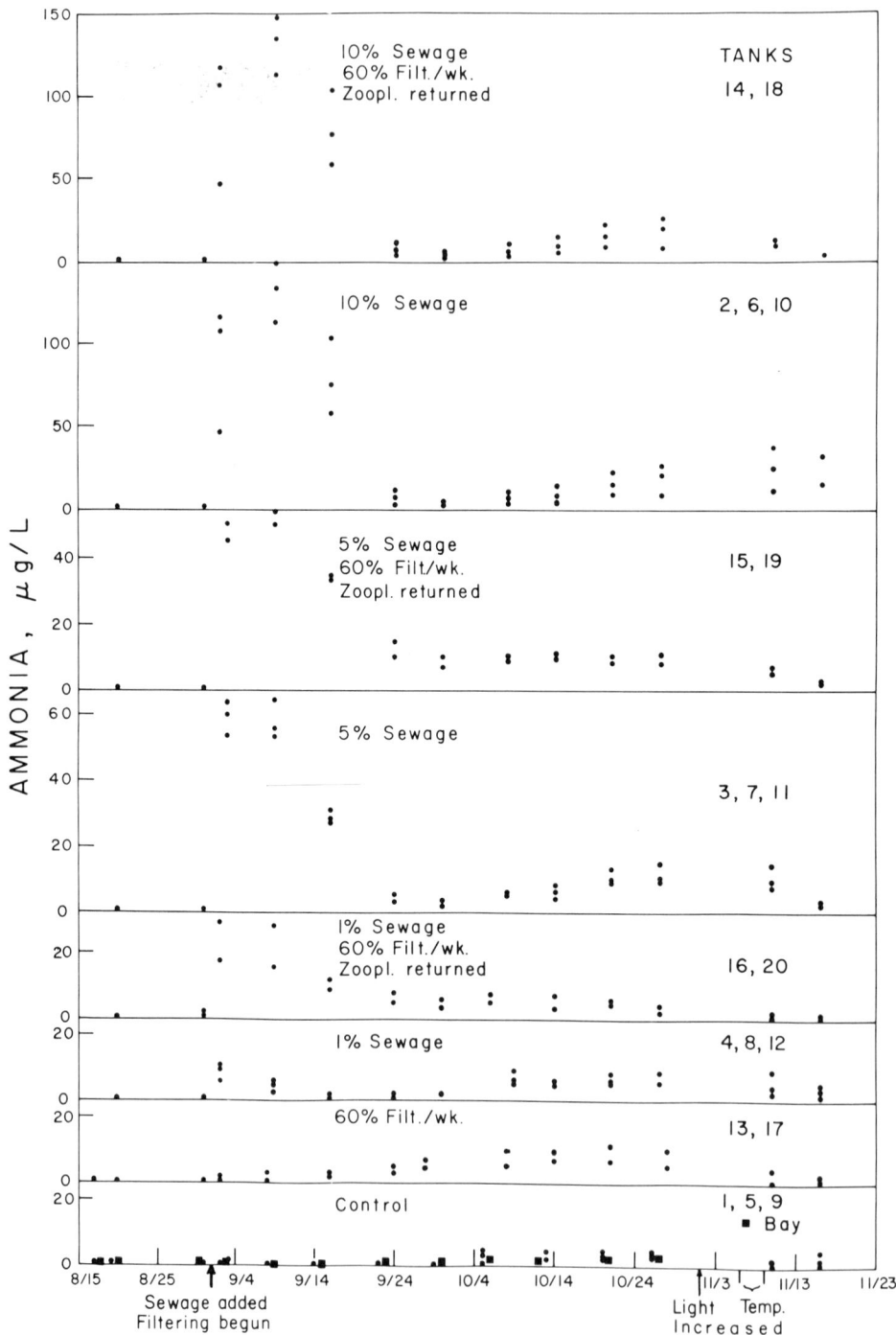

Fig. 2. Ammonia concentrations in microcosms receiving different levels of secondary treated sewage and a filtration level of 60% per week.

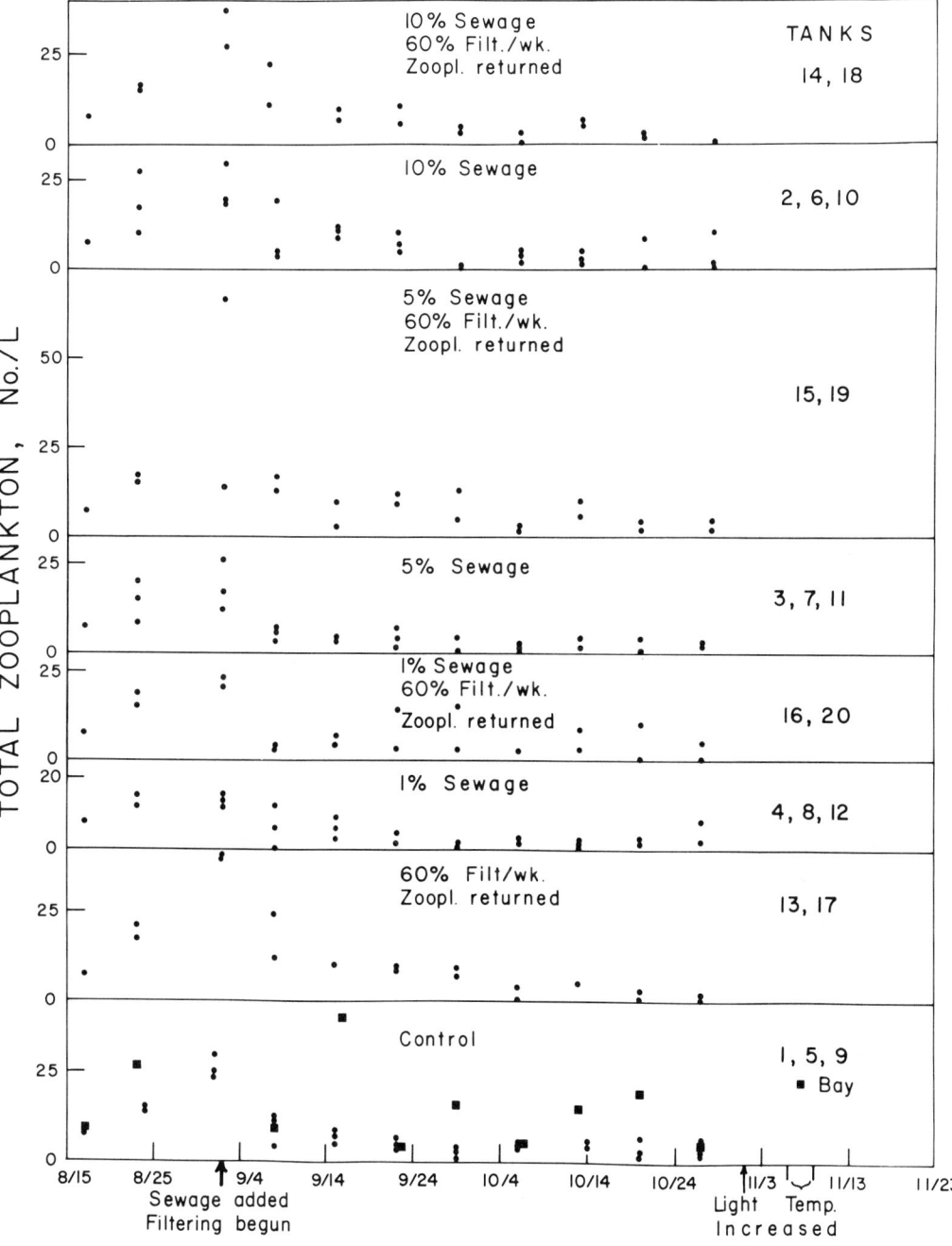

Fig. 3. Zooplankton numbers in microcosms receiving different levels of secondary treated sewage and a filtration level of 60% per week.

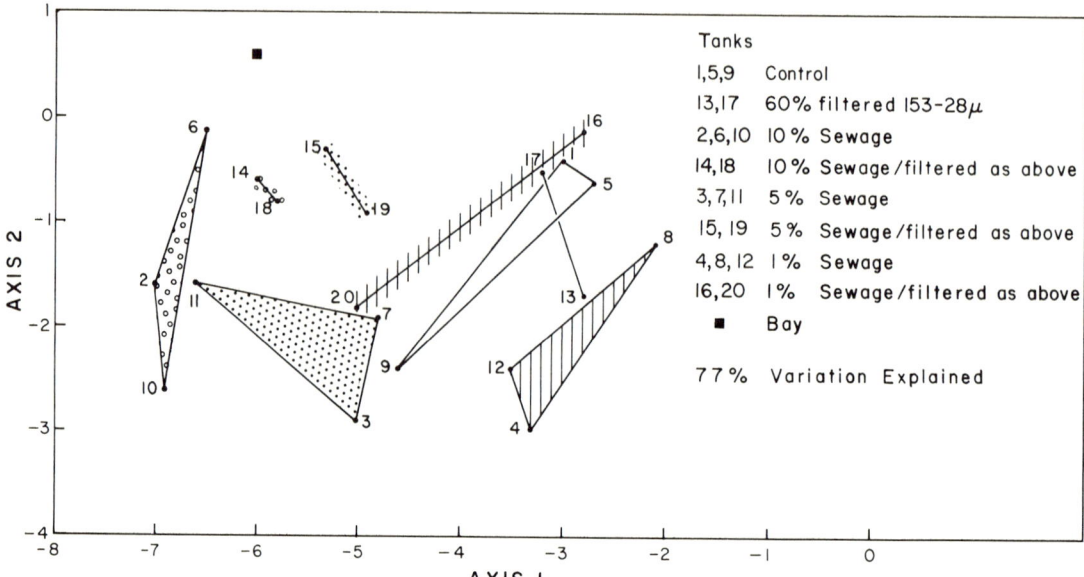

Fig. 4. Canonical variates analysis of microcosms receiving different levels of secondary treated sewage and a filtration level of 60% per week. The data set for this analysis included chlorophyll concentration, ammonia concentration, 14°C productivity, benthic oxygen flux, benthic ammonia flux, phytoplankton counts, and zooplankton counts.

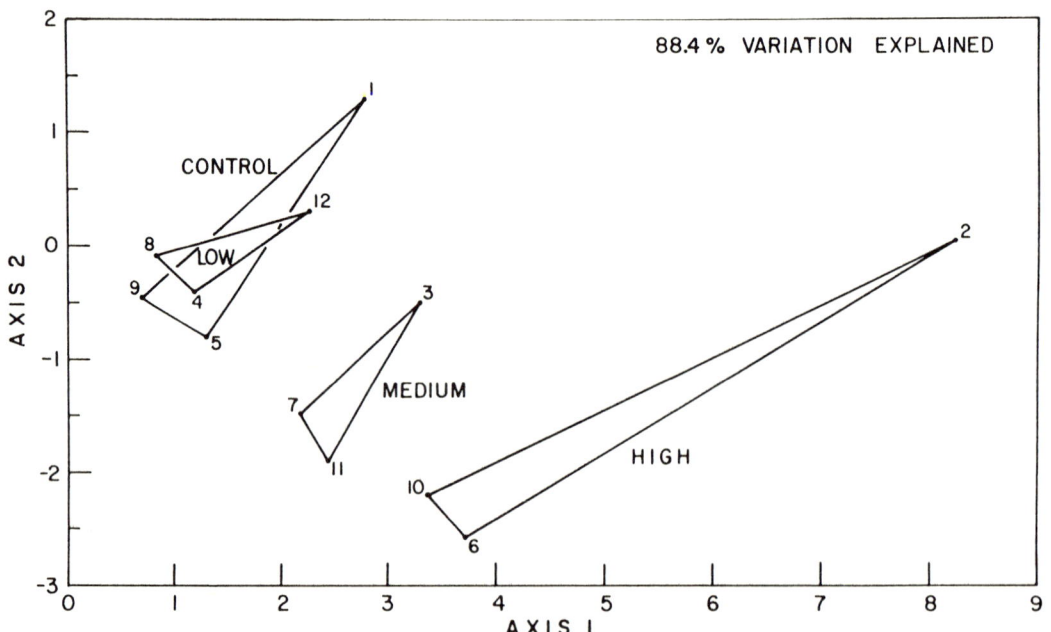

Fig. 5. Canonical variates analysis of microcosms stressed with different levels of secondary treated sewage during the spring-summer experiment. The data set which included chlorophyll concentrations, number of diatoms, number of flagellates, pelagic ATP, suspended particles, and ammonia concentrations was collected during the first month (April-March) with sewage input (from Oviatt et al., in press).

effects and compared well with the eutrophication gradient resulting from sewage inputs from the Providence River into Narragansett Bay (Oviatt *et al.*, 1977). A comparison of the two experiments indicates that while sewage was a primary factor in controlling productivity of the microcosms in the first experiment, it had only a minor effect in the fall experiment.

The lack of response to such strong sewage treatments was puzzling, and we began to wonder if improper scaling of some physical factor was responsible. As a consequence, we began to change light and temperature inputs to the microcosms.

## MANIPULATION OF SEWAGE AND PREDATED MICROCOSMS

Since there was still some uncertainty about how to scale light properly, this was the first suspect parameter (Perez *et al.*, 1977; Nixon *et al.*, this volume). On November 2, 1976, light levels were increased to 25 ly/day for all microcosms. In all past experiments this level had resulted in phytoplankton blooms.

No response to the increased light occurred. After 7 days, chlorophyll values were still less than 1 µg/l in microcosms treated with sewage; those in filtered tanks were slightly greater than 1 ug/l, whereas the bay had risen to 4.5 µg/l (Table 4A).

Water temperatures had dropped more rapidly than in average years, and it was felt possible that the bay and microcosms had undergone thermal shock as a result. Beginning November 8, the temperatures of the microcosms were brought up slowly over a period of several days to simulate summer conditions. Only filtered tanks responded by increased chlorophyll levels. There was no apparent response to the sewage gradient. After 3 days at high temperatures, microcosms treated with sewage had chlorophyll levels between 1 and 2 µg/l, and filtered tanks had levels between 4 and 6 µg/l (Table 4B). Bay levels remained about the same at 4-5 µg/l. After 10 days at higher temperatures, sewage microcosms still had concentrations

Table 4. Mean chlorophyll concentration (µg/l) in sewage/ predation microcosm after light and temperature increases. A. after 7 days at high light. B. after 3 days at higher temperature and light. C. after 10 days at higher temperature and light.

| Date | A. | Sewage Level | 0 filtering | 60% filtering |
|------|----|-----|-----|-----|
| | | Control | 0.86 | 1.17 |
| | | High | 0.61 | 1.17 |
| 11/8/76 | | Medium | 0.61 | 1.23 |
| | | Low | 0.54 | 1.29 |
| | | Bay | 4.54 | |
| | B. | Sewage Level | 0 filtering | 60% filtering |
| | | Control | 2.68 | 4.35 |
| | | High | 1.30 | 4.79 |
| 11/12/76 | | Medium | 1.32 | 5.67 |
| | | Low | 1.28 | 5.23 |
| | | Bay | 4.79 | |
| | C. | Sewage Level | 0 filtering | 60% filtering |
| | | Control | 2.20 | 8.06 |
| | | High | 2.31 | 11.08 |
| 11/19/76 | | Medium | 1.64 | 9.83 |
| | | Low | 2.02 | 12.10 |
| | | Bay | 12.60 | |

of about 2 µg/1, whereas filtered microcosms and the bay were between 8 and 13 µg/1 (Table 4C).

It seemed possible that the filtering had stimulated the growth of phytoplankton less than 28µ in size. However, a size and species check indicated similar phytoplankton species in the bay and in all of the microcosms, with *Thallassiothrix* sp., *Cosinodiscus* sp., *Thallasiosira* sp., and *Skeletonema* sp. dominating.

Nutrients, light, and temperature could not have been limiting factors during the second eutrophication experiment. Zooplankton biomass was low and did not appear to affect phytoplankton levels. An earlier experiment at a different time of year had shown strongly increased levels in response to sewage treatment. However, a strong response when it appeared in November was for filtered microcosms to track bay levels of chlorophyll rather than to show a depression in chlorophyll levels and for sewage microcosms to show no response at all. As a consequence of these results, it was hypothesized that some inhibitor or stimulator of phytoplankton growth was present or absent in the water column at this time of year.

### TEST TUBE EXPERIMENTS

A series of short-term tube experiments designed to discover the factors responsible for bay and microcosm behavior was carried out during the latter part of the eutrophication experiment. In all cases, 16 ml tubes were inoculated with 10 ml of water and a 1.5 ml culture of the diatom, *Skeletonema costatum*, and then incubated at $14^{o}$C and 13.9 ly/day. All treatments, which were carried out in duplicate or triplicate, are summarized in Table 5. Phytoplankton counts on a Coulter counter Model TA 11 were made to assess treatment effects.

Table 5. Summary of treatments in test tube experiments.

A. Water Source: Control microcosm, bay water, offshore water.

B. Heated Water: $40^{o}$C., $70^{o}$C., $100^{o}$C. (for 5 minutes). Titration with different percentages of boiled water.

C. Filtration: 0.2µ, charcoal, Dialysis (<8000mw), Amicon filter (<500mw), Chelex column (trace metals).

D. Treated Water: Bacteriocide, indolacetic acid, EDTA, nutrients (f/2), pH.

### Results and discussion

A. Water Source. During the second week of November, best growth occurred in the order of bay water, microcosm water, and offshore water, thus suggesting that the factor was becoming less potent in bay water (Table 6A). In fact, bay levels of chlorophyll were slowly increasing during this period. (Fig. 1, Table 4).

B. Heated water. With one exception, the best growth always occurred in water which had been boiled and cooled before the *Skeletonema* culture was added (Table 6B). Growth was also somewhat a function of how high the water was heated up to boiling (Table 6B a).

Table 6. Phytoplankton counts in treated test tube experiments.
Counts represent means of 2 or 3 replicates. A. effects of water
source after 8 days. B. effects of heated water after 5 days.
C. effect of filtration. D. effects of treated water.

**A. Effect of water source after 8 days.**

| Source | cells/ml (x1000) |
|---|---|
| Bay | 100 |
| Microcosm | 65 |
| Offshore | 45 |

**B. Effect of heated water after 5 days.**

a. Thermal Stress

| | |
|---|---|
| unheated | 119 |
| 40°C | 128 |
| 70°C | 187 |
| boiled | 311 |

b. Boiled Water

| | |
|---|---|
| 0% boiled | 72 |
| 1% boiled | 78 |
| 5% boiled | 85 |
| 10% boiled | 87 |
| 50% boiled | 137 |
| 75% boiled | 219 |
| 100% boiled | 281 |

**C. Effects of Filtration.**

| | Cells/ml (x1000) | | | |
|---|---|---|---|---|
| | after 5 days | after 5 days | after 5 days | after 8 days |
| Microcosm water | 72 | 119 | 129 | 100 |
| 0.2µ | | | | 100 |
| Dialysis tubing in microcosm water | | 94 | | |
| Amicon filter | | | | |
| filtrate | | | 175 | |
| concentrate | | | 174 | |
| Chelex column | 63 | | | |
| Charcoal | 336 | | | |
| With boiling | 273 | 311 | 302 | 190 |

**D. Effects of Treated Water.**

| | after 5 days | after 5 days | after 6 days | after 8 days |
|---|---|---|---|---|
| Microcosm water | 129 | 78 | 19 | 100 |
| With bacteriocide | 200 | | | |
| With $10^{-5}$M indolacetic acid | 143 | | | |
| With EDTA (0.01-0.60 µM) | | | | 81 |
| With pH 2.6[*] | | | 52 | |
| With boiling | 302 | 273 | 52 | 190 |

[*]In repeat experiments pH 2.3 and 2.8 resulted in lower cell counts
than pH 2.6 which again had counts similar to boiling. However,
adding neutralized acid solutions also had counts similar to boiling.

Titration of boiled to unboiled water resulted in more growth as the
percent of boiled water was increased (Table 6B b).

A partial explanation for the increase on chlorophyll concentrations in the filtered microcosms in early November emerges from
these results. If boiling water increased growth, so did pasteurization of the filtrate from the microcosms. It remains a mystery
why this pasteurization process was ineffective earlier in the fall.
Chlorophyll concentrations in filtered microcosms did not begin to
rise until the bay values of chlorophyll concentration started to
increase (Table 4).

C. Filtration. Filtration with 0.2µ filter had no effect on
phytoplankton growth, nor did filtration through a Chelex column to
remove trace metals (Table 6C). When boiled water was placed in
dialysis tubing in microcosm water, growth was inhibited, indicating
that the factor could move through the tubing and was therefore less
than 8000 mw. Similarly, the filtrate and concentrate of an Amicon
filter had equal growth rates, indicating the factor could be less
than 500 mw (Table 6C). By contrast, water which had been charcoal-
filtered resulted in excellent growth, suggesting the factor was an

organic compound.

D. Treatment. None of the treatments with indolacetic acid (a growth stimulator), EDTA (a metal chelator necessary for growth), and nutrients improved phytoplankton growth over plain microcosm water (Table 6D). Treatment with a bacteriocide appeared to improve growth but not as well as boiling or charcoal-filtering. Lowering the pH to 2.6 with hydrochloric acid and then bringing it back up to 8 with sodium hydroxide did result in growth comparable to boiling, but so did addition of neutralized acid and base reagents, suggesting that impurities in the reagents were responsible for enhanced growth.

These test tube experiments encourage the belief that there is some substance, probably an inhibitor, which controls the growth of phytoplankton during the late summer-early fall. Neither nutrients, either from sewage or artificial media, nor light, nor water temperature affect the impact of this substance. The substance, which appears to be of low molecular weight, could be inactivated in November, when it became less potent in the bay, by charcoal filtration or boiling. R. Ferguson (personal communication) has indicated that filtration through a Chelex column may not be a valid test for trace metals and suggests that the observed effects on phytoplankton growth may be due to trace metal chelators. The fact that impurities in acid and base reagents caused growth comparable to boiling supports his contention. Experiments continue in an attempt to reveal the nature of this substance.

### *Growth-controlling substances*

Recognition of growth-controlling substances in seawater is not new, although the extreme potence as illustrated by this experiment is not generally known. The alternate dominance of phytoplankton species *Skeletonema costatum* and *Olisthodiscus luteus* during the warmer months in Narragansett Bay has been attributed to the production of an inhibitory ectocrine by *Olisthodiscus* (Pratt, 1966). While even the presence of this ectocrine is still speculative, it was suspected to be a tannoid substance (polyphenol) important for metal chelation (Pratt, 1966). A series of papers report on the excretion of organic compounds by macroalgae and their importance in the formation of *gelbstoff* (Sieburth, 1968; Sieburth and Jensen, 1969; Sieburth, 1969; and Sieburth and Jensen, 1970). Of interest is the fact that seaweeds exude substances of a phenolic nature which complexes with carbohydrates and proteins in the water to become gelbstoff. These phenolic substances are toxic to some larval fish (Sieburth and Jensen, 1969) and probably also to phytoplankton species (Sieburth, 1968). While the nature and source of the substance which suppresses the autumn phytoplankton bloom are unknown, the reality of its presence seems increasingly clear.

## Section II. Plankton Predation Experiments

### 1. THE IMPORTANCE OF VARIOUS SIZE FRACTIONS

Microcosms, because of their small size, do not usually include members from higher trophic levels. This spring experiment was designed to simulate size-selective predation by higher trophic levels. Predators such as fish and zooplankton are recognized to be size-selective for their prey (Durban, 1976; Frost, 1972). Thus, the impact of removing different size fractions of the plankton on microcosm behavior was explored while the intensity of the predation was held constant. The intensity level chosen (20% per week) was close to the maximum level reported (3% per day) by Vargo (1976) for ingestion of diatoms by zooplankton in late winter.

### Design and methods

A size-selective filtration experiment was carried out in early spring (2/25-3/29/76) with 12 microcosms. In microcosms 2, 6, and 10, organisms greater than 153 μ in size were removed; in microcosms 3, 7, and 11, organisms greater than 80 μ were removed; in microcosms 4, 8, and 12, organisms greater than 0.2 μ in size were removed. Microcosms 1, 5, and 9 served as controls with no filtration.

The parameters measured and the methods used during the experiments are summarized in Table 2. Microcosms were filtered (15 1) Tuesdays and Fridays of each week. A number 10 net was used to remove zooplankton from 3 tanks and a number 20 net was used to remove zooplankton and phytoplankton from another 3 tanks. In 3 other tanks, two reverse flow filters in series (1260 cm$^2$ and 156 cm$^2$) were used to concentrate particles larger than 0.2 μ in 250-400 ml of water. All materials filtered from the microcosms were killed by heating to 80$^\circ$C for 5 minutes and then returned to the appropriate microcosm.

### Results and discussion

This level of size-selection predation had no discernable effect on phytoplankton concentrations or zooplankton numbers. During this experiment, the winter-spring diatom bloom decreased in the bay and in the microcosms as indicated by chlorophyll concentrations (Fig. 6). After mid-March, a second bloom occurred in the microcosms but not in the bay. Zooplankton, primarily *Acartia clausi* juveniles and nauplii and rotifers, behaved in an inverse pattern in the microcosms, with peak numbers occurring when chlorophyll concentrations were lowest (Fig. 7). Bay counts of zooplankton were more variable than the replicate microcosms but appeared to follow a similar pattern. Microcosm behavior was very similar to the bay, and zooplankton appeared to be controlling chlorophyll concentrations.

Canonical variates analysis (Blackith and Reyment, 1971) with a data set that included chlorophyll concentrations, ammonia concentrations, total phytoplankton counts, and total zooplankton counts showed no differences between treatment levels (Fig. 8). All

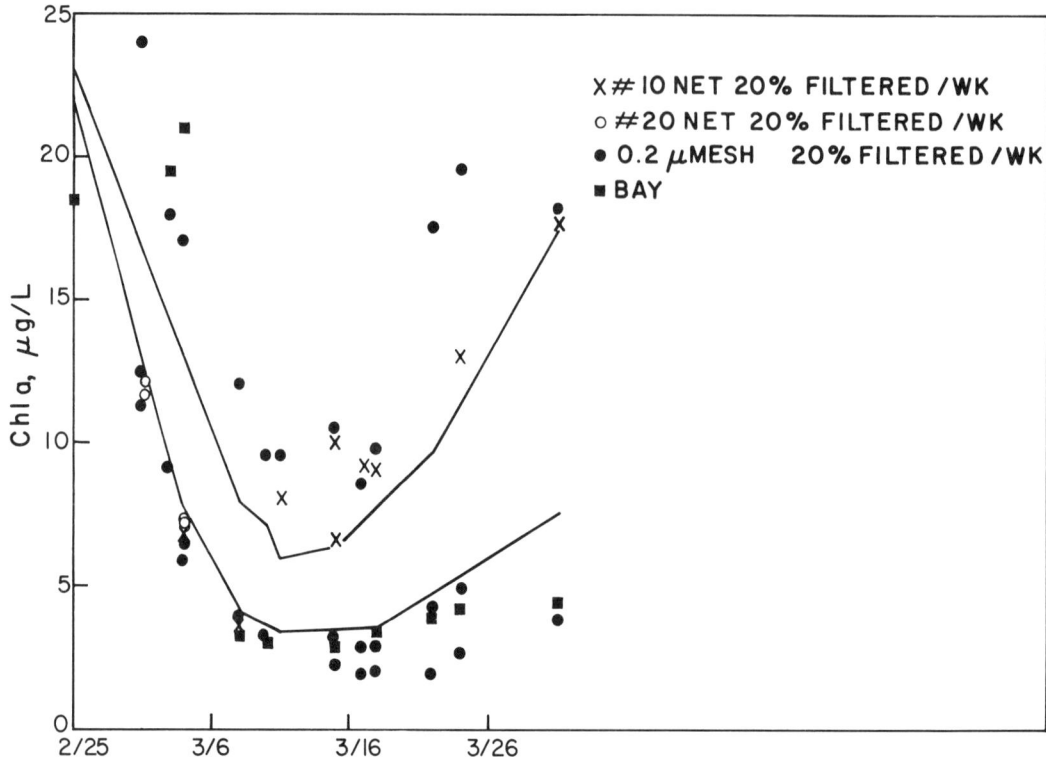

Fig. 6. Chlorophyll concentrations in microcosms with size-selective filtration levels of 20% per week. The envelope indicated by solid lines represents the variability of control microcosms. Treatment values falling within this envelope are omitted.

treatments showed a high degree of overlap, with control and predaceous zooplankton microcosms having the greatest degree of variability. The fact that the mean position of the bay variate is displaced to the left, well away from all microcosm variate positions, is probably due to the fact that a second phytoplankton bloom appeared in the microcosms but not in the bay (Fig. 6).

It is quite apparent that the imposed level of filtration stress was not effective in causing the systems to diverge in terms of the parameters measured. In contrast, it appeared that the zooplankton with their continuous filtration were much more effective in controlling phytoplankton concentrations.

## 2. THE EFFECT OF PREDATION INTENSITY

Since no changes in the systems behavior were observed with size-selective predation, the decision was made to remove all organisms and impose differing levels of predation intensity. The levels chosen this time were well above the mean phytoplankton percent eaten per day by zooplankton (2.7%) reported by Vargo (1976) for summer. It was thought that this would result in severe perturbations which would damp high natural oscillations of the plankton which are common in summertime.

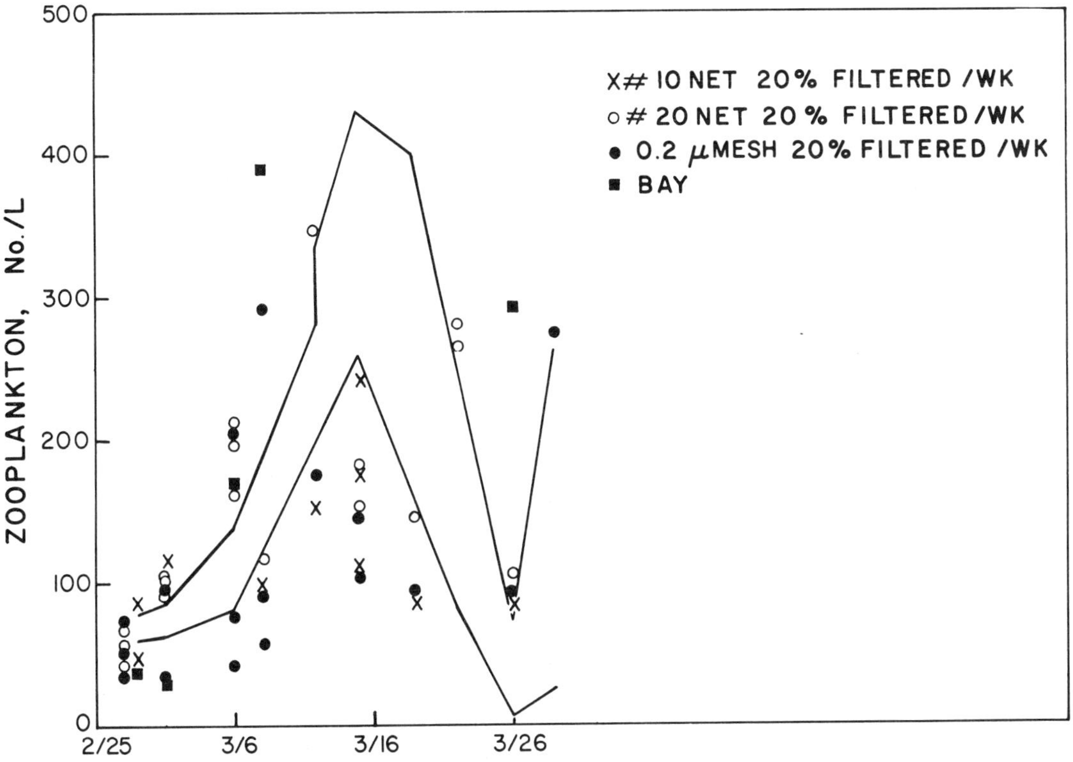

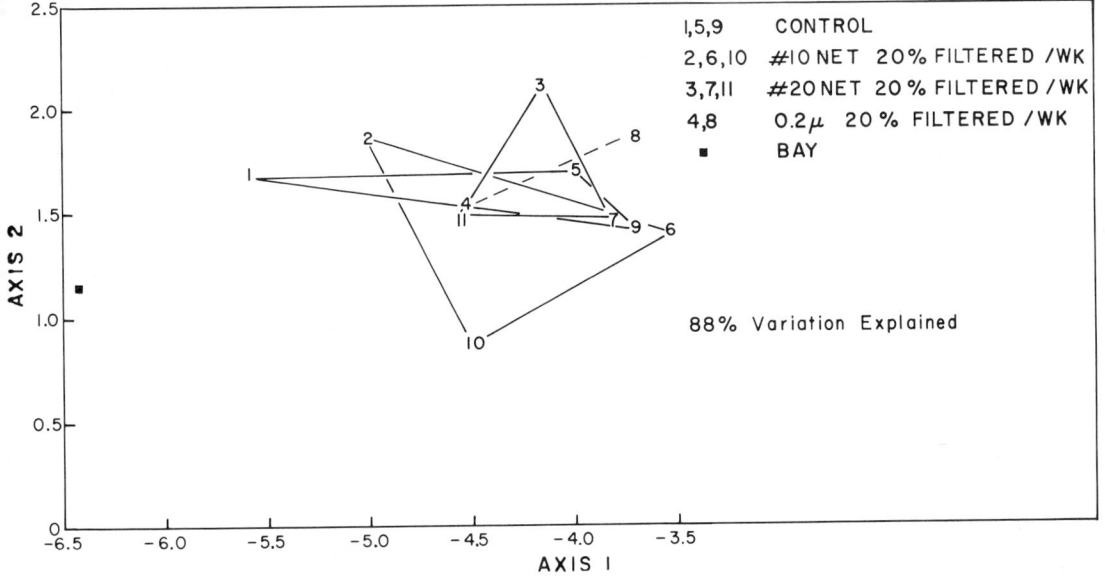

Fig. 7.  Zooplankton numbers in microcosms with size-selective
filtration levels of 20% per week.  The envelope indicated by the
solid lines represents the variability of control microcosms.
Treatment values falling within this envelope are omitted.

Fig. 8.  Canonical variates analysis of microcosms with size-
selective filtration levels of 20% per week.  The data set for this
analysis included weekly chlorophyll concentrations, ammonia
concentrations, total phytoplankton counts, and total zooplankton
counts.  Data from microcosm 12 was omitted because the benthos was
not collected at the same location as the rest of the microcosms.

*Design and methods*

In mid-summer (7/19-8/3/76) a second simulated predation experiment was carried out with differing levels of filtration. A 30% level and a 60% level (15 1 or 30 1 filtered Mondays, Wednesday, and Fridays through a 0.2 μ filter) were imposed with three different light levels on 18 microcosms. The three light levels were chosen to box light intensity effects in the bay water column and will be discussed more extensively elsewhere (Nixon *et al.*, this volume). Microcosm treatment was designed as shown in Table 7. Characteristics of the microcosms and analysis procedures were the same as in experiment 1. The filtrate was pasteurized as in experiment 1 and returned to the microcosms.

Table 7. Microcosm treatment design in the summer predation experiment.

| Filtration/week | Light Level | | |
|---|---|---|---|
| | 25 ly/dy | 19 ly/dy | 16 ly/dy |
| | Microcosm Numbers | | |
| 0% | 1,10 | 8,17 | 6,15 |
| 30% | 4,13 | 2,11 | 9,18 |
| 60% | 7,16 | 5,14 | 3,12 |

*Results and discussion*

It was puzzling to observe again no effect from the filtration. Microcosm and bay measurements showed chlorophyll concentrations dropping from about 10 μg/l to approximately 3 μg/l and then rising to higher levels. Zooplankton during the same period declined from 15/1 to 1-3/1 during the last two weeks of the experiment. No significant differences (0.05 level) between chlorophyll concentrations or between zooplankton biomass and filtration level were observed using factor analysis of variance (Table 8A, 8B). A test for interaction prior to main effects was also found to be insignificant.

Time of sampling was significant (0.05 level) in both analyses, suggesting that the high initial zooplankton numbers were responsible

Table 8A. Factor analysis of variance on chlorophyll concentration (ug/l) in 18 microcosms during summer predation/light experiment.

| Filtration | Light Level (ly/dy) | | | F | $F_{0.05}$ |
|---|---|---|---|---|---|
| Percent/week | 25 | 19 | 16 | | |
| 0 | 5.80 | 4.41 | 3.35 | | |
| 30 | 9.62 | 6.20 | 1.85 | 0.69 | 3.26 |
| 60 | 6.45 | 8.87 | 2.07 | | |
| | 5.87* | 10.83** | | | |
| | | 14.75** | | | |

Significance: $^*F_{0.05} = 4.11$; $^{**}F_{0.01} = 7.39$

Table 8B. Factor analysis of variance on zooplankton numbers (no./1) in 18 microcosms during summer predation/light experiment.

| Filtration | Light Level (ly/dy) | | | F | $F_{0.05}$ |
|---|---|---|---|---|---|
| Percent/week | 25 | 19 | 16 | | |
| 0 | 16 | 24 | 29 | | |
| 30 | 14 | 19 | 17 | 1.41 | 3.35 |
| 60 | 30 | 8 | 23 | | |
| F | | 1.25 | | | |

for the decrease in phytoplankton and that low zooplankton numbers at the end of the experiment allowed phytoplankton to increase again. Although significant differences (0.05 level) occurred between the three light levels for chlorophyll (Table 8A), there were no significant differences for interaction between time of sampling and light, thus supporting the suggested interaction between phytoplankton and zooplankton. In both experiments, when phytoplankton levels were high, zooplankton increased and caused phytoplankton to decrease.

### 3. THE EFFECT OF PREDATION ALONG A EUTROPHICATION GRADIENT

In this fall experiment we had thought that blooms would develop according to sewage treatment level and be sensitive to an extreme predation stress, even though no effects were seen in the earlier two experiments. The manner in which predation was imposed was made size-selective for phytoplankton by leaving in zooplankton and imposing the highest intensity (60%/week) for experiment 2 on phytoplankton-size particles.

#### Design and methods

Four levels of sewage and two filtration levels on phytoplankton-size particles (153µ to 80µ) were established in 20 microcosms (Table 9). Characteristics of the microcosms and analysis procedures have already been described in Section I. The filtrate was pasteurized as in the two previous predation experiments and returned to the microcosms.

Table 9. Microcosm treatment design in the fall sewage and predation experiment.

| Filtration/week | Sewage Level | | | |
| | 0% | 1% | 5% | 10% |
| | Microcosm Number | | | |
| 0% | 1,5,9 | 4,8,12 | 3,7,11 | 2,6,10 |
| 60% | 13,17 | 16,20 | 15,19 | 14,18 |

#### Results and discussion

As outlined in Section I, phytoplankton blooms failed to develop in the fall sewage experiment (Fig. 1). As a consequence, no effect due to predation stress was observable in time-series plots (Fig. 1). A slight, and variable, effect due to filtration was apparent in a canonical variates analysis. The positions of some filtered microcosms are shifted to the right, away from respective sewage microcosms in the direction of control microcosms along the first axis (Fig. 4). Thus, some of the effect of filtration was to reduce the increased biomass of phytoplankton resulting from sewage treatment.

#### ARTIFICIAL PREDATION

The microcosms proved insensitive to filtration stress when applied either twice a week or three times a week even at a 60% level. Instead, continuous feeding by zooplankton appeared to control phytoplankton levels in experiments 1 and 2. Other studies, both experimental (Vargo, 1976) and numerical modeling (Kremer and Nixon, 1978), which have studied the impact of zooplankton grazing on phytoplankton standing crop, find minimal effects, but the considerable variability in measured ingestion rates may mask the effects in specific instances. In experiment 3, during the treatment period when zooplankton numbers were low, phytoplankton growth did increase slightly according to degree of eutrophication. The effect of filtration in this experiment was to decrease phytoplankton biomass and to make the levels more like the controls, but the shift was small and questionable (Fig. 4). Apparently, the manner in which stress is imposed can be important and the microcosms were extremely stable with regard to the filtration stress episodically during the three seasons.

## Section III. Seasonality in Natural and Experimental Ecosystems

The two-factor stress experiment was confounded by a natural seasonal pattern of the plankton. In many years, phytoplankton, and consequently zooplankton, drop to very low levels in the late summer and fall in Narragansett Bay in contrast to those temperate estuaries which have a traditional fall bloom (Cushing, 1975). Since plankton levels and production were very low during the experimental period, only a slight response to either factor was observed. All microcosms behaved in almost the same manner, and similar to the bay (Figs. 1, 3).

In these microcosm experiments we were attempting to determine empirical values for resiliency and stability (Holling, 1973) in marine systems under different kinds of controlled perturbation operation singly and in combination. If resilience is the ability of a system to absorb oscillations of state variables or driving variables and still persist, and stability is the ability of a system to return to equilibrium state after a temporary disturbance (Holling, 1973), the Narragansett Bay has varying degrees of resilience and stability during an annual cycle. Systems state of the bay changes dramatically and cyclically. A time comparison of data from 13 stations on zooplankton, ammonia, nitrate, phosphate, silicate and chlorophyll, using canonical variates, illustrates this change of state (Fig. 9). Large oscillations (high resilience-low stability) occur from June to September. High stability with little oscillation occurs during January-February. Both spring and fall appear to be transitional systems with little oscillation and no equilibrium of their own.

One year of data does not define a system's boundaries. Nevertheless, if the pattern persists, then microcosm experiments must either be done on an annual basis or repeated seasonally. As the eutrophication experiments illustrate, quite different results occur at different seasons of the year. In either case, the challenge of how to quantify the resilience or stability to a particular stress is increased. One approach with annual data sets might be to perform multivariate discriminant analyses which measure distances

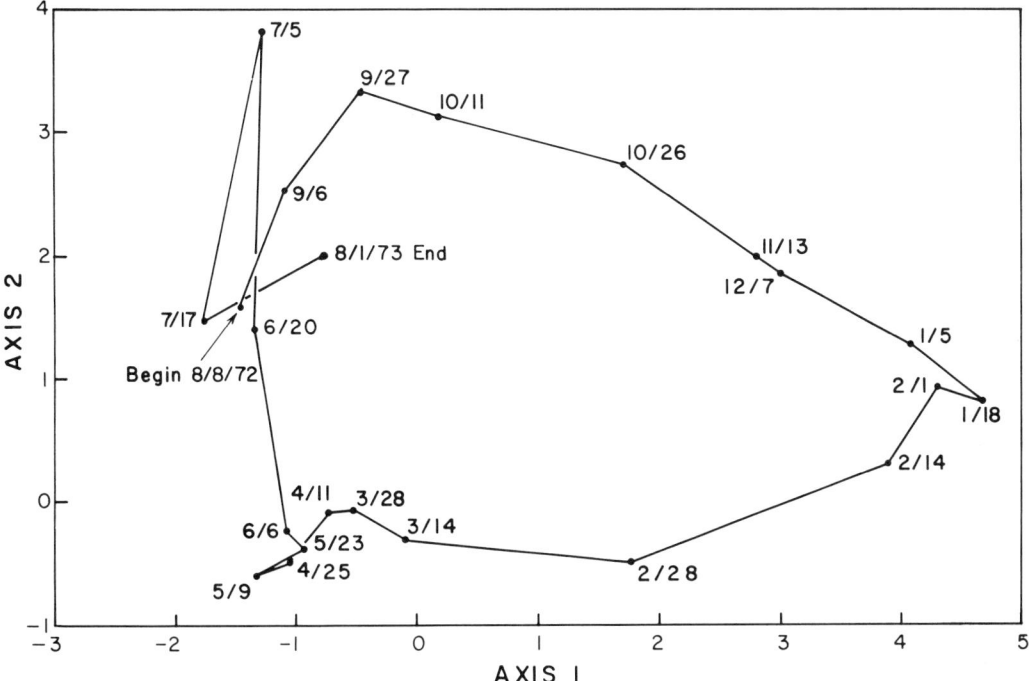

Fig. 9. Canonical variates analysis by date of data from 13 stations around Narragansett Bay taken biweekly in 1972-1973. The data set includes zooplankton biomass, ammonia, nitrate plus nitrite, phosphate, silicate and chlorophyll concentrations.

apart along multivariate axes, assuming the different treatment levels result in displaced cyclical patterns. This approach presumes long-term experiments and large data sets which decrease the rate at which questions may be posed.

## Acknowledgments

We are grateful to Jack Kelley and Patrick Roques and our EPA colleagues, George Morrison, Neal Lackie, Don Winslow, Peter Murphy, and David Dillman for help with manipulation, sampling and laboratory analyses. This research was supported by a grant from the U.S. Environmental Protection Agency, R-803143.

## References

Blackith, R.E. and R. A. Reyment. 1971. *Multivariate Morphometrics*. Academic Press, New York.

Cushing, D.H. 1975. *Marine Ecology and Fisheries*. Cambridge University Press, Cambridge, MA.

Davies, S.M., J.C. Gamble, and J.H. Steele. 1975. Preliminary studies with a large plastic enclosure. In: *Estuarine Research* (L.E. Cronin, ed.). Vol. 1. Academic Press, New York.

Durban, A.G. 1976. The role of fish migration in two coastal ecosystems. Ph.D. thesis, University of Rhode Island, Kingston, RI.

Frost, B.W. 1972. Effects of size and concentration of food particles on the feeding behavior of the marine planktonic copepod *Calanus pacificus*. Limnol. Oceanogr. 17: 805-815.

Holling, C.S. 1973. Resilience and stability of ecological systems. In: *Annual Review of Ecology and Systematics* (R.F. Johnston, P.W. Frank, and C.D. Michener, eds.), Vol. 4, pp. 1-24. Annual Review, Inc., Palo Alto, CA.

Kremer, J. and S.W. Nixon. 1977. *A Coastal Marine Ecosystem: Simulation and Analysis*. Springer-Verlag, New York.

McAllister, C.D., T. R. Parsons, K. Stephens, and J.D.H. Strickland. 1961. Measurements of primary production in coastal sea water using a large-volume-plastic sphere. Limnol. and Oceanogr. 6 (3): 237-258.

Miller, Gary. 1968. The influence of season on the radiation sensitivity of an old field community. Ph.D. thesis, Univ. of N.C., Chapel Hill.

Nixon, S.W., C.A. Oviatt, J.N. Kremer, and K.T. Perez. 1978. The use of numerical models and laboratory microcosms in estuarine ecosystem analysis -- simulations of a winter phytoplankton bloom. In: *Marsh-Estuarine Systems Simulation* (R. Dame, ed.). Vol. 8, Univ. of South Carolina Press, Columbia.

Oviatt, C. A., K. T. Perez, and S. W. Nixon. 1977. Multivariate analysis of experimental marine ecosystems. Helgolander wiss. Meeresunters.30(1-4): 30-46.

Perez, K.T., G.M. Morrison, N.F. Lackie, C.A. Oviatt, S.W. Nixon, B. Buckley, and J.F. Heltshe. The importance of physical and biotic scaling to experimental simulation of a coastal marine ecosystem. Helgolander wis. Meeresunters. (in press).

Pratt, David M. 1966. Competition between *Skeletonema costatum* and *Olisthodiscus luteus* in Narragansett Bay and in culture. Limnol. Oceanogr. 11(4): 447-455.

Sieburth, J.M. 1968. The influence of algal antibiosis on the ecology of marine microorganisms. In: *Advances in Microbiology of the Sea* (M.R. Droop and E.J. Wood, eds.). Vol. 1, pp. 63-94. Academic Press, N.Y.

Sieburth, J.M. and A. Jensen. 1968. Studies on algal substances in the sea. I. Galbstoff (humic material) in terrestrial and marine waters. J. Exp. Mar. Biol. Ecol. 2: 174-189.

Sieburth, J.M. 1969. Studies on algal substances in the sea. III. The production of extra cellular organic matter by littoral marine algae. J. Exp. Mar. Biol. Ecol. 3: 290-309.

Sieburth, J.M. and A. Jensen. 1969. Studies on algal substances in the sea. II. The formation of gelbstoff (humic material) by exudates of phaeophyta. J. Exp. Mar. Biol. Ecol. 3: 275-289.

Sieburth, J.M. and A. Jensen. 1970. Production and transformation of extra cellular organic matter from littoral marine algae: A resume. In: *Organic Matter in Natural Waters* (D.W. Hood, ed.), pp. 203-223. University of Alaska, College, Alaska.

Solorzano, L. 1969. Determination of ammonia in natural waters by the phenolhypochlorite methods. Limnol. Oceanogr. 14: 799-801.

Strickland, J.D.H. and J.T. Parsons. 1968. A practical handbook of sea water analysis. Fish. Res. Bd. Can., Bull. 167.

Takahashi, M., W.H. Thomas, D.L.R. Siebert, J. Beers, P. Koeller, and T.R. Parsons. 1975. The replication of biological events in enclosed water columns. Arch. Hydrobiol. 76(1): 5-23.

Ulanowicz, R.E., D.A. Flemer, D.R. Heinle, C.D. Mobley. 1975. The *a posteriori* aspects of estuarine modeling. In: *Estuarine Research* (L.E. Cronin, ed.), Vol. 1. Academic Press, New York.

Vargo, G.A. 1976. The influence of grazing and nutrient excretion by zooplankton on the growth and production of the marine diatom, *Skeletonema costatum* (Greville) Cleve, in Narragansett Bay. Ph.D. thesis, University of Rhode Island, Kingston, RI.

# The Use of Numerical Models and Laboratory Microcosms in Estuarine Ecosystem Analysis— Simulations of a Winter Phytoplankton Bloom

**Scott W. Nixon**
**Candace A. Oviatt**
Graduate School of Oceanography
University of Rhode Island
Kingston, Rhode Island 02881

**James N. Kremer**
Department of Biology
University of Southern California
Los Angeles, California 90007

**Kenneth Perez**
Environmental Research Laboratory
U.S. Environmental Protection Agency
Narragansett, Rhode Island 02882

### Abstract

Narragansett Bay, R.I., is characterized by an unusually early and intense mid-winter phytoplankton bloom. The timing of the bloom varies somewhat from year to year in response to circumstances that are not well understood, despite some 20 years of study. This paper reports the results of an analysis of the factors involved in the initiation of the bloom during two years in which the period of rapid growth was delayed from its usual starting time in December until late January and February. On the basis of numerical simulations using a mechanistic model, experiments with laboratory microcosms, and an analysis of field data, it seems clear that light intensity is the critical factor in the initiation of the bloom when it is delayed until this late in the year. Our experience also suggests that all three of these approaches are necessary in ecosystem analysis and that numerical simulations may be required in the design of microcosms. The interactions of living systems with the environment are often governed by non-linear processes that make it difficult to scale microcosms appropriately. For example, the work described here demonstrates that the effect on phytoplankton growth of the relationship between light input and depth is not straight forward and that a more elaborate numerical analysis than the calculation of depth-averaged light is required to compare systems adequately.

## Introduction

*It must be made clear at once that in Nature the investigation of the dependence of plankton photosynthesis on light is an extremely intricate problem. An accurate systematic study would*

*involve so many different single measurements and*
*experiments that any practical treatment of the*
*problem would be very difficult.*

*Steeman-Nielsen and Hansen (1961)*

This symposium brings together two groups of people who are trying to model coastal marine ecosystems in different ways.  One group
is developing computer simulation models that provide a mathematical
synthesis of our knowledge about the functioning of individual parts
of the system.  The other is exploring the behavior of small-scale
living models or microcosms which we hope will integrate the behavior of their parts in a manner that is ecologically meaningful.
While the computer models can take advantage of a large amount of
high quality data from environmental physiology, the microcosms make
it possible to include much more complexity and biological realism
in the modeling effort, including adaptation and self-design properties that are far beyond the state of the art in computer simulation.

The numerical models are constructed to provide information
about the mechanisms and interactions conceived to lie behind the
emergent properties of the system as they are observed in the field.
However, it is still not clear that the reductionist data on which
they are based is appropriate for the task (Patten, 1971; Mann, 1972,
1975).  This is an important point that must eventually be resolved.
A great deal of ecological research is heavily analytical, and much
of the literature is devoted to reporting rate measurements that
have been carried out on isolated parts of systems.  A basic tenet
of environmental physiology and autecology is that such measurements
are, in fact, also telling us something about the behavior of the
species or individual when it is acting as an integrated part of an
ecological system.  The computer model provides a test of this assumption by allowing us to compare predictions of the mathematical
analogues of a great many physiological responses to independent
data from the field.  This is an ambiguous test, however, since
there is no consensus for judging the success of a model.  Moreover,
any discrepancy between simulations and observed patterns in field
data may result from a fallacy in the conceptual model or from the
great variability that exists in field and laboratory data.

Microcosms provide an holistic tool for carrying out sensitivity
analysis and ecosystem experiments, but their complexity makes it
difficult to understand the mechanisms that are involved.  There also
are difficult and unresolved problems concerned with microcosm
methodology, including scaling effects, that are at least as pro-
found as the uncertainties about numerical models.  For example,
microcosms tend to eliminate higher trophic levels and to minimize
spatial heterogeneity (Oviatt *et al.*, 1977; Perez *et al.*, 1977).

Because of their complementary drawbacks and advantages, numerical models and experimental microcosms may be most successful as

tools for coastal ecosystem analysis when they are combined. We have been trying such a synthesis for several years in our studies of Narragansett Bay, R.I., and while the interaction has proved to be stimulating, the interpretation of the results is not always straightforward. This paper describes our attempts to reconcile information from numerical simulations, microcosms, field observations, and the physiological literature dealing with a classical problem-- the analysis of factors related to a winter-spring phytoplankton bloom. The reason for using such an example in this paper is historical as well as convenient. The use of numerical models in marine ecology began with Fleming's(1939) analysis of the spring bloom in the English Channel, and some of the earliest applications of the microcosm approach in the sea dealt with the study of phytoplankton blooms (Edmondson and Edmondson, 1947; Pratt, 1949; McAllister *et al.*, 1961; Antia *et al.*, 1963).

## Winter-Spring Phytoplankton
## Bloom in Narragansett Bay

Narragansett Bay is a plankton-based system of some $26.5 \times 10^3$ ha with only minor inputs of organic material from marshes, macro-algae or submerged vascular plants. It is relatively shallow with a mean depth of 8-9m and has a low fresh water input that averages only 45-50m$^3$/sec. As a result, the salinity over much of the bay is high (20-30 °/oo) and relatively constant throughout the year. With the exception of the upper bay, the water column is well-mixed and there is little vertical stratification, especially during the winter. The water temperature ranges from about 1 to 20°C during the annual cycle.

One of the unique features of Narragansett Bay is a characteristic winter-spring diatom bloom that has traditionally been marked by its winter inception (December-January), its dense populations (40 or 50 x 10$^6$ cells/1), and its long duration into spring (Smayda, 1957; Pratt, 1965). Although species such as *Detonula confervacea* and *Thalassiosira nordenskioeldii* may be more or less important components of the bloom in different years, the major species involved is usually *Skeletonema costatum*. This has remained true in spite of the fact that in several of the last few years there has been a marked delay in the initiation of the bloom until February. Our field data and laboratory experiments were carried out during the "late bloom" years of 1972-1973 and 1975-1976, when the dynamics of the system may have differed from the more "normal" pattern. Nevertheless, we felt that any insight we might gain would be useful. The problem of the winter-spring bloom in Narragansett Bay has remained a challenge, despite over 20 years of study by a number of plankton ecologists. In summarizing the results of an intensive study of the bloom during 1971-1972, Smayda (1973) concluded that "the dynamics of *Skeletonema costatum* during the 1971-1972 winter-spring bloom in Narragansett Bay could be related to temperature,

light, nutrients, grazers, possibly species interactions, and hydro-
graphic disturbances."

## Numerical Model

The numerical simulation model of Narragansett Bay is shown in
Figure 1. This model has been described in preliminary form elsewhere (Kremer and Nixon, 1975; Nixon and Kremer, 1977) and has been
documented and discussed fully in a recent book (Kremer and
Nixon, 1977). While the model includes phytoplankton (two
species groups), zooplankton (adults, juveniles, nauplii, eggs),
nutrients (ammonia, nitrite plus nitrate, phosphate, silicate),
pelagic and benthic nutrient regeneration, the predation pressure of
fish larvae, ctenophores, menhaden, and clams, as well as a numerical approximation of spatial heterogeneity and hydrodynamic mixing,
the most relevant aspect of the formulation for the analysis of the
winter-spring bloom is the computation of phytoplankton growth.

PHYTOPLANKTON FORMULATION

The approach taken is, first, to evaluate the maximum "instantaneous" growth rate ($G_{max}$) of the phytoplankton (P) as an exponential function of temperature (T), according to Eppley (1972),
Goldman and Carpenter (1974), and others.

$$G_{max} = 0.59 \ e^{0.063T_{o_C}} \ (day^{-1}) \tag{1}$$

$$\text{and} \quad P_t = P_o \ e^{G_{max} T_{days}} \tag{2}$$

There are a number of limitations inherent in such an expression,
the most important being the loss of photo-period effects, thresholds, and optima characteristic of individual species groups and
the averaging involved in the relatively long-term growth measurements on which the data are based. There is also some evidence that
phytoplankton may grow for short periods at rates higher than suggested by the curve(Yoder, 1973). Nevertheless, the formulation
appears to be a reasonable approximation based on a large number of
observations, and it is compatible with measurements on natural and
cultured phytoplankton from Narragansett Bay (Smayda, 1973).

The maximum growth rate is then reduced by terms for nutrient
limitation (NUTLIM) and non-optimum light (LTLIM). Nutrient limitation is calculated using the Monod (1942) formulation for a rec-
tangular hyperbola relating growth to nutrient concentrations:

$$G = G_{max} \ \frac{[N]}{K_s \ [N]} \tag{3}$$

# PHYTOPLANKTON COMPARTMENT

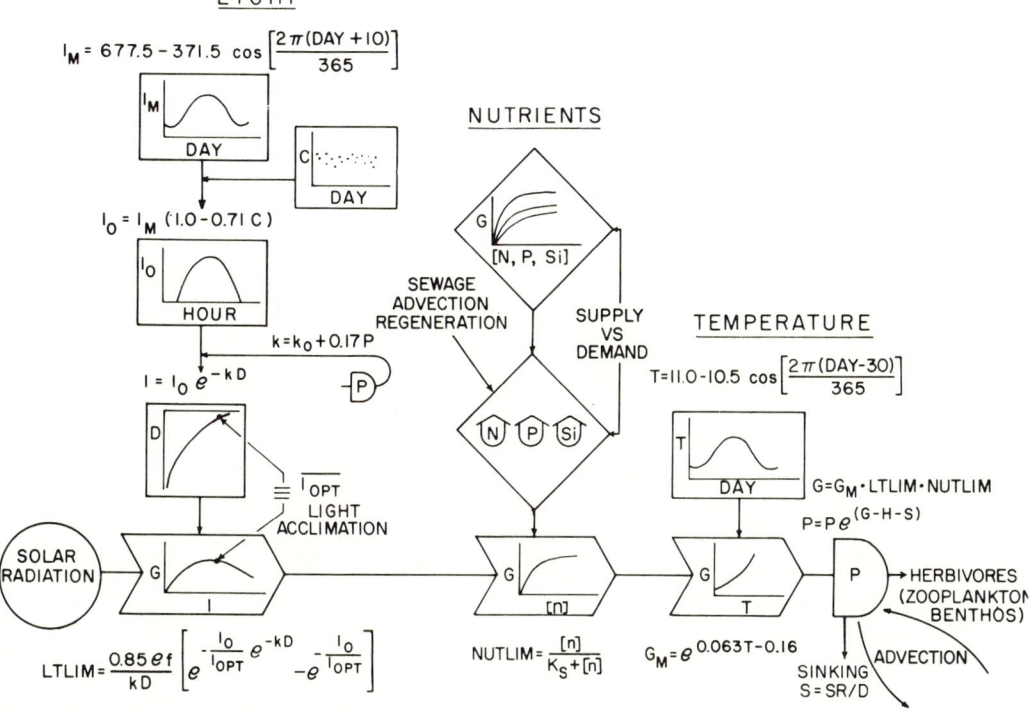

# ZOOPLANKTON COMPARTMENT

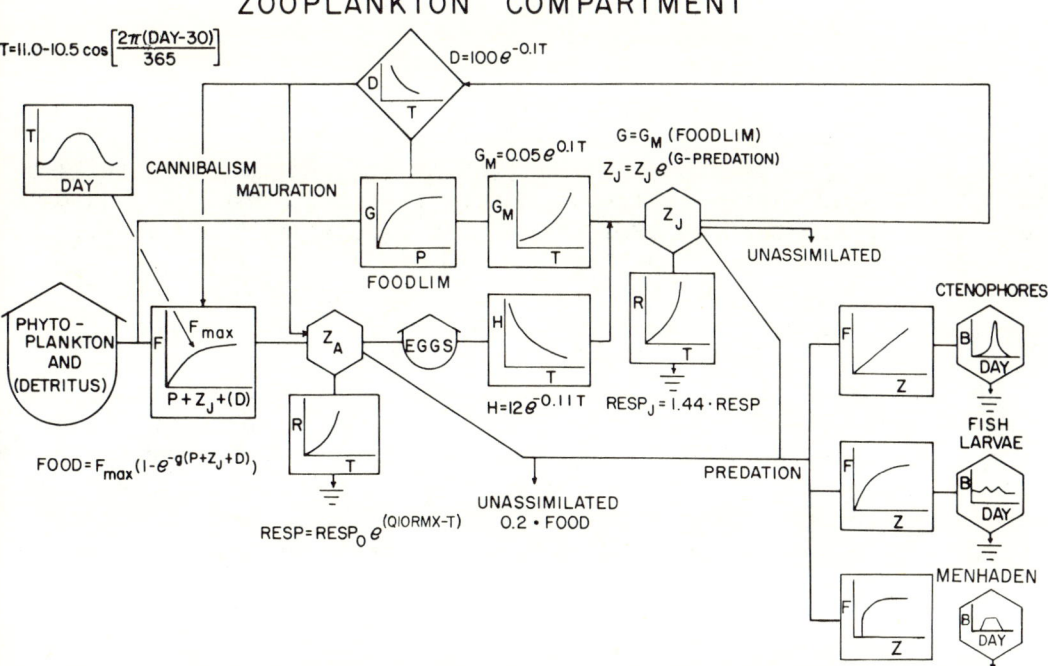

Fig. 1. Mathematical and symbolic representation of the flows and controls operating on the phytoplankton and zooplankton compartments of the model. Details in text and in Kremer and Nixon, 1975, and in Kremer and Nixon, 1977.

$$\text{NUTLIM} = \frac{G}{G_{max}} \tag{4}$$

where $G_{max}$ is the maximum growth rate from eq. (1), [N] is the ambient steady-state concentration of nutrient, and $K_s$ is a characteristic half-saturation constant defined as the concentration at which the growth rate is one half the maximum. An identical formulation was suggested by Michaelis and Menten (1913) for the dynamics of enzyme reactions, and Odum (1972) has discussed the equation as descriptive of general processes involving "double-flux" reactions. The empirical agreement of the expression is well-established (Dugdale, 1967; Eppley and Thomas, 1969; Eppley *et al.*, 1969; Hanton, 1969; MacIsaac and Dugdale, 1969; Paasche, 1973, 1973a and many others). It is important to note, however, that the calculation of nutrient limitation using this expression does not explicitly take the standing stock of phytoplankton into account. The model compensates for this by calculating the amount of nutrient that would be required for the projected growth and comparing it with the amount available. The growth rate is then adjusted appropriately so that demand cannot exceed supply during the one-day time step of the simulations. The decision as to which nutrient is most important is made by calculating NUTLIM for each and by using the "most limiting" to reduce growth.

$$P_t = P_o \, e^{(G_{max} \cdot \text{NUTLIM}) \, T_{days}} \tag{5}$$

There are a number of additional aspects of the nutrient formulation that could be described and argued at length (see Kremer, this volume), but for the time period involved in this discussion (initiation of the winter-spring bloom), nutrients are generally abundant and do not appear to play a major role in regulating the plankton dynamics.

The calculation of light limitation is substantially more complicated, and we have given a detailed treatment of the subject elsewhere (Kremer and Nixon, 1977). It is also critical to much of the discussion that follows. In essence, the formulation includes a light input with stochastic day-to-day variation based on cloud cover statistics (options include observed data or long-term daily means), an exponential attenuation of light with depth, a feedback between chlorophyll concentrations and the extinction coefficient (Riley, 1956), and a relationship between phytoplankton growth and light intensity that is described by the Steele (1962) equation:

$$G = G_{max} \frac{I}{I_{opt}} \, e^{\left(1 - \frac{I}{I_{opt}}\right)} \tag{6}$$

$$\text{LTLIM} = \frac{G}{G_{max}} = \frac{I}{I_{opt}} e^{\left(1 - \frac{I}{I_{opt}}\right)} \qquad \text{(6 cont'd)}$$

where I is the incident light intensity, and $I_{opt}$ is the light at which maximum production occurs (Fig. 2).

The development and application of this and other light-photo-synthesis formulations have been critically reviewed by Vollenweider (1965), Fee (1969) and others. The equation proposed by Steele (1962) remains one of the simpler and more remarkably versatile, and the degree to which it approximates a wide range of light-photo-synthesis observations is quite satisfactory (Ryther and Menzel, 1959; Steele, 1962, 1965; Fee, 1973; Bannister, 1974; Goldman and Carpenter, 1974; Nixon *et al.*, in prep.). It also has the additional advantage in normalized form of requiring only one parameter, $I_{opt}$, which has a biologically meaningful interpretation. Moreover, the parameter, $I_{opt}$, can be used in the model to simulate light acclimation by the phytoplankton.

### $I_{opt}$ AND LIGHT ACCLIMATION

Fluctuations in the light response of natural phytoplankton populations over short spans of time and space have been widely recognized for many years, and numerous investigations have documented variations within and between species in the laboratory (Ryther, 1956; Jitts *et al.*, 1964; McAllister *et al.*, 1964; Smayda, 1969; Ignatiades and Smayda, 1970), as well as in many mixed natural populations (Ryther and Menzel, 1959; Steeman-Nielsen

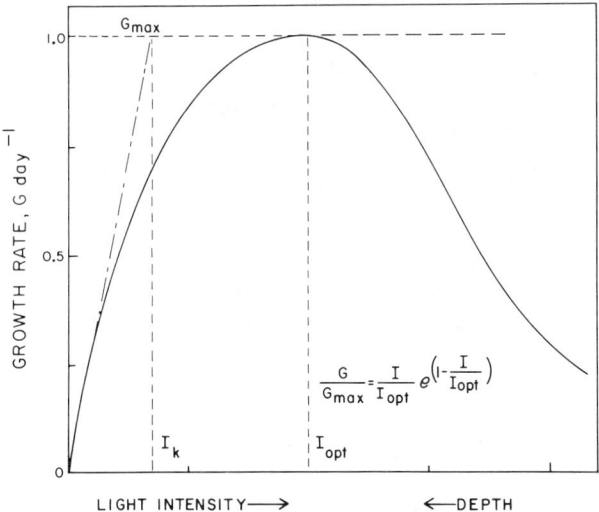

Fig. 2. The instantaneous photosynthesis-light response used in the model for calculating phytoplankton growth (Steele, 1962). The normalized equation predicts growth relative to the maximum as a function of the ratio of the incident light (I) and the intensity at which production is maximum ($I_{opt}$). In cases where there is a plateau rather than a clear optimum, another parameter ($I_k$) is often used (Talling, 1957).

and Hansen, 1959, 1961; Nixon *et al.*, in preparation; Smayda and
Hitchcock, URI, personal communication). In some cases, the photosynthetic rate may show a hyperbolic response to increasing light
intensity rather than a clear optimum. Under these circumstances a
parameter, $I_k$, defined by Talling (1957) as the intersection of the
initial slope of the hyperbola and its asymptote (Fig. 2), is often
used to characterize the changing light-photosynthesis response.
For example, Yentsch and Lee (1966) present experimental data and a
theoretical discussion of the importance of changes in $I_k$. With
the Steele equation the initial slope can be shown to be $I_{opt}/e$,
thus allowing a direct comparison of $I_k$ and $I_{opt}$.

Changes in $I_k$ or $I_{opt}$ may be due to a variety of environmental
factors, only some of which actually reflect changes in the light
reactions of photosynthesis (Yentsch and Lee, 1966). However, it
now appears well-established that the photosynthetic capacity reflects to a significant degree the past light history of the species
or community. Accordingly, the basic assumption in the model is
that $I_{opt}$ tracks the previous light history to which the phytoplankton have been exposed. Experimental data for *Chlorella* (Steeman-Nielsen *et al.*, 1962) first suggested an approximate rate for
acclimation of 2 or 3 days with the most rapid change during the
first day. Further work (Steeman-Nielsen and Park, 1964) continues
to suggest this time-frame for changes in the response pattern of
cultures and communities.

Another consideration in the specification of a changing $I_{opt}$
is the light level to which acclimation is coupled. Our field
vertical productivity profiles suggest that fully acclimated natural
bay populations frequently demonstrate maximum production at a depth
of about one meter, or approximately 40-70% of the surface light
with the extinction coefficients found in the bay. Over a wide
range of locations and seasons, the tendency was remarkably consistent. Considering the ratio $I_{opt}:I_{surface}$, Steele reports a number of observations which agree with the assumption that $I_{opt} \cong$
50% $I_{sfc}$ (Steeman-Nielsen and Jensen, 1957, 33-50%; Sorokin *et al.*,
1959, 30-50%; Steele and Baird, 1961, 50-100%; Steele, 1962, 50%).
Based on these considerations, the model determines $I_{opt}$ for each
day as a weighted, moving average of the light intensity at the 1m
depth for the previous three days. The weighting factors were chosen
to approximate the observations of Steeman-Nielsen *et al.* (1962) with

$$I_{opt} = 0.7 \ I'_1 + 0.2 \ I'_2 + 0.1 \ I'_3 \qquad (7)$$

for $I'_j$, the average light level at 1m, j days earlier. Under a
constant light regime, the population will be 70% acclimated after
1 day, 90% after 2 days and 100% by the third day. Under the scheme
of stochastic cloudiness used in the model, continuous acclimation
occurs which will, on the average, reflect a seasonal trend of mean
insolation.

A lower limit to light acclimation is usually also specified in
the model. As a first approximation, it seemed reasonable to set

this lower limit to $I_{opt}$ as the light intensity found at a depth of one meter when the average light in the water column equaled 40 ly/day.

$$\bar{I} = \frac{I_o}{kz} (1 - e^{-kz}) \tag{8}$$

where $\bar{I}$ = mean light in the water column (depth-averaged), $I_o$ = non-reflected incident light, k = extinction coefficient (m$^{-1}$), z = depth, and $I_{1m} = I_o e^{-k}$. This value was chosen on the basis of Riley's (1967) observation that an average light intensity of 40 ly/day in the water column seemed to be a critical condition for initiation of the winter-spring bloom in Long Island Sound.

Once $I_{opt}$ is specified, however, the use of the Steele equation still requires a transformation to account for the integrated effects of changing light regime both throughout the day and with depth in the water column.

### TIME- AND DEPTH-INTEGRATION OF THE STEELE CURVE

The non-linear pattern of diurnal irradiance in the simplest case of a cloudless sky makes analytical integration of even the relatively simple Steele equation impossible. Steele (1962) used an assumption of a linear increase from dawn to mid-day and subsequent linear decrease to dusk, allowing a time-averaged production estimate in terms of the time-averaged light, $\bar{I}$:

$$\frac{\bar{G}}{G_{max}} = \frac{e \cdot I_{opt}}{2\bar{I}} \left[ 1 - \left( 1 + \frac{2\bar{I}}{I_{opt}} \right) e^{-\frac{2\bar{I}}{I_{opt}}} \right] \tag{9}$$

The depth integral of equation 9 under the exponential extinction of light becomes formidable due to the double exponential terms ($e^{e-kz}$), and Steele does not present a solution.

A somewhat different, simplifying assumption invoked by DiToro *et al.* (1971) permits analytical solution of the depth-time integral but not without some loss of biological validity. If the incident radiation is taken to be constant through the entire photo-period, i.e. $I = \bar{I}$ (time-averaged) at all times, the exact double integral becomes:

$$\frac{\bar{G}}{G_{max}} = \frac{e \cdot f}{k \cdot z} \left( e^{-\frac{\bar{I}}{I_{opt}} e^{-kz}} - e^{-\frac{\bar{I}}{I_{opt}}} \right) \tag{10}$$

where k = extinction coefficient (m$^{-1}$), z = depth (m), f = the photoperiod as a fraction (e.g. 0.5), $\bar{I}$, $I_{opt}$ = incident time-averaged light and the optimal light (ly/day).

The troublesome theoretical consideration concerning the DiToro double integral of Steele's equation is the weakness of the constant light assumption. The power of the Steele equation to represent surface inhibition is substantially diminished when mid-day brightness is lost in a daily average. To evaluate this error, a program was written to integrate numerically equation 6 over 0.25 meter depth-intervals every 10 minutes during the day. Radiation was assumed to follow a half-sine wave peaking at noon.

The discrepancy between equations 6 and 10 is non-linear and is increasingly severe for clear waters and bright days. While a curve could be fitted to provide a correction factor depending on $I/I_{opt}$, we did not feel that such detail was justified, and a constant overestimate of 15% by equation 10 has been assumed to be representative of the range of $I/I_{opt}$ between 1 and 4 (Kremer and Nixon, 1977). The resulting corrected expression used in the model estimates the accumulated effects of non-optimum light throughout the water column during a 24 hr. day. After addition of the terms for advection, A (from a mixing model), sinking, S (specified as a % day$^{-1}$) and grazing, G (calculated in other sub-routines), the final equation for phytoplankton growth becomes:

$$P_t = P_o \, e^{[(G_{max} \cdot NUTLIM \cdot LTLIM) \pm A - S - G] \, T_{days}}$$

(11)

### SIMULATED PHYTOPLANKTON DYNAMICS

While there are times during the summer and fall when the computed and observed phytoplankton standing crops differ, the overall seasonal and spatial patterns are similar, and the agreement during the winter-spring bloom is particularly strong (Fig. 3). The output shown here is from the "standard run" of the model, in which all of the many coefficients have been assigned values well within the mid-range of reported values (Kremer and Nixon, 1977). It is, of course, possible to improve the overall correspondence between the observed and computed values by shifting various coefficients toward one or the other of the extreme values that have also been reported. However, the purpose of the model is not to get the "best fit," but to explore some of the possible consequences of assigning various values to processes in the system.

We used the model to carry out an analysis of what may have been important in regulating the initiation of the winter-spring bloom during 1972-1973. A traditional explanation for the timing of the bloom in Narragansett Bay has been a reduction in grazing pressure from zooplankton (Martin, 1965; Pratt, 1965). However, an analysis of the simulated phytoplankton-zooplankton interactions for the period preceding and during the bloom indicates that grazing was not significantly involved during this year (Fig. 4). Similar analysis also suggests that neither nutrients nor temperature was exerting a major controlling effect on the initiation of the bloom. In fact, the basic pattern of the bloom remains even if the seasonal temperature cycle is entirely removed and if the simulation is run at 11°C,

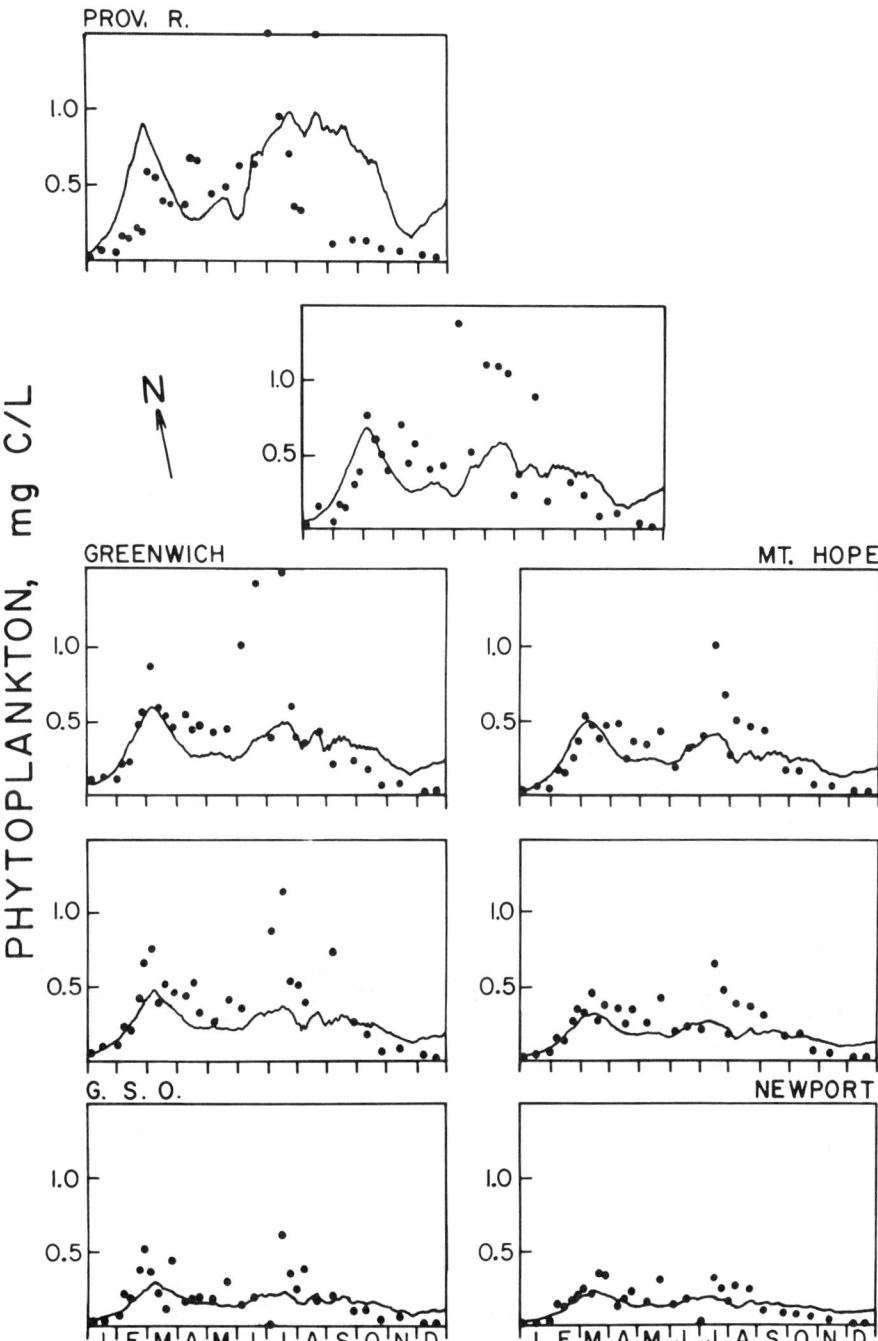

Fig. 3. Comparison of observed (•) and computed (—) phytoplankton abundance for eight spatial elements of Narragansett Bay
during 1972-1973. Observed data were converted to carbon using a
constant ratio of C/Chl a of 30:1. Upper elements represent the
Providence River and the head of the bay; lower columns show the
West Passage (left) and East Passage (right).

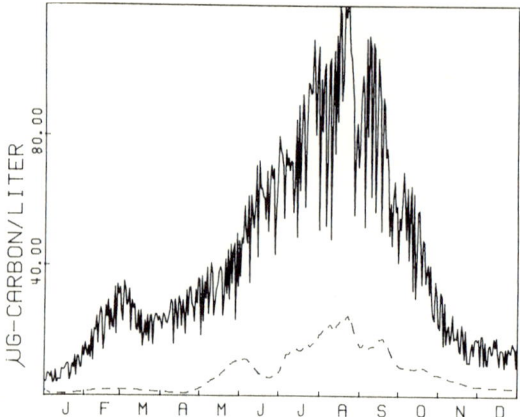

Fig. 4. Comparison of the simulated daily phytoplankton growth (solid line) and the calculated daily zooplankton ration (broken line) for the upper West Passage of Narragansett Bay during 1972-1973.

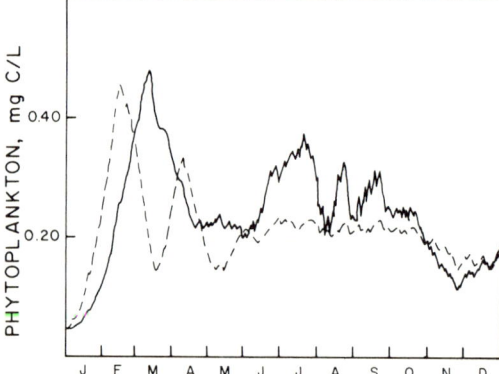

Fig. 5. Effect on phytoplankton abundance of removing the annual temperature cycle from simulations of the numerical model (---). While all other forcing functions followed their normal pattern, water temperature was held constant at 11.5C, the annual mean.

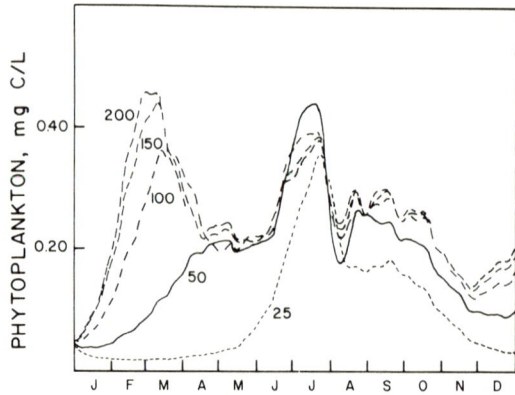

Fig. 6. Effect on phytoplankton abundance of maintaining various constant light levels (ly/day) in simulations of the numerical model. All other forcing functions followed their normal pattern.

the mean annual temperature for the bay (Fig. 5).  However, simulations run with various levels of light input have a dramatic effect on the timing of the bloom during this year (Fig. 6).  In fact, if the various light runs are compared with the observed data, it appears that there is a critical intensity for the initiation of the winter bloom of between 125-175 ly/day.

## Analysis of Field Data

If the observed chlorophyll concentrations in the upper West Passage of Narragansett Bay are plotted in time-series with the incident solar radiation for the period preceding and during the initiation of the 1972-1973 winter-spring bloom, there also appears to be a relationship between increasing light and the initiation of the bloom (Fig. 7).  There is also a strong indication that the critical intensity may have been between 150-200 ly/day, or slightly higher than suggested by the numerical model.  This relationship has also been noted by Hitchcock and Smayda (1977), using a similar field data set, who point out that this value results in an average light intensity in the water column ($\bar{I}$) in this area of the bay (where the bloom always begins) of about 40 ly/day.  As mentioned earlier, this is the critical value suggested by Riley (1967) from his study of the annual bloom in Long Island Sound.  The calculation of $\bar{I}$ for the field is subject to a considerable uncertainty, however, since it depends on the extinction coefficient, $k$, (see eq. 8).  The value

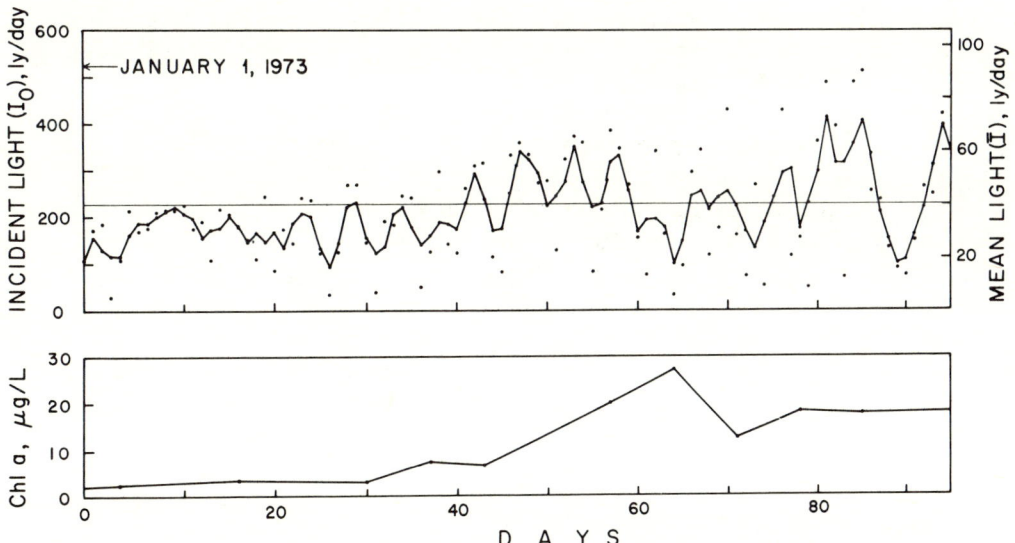

Fig. 7.  Time-series plot of incident solar radiation (I) and chlorophyll a concentrations in the upper West Passage of Narragansett Bay for the period preceding and during the winter bloom of 1973.  Solar radiation is given as daily totals ($\cdot$) and the 3-day moving average (—).  The approximate mean light in the water column ($\bar{I}$) is indicated on the right.  Solar radiation data are from the Eppley Laboratory pyrheliometer in Newport, R.I. Chlorophyll values are the mean of near-surface and near-bottom measurements.

of this parameter is often highly variable at any one location and over short time periods. Based on an extensive turbidity survey by Schenck and Davis (1975) as well as the extinction coefficients given by Hitchcock and Smayda (1977) and our own data, there may well be a coefficient of variation of some 20% associated with any particular value of $k$. Thus, an incident light energy ($I_0$) of 200 ly/day at an 8m deep station where the extinction coefficient is thought to be about 0.6 m$^{-1}$ may actually result in an average light in the water column ($\bar{I}$) between 35 and 51 ly/day, or, put the other way, we may have a mean light intensity in the water column of 40 ly/day with an incident light input between 155 and 230 ly/day. Thus, the agreement between the sensitivity analysis of the numerical model and the time-series field data is clearly within the error of the measurements.

Hitchcock and Smayda (1977) have compiled historical light data which show that December 1972 was appreciably darker than the 15 year mean for that month and which suggest that reduced light was the reason for the delay in the bloom until late January-February, 1973. The average daily light received during December 1972, was only 92 ly, with the last 3 weeks having daily averages of only 82-87 ly. In his study of the winter-spring bloom over several annual cycles some ten years earlier, Pratt (1965) noted that solar radiation was between 120-175 ly/day for the two weeks preceding the inception of the bloom in mid-December or January. If the numerical model is run with daily light input set equal to the 15 year mean instead of the values observed during 1972-1973, there is an increase in simulated phytoplankton growth rate during late November and December, in contrast to the lack of growth simulated during this period with 1972-1973 data.

## Microcosm Experiments

Since light appeared to be of such importance in regulating the inception of the winter-spring bloom during 1972-1973, the next step seemed to be to carry out some simple light experiments with laboratory microcosms. The first opportunity to do this came during the winter of 1975-1976, which again turned out to be a "late bloom" year very similar to 1972-1973. The microcosm laboratory has been described elsewhere (Oviatt *et al.*, 1977; Perez *et al.*, 1977; Oviatt *et al.*, this volume). Basically, it consists of 12 plastic tanks of 150 liters each that can be maintained under controlled temperature and light conditions. Each tank has a heterotrophic bottom community associated with it, and each is operated as an open system that is coupled to the larger natural system through regular inputs of Narragansett Bay water that are collected by bucket to prevent any damage to the plankton. Mixing is provided by reversing plastic mesh paddles. The systems appear to replicate well, are self-maintaining over periods at least as long as six months, have been shown to respond to eutrophication in a manner similar to that of the natural system, and by manipulating environmental conditions can be made to track the bay reasonably well at various times during the year in terms of species composition, biomass, primary production, and benthic metabolism (Oviatt *et al.*, 1977; Perez *et al.*, 1977; Oviatt *et al.*, this volume). It seemed a relatively straightforward matter to set up the microcosms in late January when ambient light levels were low and the bay had not yet bloomed and to expose them to a range of light levels in the laboratory to see if a bloom could be induced. Our hope was that the critical intensity might be similar to that indicated by the numerical model simulations and field data analysis for this period in 1973.

### SCALING LIGHT INPUTS

One of the first problems in designing the microcosms was to choose a light input that was appropriately scaled to Narragansett Bay. The resolution of this problem was, of course, particularly important for the experiments reported here. It did not seem right to use ambient light levels from the field because the depth of the microcosms is only 0.7m, compared to a depth of about 7m in the upper West Passage of the bay. Moreover, as soon as water from Narragansett Bay is placed in the tanks, its apparent extinction coefficient increases from about $0.65$ $m^{-1}$ to about $2.5$ $m^{-1}$, simply as a result of changes in the effective light field and the proportion of direct and scattered light.

As a first approximation, it might seem that the microcosms should be scaled to the bay by adjusting the light input in such a way that the mean light intensity in the tanks ($\overline{I}$, eq. 8) is equal to that in the bay. This scaling is especially attractive since both systems are well-mixed, and it was, in fact, the approach we first used in some preliminary runs during the preceding year.

The problem with scaling light in this manner is that it accounts for only one of at least two major processes that need to be considered. Once the light is suitably attenuated and distributed in the tanks, the photosynthetic response to it is also non-linear (see Fig. 2). The need for evaluating this response over depth and time of day was discussed earlier as part of the light formulation in the description of the numerical model. Thus, if all other things are equal and if one wishes to achieve the same phytoplankton growth rate in the tanks as in the bay, it is necessary to adjust the light input to the tanks so that the parameter LTLIM is equal over both water columns (see eq. 10). This approach calls for the use of appreciably lower light inputs to the tanks than a simple matching of $\overline{I}$ and may explain why our preliminary trials resulted im blooms that were not seen at the same time in the bay.

In comparing the microcosms with the upper West Passage of the bay, however, all other things are not equal, at least in terms of flushing rate. While the turnover time of the microcosms was designed to match that of the bay as a whole ("instantaneous" flushing coefficient $\cong -0.03$ $day^{-1}$), the hydrodynamic mixing model used in the numerical ecosystem model suggests that the flushing rate in the area where the winter-spring bloom begins is more on the order of

$-0.12$ day$^{-1}$ (Kremer and Nixon, 1977). However, as shown in eq. 11, this is a simple correction to make in computing the expected growth rate for selection of light levels. We found that it seemed to have been necessary to go through at least a part of the numerical modeling effort in order to design the microcosm experiment.

For the period of the light experiment (Jan. 20-Feb. 23, 1975), the calculations described above (using $z = 0.7$ m, $k = 2.5$ m$^{-1}$) indicated that light inputs to the microcosms of 25, 16, 9 and 5 ly/day might provide a range of expected phytoplankton growth rates (logarithm base e) of 0.05, 0.02, 0.002, and $-0.01$ day$^{-1}$, respectively. Such a range would bracket the calculated rate of 0.01-0.03 for the upper West Passage during the initiation of the winter-spring bloom ($\bar{I} \cong 40$ ly/day, $z \cong 7$ m, $k \cong -0.65$ m, $I_{opt} \cong 100$ ly/day $= I_{1m}$ when $\bar{I} = 40$) and would provide a good test of the hypothesis that increased light initiates the bloom when it has not occurred earlier in the year. Additional tanks with very low light (1.5 ly/day) and with a light input that switched on alternate days (25-5-25...ly/day) were also included to observe the effects of extreme conditions and daily variation. In all cases, the light levels were imposed by adding layers of screen over duplicate tanks. Light measurements were made using a Li-COR model 185 quantum meter (400-725 nm) that was intercalibrated with an Eppley laboratory pyranometer, model 8-48 (300-3000 nm) under the "cool white" fluorescent lights used with the microcosm and in sun light. The spectral composition of the "cool white" bulbs was measured with an ISCO spectral radiometer and closely matched that which might be found at about 4m (near mid-depth) in Narragansett Bay (Fig. 8).

December of 1974 was not as "dark" as that of 1972, but the incident radiation during January 1976 was considerably lower than in January 1973 (Table 1), and the bloom had showed no signs of appearing in the bay when the microcosm tanks were filled in late January.

## Results

The effect of the different light levels was apparent after only one or two days. Dramatic blooms developed in the tanks receiving 25, 16, and 9 ly/day, while those at 5 ly/day showed stable levels of chlorophyll a and those at 1.5 ly/day declined (Fig. 9). The microcosms with alternating days at 25 and 5 ly appeared to average the inputs and behaved almost exactly as the tanks with 16 ly/day. The chlorophyll levels in tanks with 5 ly/day input remained almost identical to the values in Narragansett Bay for 15-20 days (until about Feb. 4-9). After that time the average light in the water column ($\bar{I}$) in the bay exceeded 40 ly/day, and the bloom began (Fig. 9) just as it had during the February bloom of 1972-1973 (Fig. 6). It is clear from these results that blooms developed in the microcosms with an average light in the water column of much less than 40 ly/day. For example, at $I_o = 9$ ly/day, $\bar{I} = 4.2$ ly/day, almost

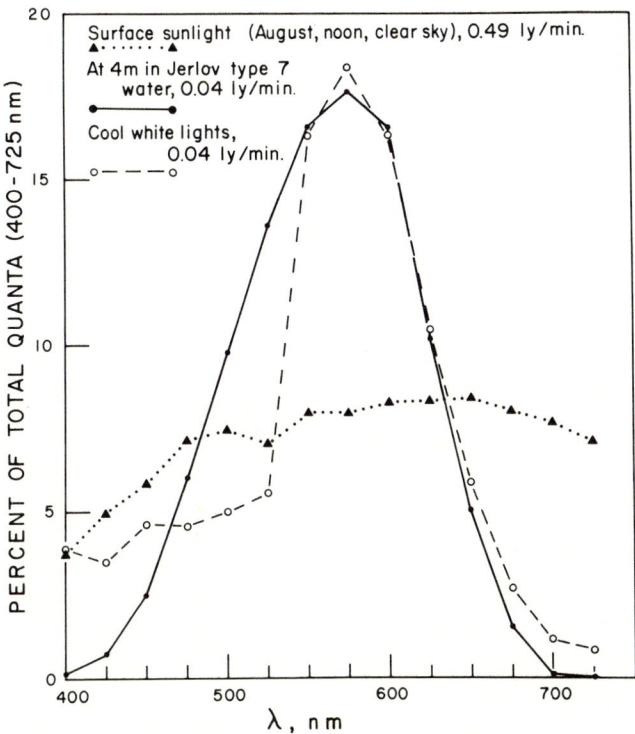

Fig. 8.  Comparative spectral composition of the "cool white" lights used with the microcosms, solar radiation incident on Narragansett Bay, and the light available at about 4m in coastal water similar to that of Narragansett Bay (Jerlov, 1968, Type 7). The measurements of solar radiation and the computation of the distribution at 4m were provided by James Yoder at the Graduate School of Oceangraphy, URI.

Table 1.  Mean daily incident solar radiation* (ly/day) on Narragansett Bay during each week in December and January of the "late bloom" years of 1972-1973 and 1975-1976.  Departure of the values from the comparable mean 15 year value is also shown.

| | | 1972-1973** | | 1975-1976 | |
|---|---|---|---|---|---|
| | | $I_o$ | $\Delta$ | $I_o$ | $\Delta$ |
| Dec. | I | 110 | −24 | 139 | + 4 |
| | II | 87 | −39 | 68 | −57 |
| | III | 85 | −36 | 123 | + 2 |
| | IV | 82 | −48 | 122 | − 8 |
| Jan. | I | 141 | −20 | 139 | −22 |
| | II | 201 | +33 | 137 | −31 |
| | III | 174 | + 2 | 163 | − 9 |
| | IV | 148 | −17 | 133 | −32 |

*From Eppley Laboratory roof top pyrheliometer
**From Hitchcock and Smayda (1977)

an order of magnitude lower than the "critical value" in the bay. A
portion of this discrepancy may be explained by the differences in
spectral distributions  being measured in the laboratory and in the
field, but this cannot by itself account for the difference.  Our
intercalibration measurements indicate that virtually all of the
energy from the fluorescent "cool white" lights is in the range of
400-725 nm, while only some 55% of the solar radiation falls in this
range.  It is not possible to correct the $\bar{I}$ values further even
though the $k$ values for the microcosms were obtained with a quantum
meter  and those for the bay were taken from Secchi disk and submarine photometer profiles.  While Morel and Smith (1974) suggest
that the discrepancy of quanta and energy with depth is on the order
of 10% in the 400-700 nm range, field comparisons in Narragansett
Bay by Yoder and Hargraves (unpublished ms.) show that while this
magnitude of correction appears valid, it is not always true that
the attenuation of quanta is more rapid.

   As discussed in the section on light scaling, the non-linear
response of photosynthesis to light intensity is very much involved
in producing growth at lower levels of $\bar{I}$ in the tanks.  However,
while the preliminary calculations of LTLIM indicated that blooms
should have been expected in the tanks at about 16 ly/day input,
they did not suggest that the rate of growth during the bloom would
be as high as that observed.  While the average "instantaneous"
growth rate (base e) during the bloom in the microcosm was 0.23 day$^{-1}$
at 25 ly, 0.16 day$^{-1}$ at 16 ly, and 0.07 day$^{-1}$ at 9 ly/day, it was
only on the order of 0.03-0.04 day$^{-1}$ in the bay during the blooms
of 1973 and 1975.  Moreover, the high growth rates in the microcosms
did not represent short-term  or transient behavior but were sus-
tained over a period of a week or longer.

   While the microcosm experiment clearly demonstrated the importance of light in stimulating the phytoplankton bloom at this
time of year, we have puzzled over the "super bloom" behavior of the
microcosms a good bit in trying to reconcile their behavior with the
results from the numerical model and field studies.  The most probable explanation at this point involves a shift in the $I_{opt}$ parameter
of the Steele curve (eq. 10, Fig. 2) from the higher values found
in the field to lower values more characteristic of laboratory cultures.  It is also possible, of course, that other factors may be
involved, including the absence of any exposure to the high intensity
light and certain wavelengths that field populations receive from
sunlight near the surface.

## LIGHT ACCLIMATION

   As we discussed earlier, the numerical ecosystem model was designed to allow for light acclimation in field populations by computing $I_{opt}$ as a moving, weighted average of light intensity at the
one meter depth (eq. 7).  During mid-winter the value of $I_{opt}$ would
decline to about 50 ly/day if it were not constrained above about
75-100 ly/day by the $\bar{I}$ = 40 ly/day limit described in the light-
production section.  When the model was used to calculate LTLIM for

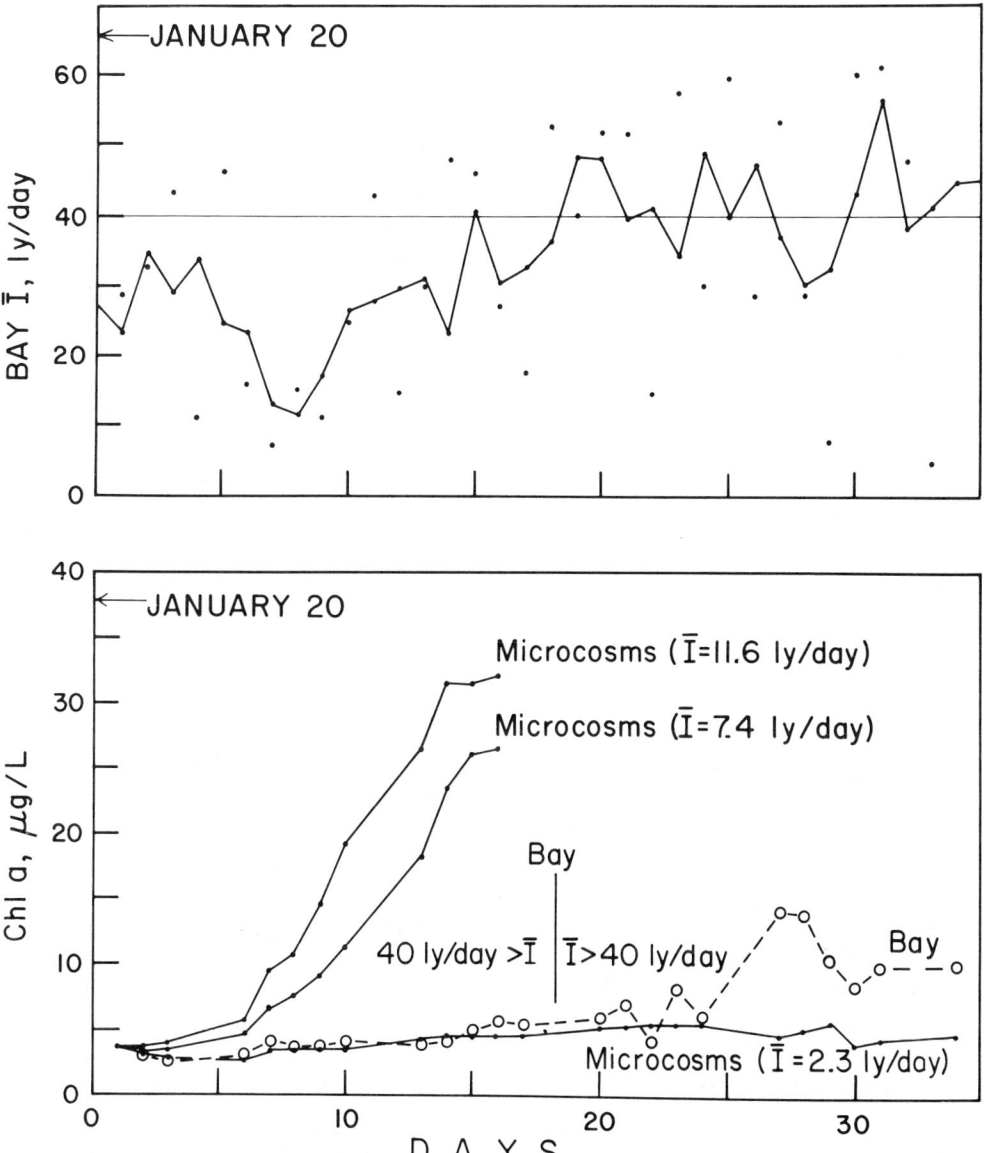

Fig. 9. Chlorophyll a concentrations over time in the lower West Passage of Narragansett Bay and in laboratory microcosms maintained at different light levels but with the same photo period as the bay. The microcosms were also kept within 1-2°C of the bay temperature. Only the results for the microcosms at 25, 16, and 5 ly/day are shown here for clarity. The tanks at 9 and 1.5 ly/day behaved proportionally as shown in Fig. 10. Values plotted here are the means for duplicate tanks. Daily values (·) for the approximate mean light in the water column in the bay are also shown, along with the 3 day moving average (—) for the period preceding and during the bay bloom in February 1975.

scaling light in the microcosms, an $I_{opt}$ of 100 ly/day was assumed, since this is about the value for the bay at the inception of the bloom. However, there is considerable evidence that marine phytoplankton may display optimum or light-saturated production at much lower intensities (Ryther, 1956; Jitts *et al.*, 1964; McAllister *et al.*, 1964; Smayda, 1969), and Rodhe *et al.* (1958) have found light adaptation down to an $I_{opt}$ of 15 ly/day in the winter phytoplankton of lakes. Since the microcosms in the model were exposed to much lower light intensities than in the field, it is not unreasonable to suppose that the $I_{opt}$ values for the microcosm phytoplankton were much lower than 100 ly/day as a result of shade adaptation.

If we assume that $I_{opt}$ for the microcosm phytoplankton was on the order of 20 ly/day with the "cool white" lights (equivalent to about 35-40 ly/day in sunlight), a comparative plot of the LTLIM parameter for the tanks and Narragansett Bay shows a dramatic difference in the response of the two systems to light input (Fig. 10). The combined effects of the differences in depth, apparent extinction coefficient, and $I_{opt}$ result in equivalent light limitation at much lower light inputs for the microcosms. It is also clear that the microcosms should be very sensitive to light input, especially at low levels, and that it is possible to reduce the amount of light limitation in the microcosms far below that which might be attained in the bay. The latter effect is apparent at low light intensity in the microcosms and would result in substantially higher growth rates in the tanks than in the field. In fact, if the numerical model is applied directly to the microcosms with $I_{opt}$ set at 20 ly/day, the simulated bloom at each light intensity falls closely along the observed data during the experiment (Fig. 11).

Since the use of a lower value for $I_{opt}$ has such a dramatic effect on the computed phytoplankton dynamics of the microcosms, it is reasonable to wonder what the effect might be on the simulations of Narragansett Bay if the lower constraint on $I_{opt}$ were removed. When such a simulation is run, the only apparent difference is a slightly earlier and more rapid winter-spring bloom (Fig. 12). Even a very high constant $I_{opt}$ of about 250 ly/day produces a marked bloom with only slightly lower growth rates, although its inception is considerably delayed (Fig. 12). The larger system appears much less sensitive to the $I_{opt}$ parameter (and to I) because of the integration of the Steele curve through a greater depth (7m instead of 0.7m). During the summer this feature makes the numerical model particularly insensitive to the choice of $I_{opt}$ when simulations are run for the bay (Fig. 12). Even if a low $I_{opt}$ is used, it makes little difference because the inhibition of photosynthesis that this implies is only of importance in the surface one or two meters.

### SHALLOW TANKS, DEEP TANKS, MECHANISTIC MODELS AND NATURAL SYSTEMS

A final observation is that if one makes his microcosms deep enough and maintains them outside in natural daylight, some of these complications ought to disappear. At the Marine Ecosystems Research

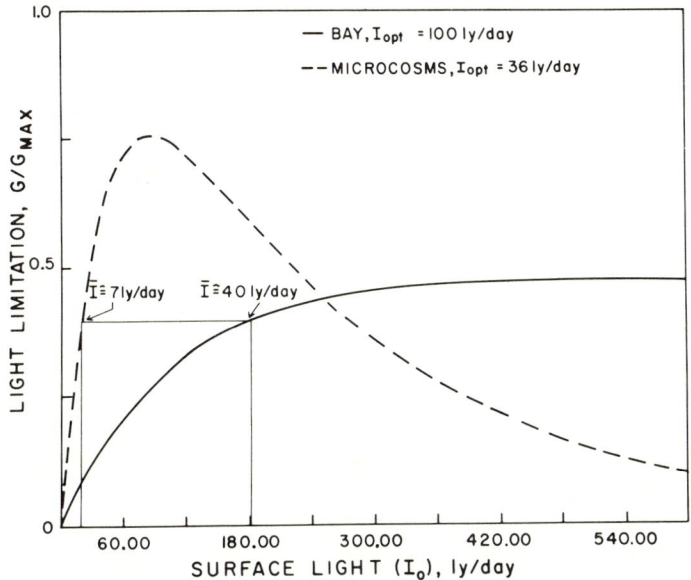

Fig. 10. The parameter LTLIM as a function of incident light $(I_o)$ for the experimental microcosms and for the upper West Passage of Narragansett Bay (see eqn. 10). Note that as LTLIM increases to 1.0, light becomes less limiting.

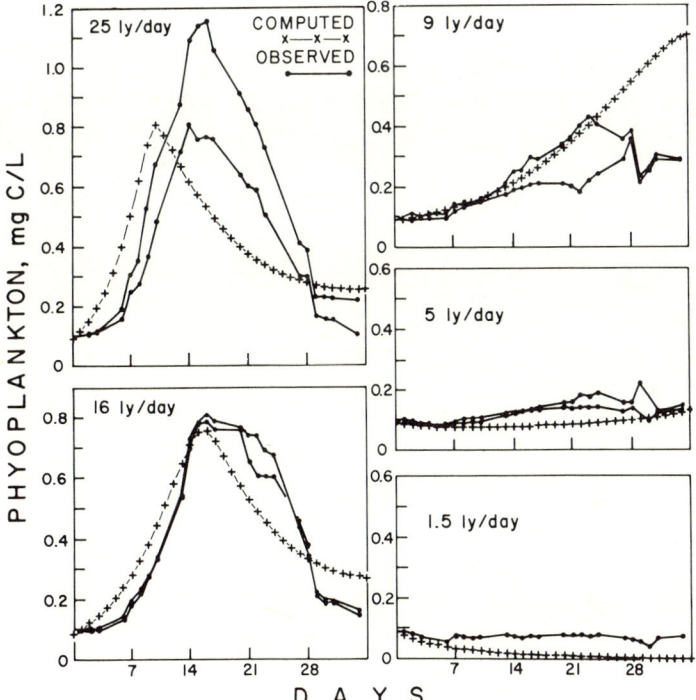

Fig. 11. Simulated (---) and observed (—) chlorophyll concentrations over time in the experimental microcosms with various levels of light. Computations were made as in the "standard run" of the Narragansett Bay model except for appropriate changes in depth, column flushing rate, and initial conditions, and the setting of $I_{opt}$ to 20 ly/day (equivalent to 36 ly/day of solar radiation).

Laboratory (MERL) operated by the University of Rhode Island, we have been involved in a large cooperative study employing 12 such large replicate "mesocosms" containing some 13m$^3$ each of Narragansett Bay water and 2.4m$^2$ of natural bay sediment. The water in the tanks is almost 5m deep, and the tanks are maintained out of doors adjacent to the bay. While the facility has only been in operation since September 1976, the winter phytoplankton bloom developed during November this year in both the tanks and the bay, and the growth rates were almost identical (G. Vargo, MERL, URI, personal communication).

Of course, as one set of problems is eliminated, others appear, and the design of experimental ecosystems is very much an exercise in compromise. The major point is that while microcosms *may* be simpler than nature, they are far from simple. The physical and biological scaling of such living models is a subtle and poorly understood process that may have dramatic influences on the experimental results that are obtained from them. Moreover, the scaling process itself may require the use of mechanistic numerical models, at least for some parameters. On the other hand, the microcosms provide an experimental capability that can play an important part in the verification and refinement of numerical models and in the interpretation of field data from natural systems.

## Acknowledgments

The field studies, microcosms, and numerical models described here all required and benefited from the help of many people. Sharon Northby, Fred Short and Eric Klos carried out much of the field sampling, and Betty Buckley, George Morrison and Neal Lackie helped with the maintenance and sampling of the experimental micro-

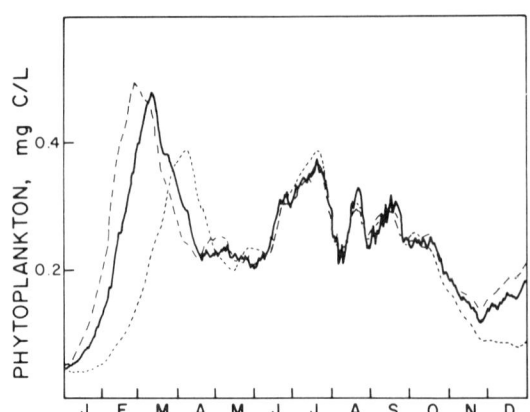

Fig. 12.   Effect in the numerical model of changes in the arbitrary lower limit to light acclimation. The standard run ($I_{opt} > 75-100$ ly/day) is shown with a solid line while the option with no lower limit to $I_{opt}$ is shown with a broken line, and an unusually high lower limit ($I_{opt} > 200-250$ ly/day) is represented by the dotted line. Stochastic light input has been removed from the two extreme cases for clarity.

cosms. Wendell Hahm assisted with the computer simulations. We are grateful to Gary Hitchcock, Jom Yoder and Gabe Vargo for the use of their unpublished data. Solar radiation measurements were provided by the Eppley Laboratory in Newport, R.I. The research was supported by the Office of Sea Grant Programs, U.S. Dept. of Commerce, NOAA and by grant No. R-803143 from the U.S. Environmental Protection Agency.

# References

Antia, N. J., C. D. McAllister, T. R. Parsons, K. Stephens, and J. D. H. Strickland. 1963. Further measurements of primary production using a large-volume plastic sphere. Limnol. Oceanogr. 8: 166-183.

Bannister, T. T. 1974. Production equations in terms of chlorophyll concentration, quantum yield, and upper limit to production. Limnol. Oceanogr. 19: 1-12.

DiToro, D. M., D. J. O'Connor and R. B. Thomann. 1971. A dynamic model of phytoplankton populations in the Sacramento-San Joaquin delta. Adv. in Chem. Series 106: 131-180.

Dugdale, R. C. 1967. Nutrient limitations in the sea: dynamics, identification, and significance. Limnol. Oceanogr. 12: 685-695.

Edmondson, W. T. and Y. H. Edmondson. 1947. Measurements of production in fertilized salt water. J. Mar. Res. 6: 228-246.

Eppley, R. W. 1972. Temperature and phytoplankton growth in the sea. Fish. Bull. 70: 1063-1085.

Eppley, R. W. and W. H. Thomas. 1969. Comparison of half-saturation constants for growth and nitrate uptake of marine phytoplankton. J. Phycol. 5: 375-379.

Eppley, R. W., J. N. Rogers and J. J. McCarthy. 1969. Half-saturation constants for uptake of nitrate and ammonium by marine phytoplankton. Limnol. Oceanogr. 14: 912-920.

Fee, E. J. 1969. A numerical model for the estimation of photosynthetic production, integrated over time and depth, in natural waters. Limnol. Oceanogr. 14: 906-911.

_____. 1973. A numerical model for determining integral primary production and its application to Lake Michigan. J. Fish. Res. Bd. Can. 30: 1447-1468.

Fleming, R. H. 1939. The control of diatom populations by grazing. J. cons. perm. Expl. Mer. 14: 210-227.

Goldman, J. C. and E. J. Carpenter. 1974. A kinetic approach to the effect of temperature of algal growth. Limnol. Oceanogr. 19: 756-766.

Hanton, J. T. 1969. Algal phosphate uptake kinetics: growth rates and limiting phosphate concentration. M. S. Thesis, Univ. of North Carolina, Chapel Hill, N. C.

Hitchcock, G. and T. J. Smayda. 1977. The importance of light in the initiation of the 1972-1973 winter-spring diatom bloom in Narragansett Bay. Limnol. Oceanogr. 22: 126-131.

Ignatiades, L. and T. J. Smayda. 1970. Autecological studies on the marine diatom *Rhizosolenia fragilissima* Bergon. I. The influence of light, temperature, and salinity. J. Phycol. 6: 332-339.

Jerlov, N. G. 1968. *Optical Oceanography*. Elsevier, Amsterdam.

Jitts, H. R., C. D. McAllister, K. Stephens, and J. D. H. Strickland. 1964. The cell division rates of some marine phytoplankters as a function of light and temperature. J. Fish. Res. Bd. Can. 21: 139-157.

Kremer, J. N. 1978. An analysis of the stability characteristics of an estuarine ecosystem model. In: *Marsh-Estuarine Systems Simulation* (R. Dame, ed.). Vol. 8., Univ. of South Carolina Press, Columbia.

Kremer, J. N. and S. W. Nixon. 1975. An ecological simulation model of Narragansett Bay--the plankton community. In: *Recent Advances in Estuarine Research* (J. Cronin, ed.). Academic Press, N. Y.

Kremer, J. N. and S. W. Nixon. 1977. *A Coastal Marine Ecosystem: Simulation and Analysis*. Ecological Studies, Vol. 24. Springer-Verlag, N. Y.

Mann, K. H. 1972. The analysis of aquatic ecosystems. In: *Essays in Hydrobiology* (R. B. Clark and R. J. Wooton, eds.), pp. 1-14. Univ. of Exeter.

Mann, K. H. 1975. Relationship between morphometry and biological functioning in three coastal inlets of Nova Scotia. In: *Estuarine Research* (L. E. Cronin, ed.). Vol. 1, pp. 634-644. Academic Press, N. Y.

Martin, J. H. 1965. Phytoplankton-zooplankton relationships in Narragansett Bay. Limnol. Oceanogr. 10: 185-191.

McAllister, C. D., T. R. Parsons, K. Stephens, and J. D. H. Strickland. 1961. Measurements of primary production in coastal sea water using a large volume plastic sphere. Limnol. Oceanogr. 6: 237-258.

McAllister, C. D., N. Shah, and J. D. H. Strickland. 1964. Marine phytoplankton photosynthesis as a function of light intensity: A comparison of methods. J. Fish. Res. Bd. Can. 21: 159-181.

MacIsaac, J. J. and R. C. Dugdale. 1969. The kinetics of nitrate and ammonia uptake by natural populations of marine phytoplankton. Deep-Sea Res. 16: 45-57.

Michaelis, L. and M. L. Menten. 1913. Der Kinetic der Inverteinwirkung. Biochem. Zeitschr. 49: 333-369.

Monod, J. 1942. *Recherches sur la Croissance des Cultures Bacteriennes*. Herman et Cie, Paris.

Morel, A. and R. C. Smith. 1974. Relation between total quanta and total energy for aquatic photosynthesis. Limnol. Oceanogr. 19: 591-600.

Nixon, S. W. and J. N. Kremer. 1977. Development of a composite simulation model for a New England estuary—Narragansett Bay. In: *Models as Ecological Tools: Theory and Case Histories* (C. Hall and J. Day, eds.). Wiley Interscience, N. Y.

Nixon, S. W., J. N. Kremer and C. A. Oviatt *et al.* Systems ecology data report for Narragansett Bay 1972-73. Univ. of Rhode Island Sea Grant Tech. Rep. (in preparation).

Odum, H. T. 1972. An energy circuit language for ecological and social systems: Its physical basis. In: *Systems Analysis and Simulation in Ecology* (B. C. Patten, ed.). Vol. II, pp. 140-212. Academic Press, N. Y.

Oviatt, C. A., K. Perez and S. W. Nixon. 1977. Multivariate analysis of experimental marine ecosystems. Helgolander wiss. Meeresunters. 30: 30-46.

Oviatt, C. A., S. W. Nixon, K. T. Perez and B. Buckley. 1978. On the season and nature of perturbations in microcosm experiments. In: *Marsh-Estuarine Systems Simulation* (R. Dame, ed.). Vol. 8, B. W. B. Library in MSCI., Univ. of South Carolina Press, Columbia.

Paasche, E. 1973. Silicon and the ecology of marine plankton diatoms. I. *Thalassiosira pseudonana* grown in a chemostat with silicate as limiting nutrient. Mar. Biol. 19: 117-126.

Paasche, E. 1973a. Silicon and the ecology of marine plankton diatoms. II. Silicate uptake kinetics in five diatom species. Mar. Biol. 19: 262-269.

Patten, B. C. 1971. Coda. In: *Systems Analysis and Simulation in Ecology* (B. C. Patten, ed.). Vol. 1. Academic Press, N. Y.

Perez, K., G. Morrison, C. Oviatt, S. Nixon, and B. Buckley. 1977. The importance of physical and biotic scaling to the experimental simulation of a coastal marine ecosystem. Helgolander wiss. Meeresunters. 30: 144-162.

Pratt, D. M. 1949. Experiments in the fertilization of a salt water pond. J. Mar. Res. 8: 36-59.

Pratt, D. M. 1965. The winter-spring diatom flowering in Narragansett Bay. Limnol. Oceanogr. 10: 173-184.

Riley, G. A. 1956. Oceanography of Long Island Sound 1952-54. II. Physical Oceanography. Bull. Bingham. Ocean. Coll. 15: 15-46.

Riley, G. A. 1967. The plankton of estuaries. In: *Estuaries* (G. H. Lauff, ed.). AAAS Pub. #83.

Rodhe, W., R. A. Vollenweider, and A. Nauwerck. 1958. The primary production and standing crop of phytoplankton. In: *Perspectives in Marine Biology* (A. A. Buzzati-Traverso, ed.). Univ. of California Press, Berkeley.

Ryther, J. H. 1956. Photosynthesis in the ocean as a function of light intensity. Limnol. Oceanogr. 1: 61-70.

Ryther, J. H. and D. W. Menzel. 1959. Light adaptation by marine phytoplankton. Limnol. Oceanogr. 4: 492-497.

Schenck, H. and A. Davis. 1973. A turbidity survey of Narragansett Bay. Ocean. Eng. 2: 169-178.

Smayda, T. J. 1957. Phytoplankton studies in lower Narragansett Bay. Limnol. Oceanogr. 2: 342-359.

Smayda, T. J. 1969. Experimental observations on the influence of temperature, light and salinity on cell division of the marine diatom, *Detonula confervacea* (Cleve) Gran. J. Phycology 5: 150-157.

Smayda, T. J. 1973. The growth of *Skeletonema costatum* during a Winter-Spring bloom in Narragansett Bay, R. I. Norw. J. Bot. 20: 219-247.

Sorokin, Y. I., V. G. Snopkov, and V. M. Grinberg. 1959. The determination of the relation between phytoplankton photosynthesis and submarine illumination in the waters of the central part of the Atlantic. C. R. Acad. Sci. U.S.S.R. 124: 432-435.

Steele, J. H. 1962. Environmental control of photosynthesis in the sea. Limnol. Oceanogr. 7: 137-150.

Steele, J. H. 1965. Notes on some theoretical problems in production ecology. In: *Primary Productivity in Aquatic Environments*(C. R. Goldman, ed.), pp. 382-398. Men. Ist. Ital. Ishobiol., 18 Suppl. Univ. of Calif. Press, Berkeley.

Steele, J. H. and I. R. Baird. 1961. Relations between primary production, chlorophyll and particulate carbon. Limnol. Oceanogr. 6: 68-78.

Steeman-Nielsen, E. and V. K. Hansen. 1959. Light adaptation in marine phytoplankton populations and its interrelation with temperature. Physiol. Plant. 12: 353-370.

_____. 1961. Influence of surface illumination on plankton photosynthesis in Danish waters (56°N) throughout the year. Physiol. Plant. 14: 595-613.

Steeman-Nielsen, E. and A. Jensen. 1957. Primary oceanic production. Galathea Rep. 1: 49-136.

Steeman-Nielsen, E. and T. S. Park. 1964. On the time course of adapting to low light intensities in marine phytoplankton. J. Cons. Int. Explor. Mer. 29: 19-24.

Steeman-Nielsen, E., V. K. Hansen and E. G. Jorgensen. 1962. The adaptation to different light intensities in *Chlorella vulgaris* and the time dependence on transfer to a new intensity. Physiologia Plantarum 15: 505-517.

Talling, T. F. 1957. Photosynthetic characteristics of some freshwater plankton diatoms in relation to underwater radiation. New Phytologist 56: 29-50,

Vollenweider, R. A. 1965. Calculation models of photosynthesis depth curves and some implications regarding day rate estimates in primary production measurements. In: *Primary Productivity in Aquatic Environments* (C. R. Goldman, ed.)., Me. ist. Ital. Idrobiol., 18 Suppl. Univ. California Press, Berkeley.

Yentsch, C. S. and R. W. Lee. 1966. A study of photosynthetic light reactions and a new interpretation of sun and shade phytoplankton. J. Mar. Res. 24: 319-337.

Yoder, J. A. 1973. The interaction of temperature and nutrient concentration on the growth of *Detonula confervacea* (Cleve) Gren, *Skeletonema costatum* (Greville) Cleve and *Thallasiosira nordenskioeldii* Cleve. M. S. Thesis, Univ. Rhode Island, Kingston, R. I.

Yoder, J. A. and P. E. Hargraves. Attenuation of submarine photosynthetic radiation: its measurement and relationship to primary production. Graduate School of Oceanography, URI, unpub. MS.

# An Analysis of the Stability Characteristics of an Estuarine Ecosystem Model

**James N. Kremer**

Department of Biological Sciences
University of Southern California
Los Angeles, California 90007

A numerical simulation model for the plankton-based Narragansett Bay, R.I., ecosystem consistently exhibits persistent cyclic interactions between phytoplankton and zooplankton. The nature of the oscillations shifts seasonally through times characterized by strong stability to periods of undamped, stable cycles. The theoretical foundations for these behavioral characteristics appear analogous to the general conclusions drawn from some biologically simplified, analytical models. Thus, the strongly damped behavior occurs when phytoplankton are resource-limited, and the stable cycles occur when they are limited by herbivorous zooplankton. Further, when the zooplankton are, in turn, limited by factors other than food, the system is highly damped and stable, again in close agreement with analytical theory. While this agreement is desirable, it is far from trivial, considering the wide differences in the assumptions, methods, and goals of the two modeling approaches. In fact, additional examples drawn from the bay model reveal cases where the increased ecological detail distorts or masks the analytical expectations. In some cases, apparent agreement is even due primarily to entirely different underlying causes. In drawing inferences from analytical models to complex mechanistic models, it appears difficult to anticipate which mechanistic feature will assume the dominant role in determining the emergent stability characteristics.

## Introduction

Ecological modeling has tended to develop along two different theoretical lines. On one hand are the extensive analyses of coupled differential equations representing various types of ecological interactions in relatively simple terms. These studies are typically concerned with characterizing classes of behavior for various possible combinations of coefficients. Often such studies explore points of stability and stable or unstable cyclic patterns in the abundance of two or more populations. These models are usually, though not always, of the unforced type, considering only the interactions among the state variables themselves and ignoring, for mathematical simplicity, external influences that are so important in natural systems. I will refer to models of this general type as *analytical*, even though some versions involve complications which preclude strictly analytical solution. This approach often has the virtue of leading to precise, general solutions and of being amenable to graphical constructions explicitly summarizing fundamental properties. But analytical models have been criticized, or at least viewed with skepticism, by experimental or field biologists who deal with physiological responses and detailed functional patterns in living systems.

On the other hand are deterministic simulation models that attempt to represent the mechanistic interactions underlying the emergent behavior of a system, often with time-varying solutions and with varying degrees of biological fidelity. These models are generally of the forced type, including influences such as light, temperature, or other arbitrarily defined energy inputs from outside the specified system. I will refer to these models as *mechanistic*, although in the strictest sense many of the "mechanisms" are empirical expressions. This approach has been criticized for purporting to mimic reality with models that are crude and simple versions of complex natural systems and for claiming predictability when "success" was dependent on the fine tuning of many parameters. The predictive capability of mechanistic models has perhaps been overemphasized, but their value extends beyond satisfactory agreement of predicted and observed data. Such models are strongest, in my opinion, when they seek to evaluate critically the quantitative implications of integrating the complex network of hypotheses that compose the mechanistic formulations.

Both approaches to ecological modeling are widely used and have achieved reasonable success. While still relatively simple compared to most mechanistic models, recent progress in the analytical areas has included increasing biological detail involving numerical simulation (Maynard Smith and Slatkin, 1973; Marchessault, 1974). The analytical models, however, continue to be difficult to relate to natural situations with confidence, because the assumptions required to achieve a tractable problem are restrictive and because the highly aggregated interaction coefficients are often impossible to estimate empirically.

Mechanistic models often show pronounced seasonal patterns, times of rapid change or periodic cycling, and characteristic features of stability or persistence. Further, fundamental changes in some of the formulations or in trophic complexity may alter the nature of the model's behavior. Such results have interesting implications for our understanding of the natural community. However, they may also serve as useful numerical experiments for comparing propositions derived from simple analytical models with the simulated behavior of a system that encompasses much more ecological and biological complexity. Unfortunately, the relationship between mechanistic and analytical models often is not apparent because of differences in application, in literature audience, and even in the conventions used for output display.

This paper will examine some specific aspects of the behavior of a mechanistic model of a temperate estuarine ecosystem in light of the implications of some basic analytical models. It is important to emphasize the basic difference between the forced and unforced types of models which will be discussed. Responses of unforced analytical models are not rigorously comparable to patterns achieved in forced models, or, for that matter, in nature. Behavioral features, such as local stability, equilibrium points, limit cycles, etc., have specific meaning in the unforced case, and their interpretations relate directly to the form of the coupling of the equations for the state variables. In the forced case, the interactions are additionally complicated by influences imposed from outside the system of interest. Phase plots of two species which so graphically express the behavior of unforced models necessarily represent only part of the picture for forced models. In discussions of forced models, therefore, the use of the specific terms for classical behavioral features, and even the use of the phase plane as a basis for these descriptive terms, is not entirely consistent with the definitions for the unforced case. However, the usefulness of the idealized unforced models must rest on their implications for real forced systems, and it is with this goal in mind that the comparisons are undertaken.

## The Narragansett Bay Model

Under support from the Office of Sea Grant, Scott Nixon, of the University of Rhode Island Graduate School of Oceanography, and I developed a numerical simulation model of the mechanistic type for the Narragansett Bay ecosystem. The goal of our initial modeling effort was to simulate gross features of spatial and temporal variability for the main components of the plankton community. Detailed mechanistic formulations were developed for the following state variables: phytoplankton (two species groups), zooplankton (adults, eggs, and juveniles), nitrogen (ammonia, nitrate plus nitrite), phosphorus, and silicon. Formulations for secondary biological compartments, such as fish and bivalves, included mechanistic detail only as considered important to reflect critical feedbacks to interactions with the primary state variables. Their biomasses were not simulated but were input to the program based on empirical data. Similarly, physical forcing functions of temperature, clear-sky isolation, and stochastic cloudiness were programmed to reflect reasonable patterns for the bay region. Tidal mixing was accomplished by a fast-executing crude circulation scheme derived directly from a detailed numerical hydrodynamic model previously developed for Narragansett Bay (Hess and White, 1974). Detailed justifications of our mechanistic formulations and discussions of results have been presented elsewhere and would be inappropriate here (see Kremer and Nixon, 1975; Nixon and Kremer, 1977; Kremer and Nixon, 1978). However, any discussion of specific results and inferences is substantially weakened if not placed in the proper context with the implicit assumptions of the model. Since this paper, and the preceding one (Nixon *et al.*, this volume), will draw heavily on some of our results and interpretations, it is necessary to include here a summary of the pertinent formulations.

## CARNIVORES

Predation on the smaller, primarily herbivorous copepods results from seasonal pulses of various species in the bay. For the purposes of the model, three types of carnivores were considered.

Fish larvae of many species are in the bay in significant numbers from March through September, with peaks of abundance in April and July (Matthiessen, 1973). The biomass patterns are forced in the model, but their grazing pressure is based on a mechanistic formulation. A temperature-dependent maximum daily ration is calculated, but the actual ration achieved may be reduced by food limitation according to the hyperbolic formulation of Ivlev (1945).

Ctenophores, *Mnemiopsis leidyi*, undergo a remarkable population explosion in the summer, and recent work indicates that the growth is initiated by residual overwintering individuals which become reproductively active in the early summer (P. Kremer and Nixon, 1976). Climbing five orders of magnitude in biomass, ctenophores reach a peak in late August or September and exert significant predation pressure on the copepods. Direct observations have indicated that feeding is by passive filtration (P. Kremer, 1975), and the model formulation multiplies the scheduled biomass times a weight-specific clearing rate.

The Atlantic menhaden, *Brevoortia tyrannus*, is the third carnivore considered in our model. Experimental evidence for a lower size threshold for active feeding and the predominance of small flagellated phytoplankton in the upper bay where they are found (Durbin and Durbin, 1975; Durbin, 1976) suggest that adult menhaden are important carnivores despite their more traditional herbivorous feeding mode. A mechanistic predation formulation, similar to that of the fish larvae, was developed for menhaden based as much as possible on recent observations on the Narragansett Bay population (Durbin, 1976). A maximum ration is again specified, here using measured respiration and excretion rates and growth estimates. This metabolic requirement is then reduced by a density-dependent Ivlev food limitation term, in this case with the inclusion of a feeding threshold (Parsons *et al.*, 1967). The threshold value and Ivlev exponent were chosen to simulate the rapid switch to active feeding that has been observed.

Menhaden biomass patterns are uncertain and widely variable year-to-year. Thus, a variety of schedules was tested in the model, ranging from short, intense pulses in localized sections to broad, average abundances distributed widely in time and space

(Kremer and Nixon, 1978). A certain characteristic behavior consistently occurred in all reasonable cases, however, and the damping of trophic oscillations to be discussed later is of particular interest.

### PHYTOPLANKTON

The phytoplankton growth formulation was designed to include the interacting effects of temperature, light, and nutrients on average daily net primary production. The assumption of a multiplicative interaction of independently determined regulatory factors is widely used (Di Toro *et al.*, 1971; Goldman and Carpenter, 1974; Lehman *et al.*, 1975), but it contrasts with the explicit use of the classical Law of the Minimum in which a single factor regulates growth (Walsh, 1975).

Among those choosing the multiplicative scheme, authors differ in the extent to which it is applied. Our model assumes temperature, light, and a nutrient to interact multiplicatively, but selects a single, most-limiting nutrient. Others (Di Toro *et al.*, 1971; Lehman *et al.*, 1975) have argued that all terms should interact similarly, with chain multiplication of light and all nutrients under consideration. It is significant that these two schemes have yielded virtually identical results when they were compared in two published models (Kremer and Nixon, 1978; Lehman *et al.*, 1975). While this result is pragmatically reassuring, the fundamental question is still of interest.

Recently, a thermodynamic analysis of phytoplankton growth has suggested that these conflicting views may have a theoretical solution (Kiefer and Enns, 1976), and we are presently pursuing this possibility. The tentative indication is that temperature and nutrients may interact multiplicatively, while the relative importance of light versus this T-N term should depend on which is more limiting (Kiefer, personal communication). Interestingly, this is a combination which appears not yet to have been employed.

The equations of the phytoplankton model are summarized in Table 1. Maximum daily net growth (GMAX) is an exponential function of temperature (Eppley, 1972). The extinction coefficient (K) includes a background term and a self-shading component based on the chlorophyll concentration (Riley, 1956). The final 24-hour time- and depth-integrated light limitation term (LTLIM) is derived from the analytical solution to Steele's (1962) expression for the instantaneous photosynthesis-light response (Di Toro *et al.*, 1971). We made two innovations. First, physiological acclimation to day-to-day variations in insolation was modeled as a weighted, moving average over three days of the optimum insolation value (IOPT), subject to an arbitrary lower limit (LOIOPT). Second, a factor of 0.85 was derived to correct for the square-wave light input assumed in Di Toro's analytical solution.

Finally, limitation terms (NLIM, PLIM, SILIM) are computed for three nutrients (total nitrogen, phosphorus, and silica) using ambient concentrations predicted by the model and empirical Monod

Table 1. Basic equations of the phytoplankton formulation in the Narragansett Bay ecosystem model. See text for further explanation.

$$GMAX = 0.59 \cdot e^{(0.0633 \cdot TEMP)} \qquad (day^{-1})$$

$$CHL = PMGC / C:CHL \qquad (ug/L)$$

$$K = KO + 0.0088 \cdot CHL + 0.054 \cdot CHL^{(0.06667)} \qquad (m^{-1})$$

$$IOPT = 0.7 \cdot I1 + 0.2 \cdot I2 + 0.1 \cdot I3 \qquad (ly/day)$$

$$IF \ (IOPT < LOIOPT) \quad IOPT = LOIOPT$$

$$TERM1 = DAYRAD \cdot 0.9/IOPT$$

$$TERM2 = TERM1 \cdot e^{(-K \cdot DEPTH)}$$

$$LTLIM = 2.72 \cdot PHOTPD \cdot (e^{(-TERM2)} - e^{(-TERM1)}) \cdot 0.85 / (K \cdot DEPTH)$$

$$NTOT + NH4 + NO2NO3 \qquad (ug - at/L)$$

$$NLIM = NTOT / (KN = NTOT)$$

$$PLIM = PO4 / (DP + PO4)$$

$$SILIM = SI / (KSi + SI)$$

$$MXLIM = AMIN1 \ (NLIM, PLIM, SILIM)$$

$$GP = GMAX \cdot LTLIM \cdot MXLIM \qquad (day^{-1})$$

$$SINK = (SNKRAT/DEPTH) \cdot PMGC \qquad (mgC/L)$$

$$PMGC = PMGC \cdot e^{( GP - RLF - RLFJ - RLFBEN)} - SINK \qquad (mgC/L)$$

Where:
| | | |
|---|---|---|
| PMGC | = | Phytoplankton biomass, mgC/L. |
| I1,I2,I3 | = | Average insolation at 1 meter, one, two, and three days ago. |
| TERM1,TERM2 | = | Intermediate values for the calaulation of LTLIM. |
| DAYRAD | = | Insolation, reduced by 0.9 to correct for albedo. |
| PHOTPD | = | Photoperiod, as the daylight fraction. |
| KN,KP,KSi | = | Nutrient half-saturation constants. |
| MXLIM | = | Most limiting nutrient term, determined by the function AMIN1. |

half-saturation constants. From these, the single "most limiting" nutrient is then selected (MXLIM). The final estimate of the instantaneous rate of net phytoplankton growth is the product of the temperature, light, and nutrient terms. This rate is explicitly integrated in the exponential form in combination with the total herbivorous grazing by the zooplankton and the benthos and by sinking.

Alternative formulations were included in the phytoplankton model to extend its usefulness in a range of applications. Among these features were (1) nutrient kinetic calculations assuming internal storage pools for nitrogen and phosphorus and "luxury uptake," (2) two competing phytoplankton species groups with potentially unequal temperature and nutrient responses, (3) variations in the input scheme for insolation and in the light acclimation ability, and (4) the inhibitory or stimulatory role of any additional substance assumed to be present in or introduced into the bay. However, the implications of these additions are beyond the scope of this paper and are discussed elsewhere (Kremer and Nixon, 1978).

ZOOPLANKTON

One of the important features of our zooplankton formulation is the inclusion of realistic life history and age structure, with variable time delays for egg hatching and juvenile development. A second critical point is the assumption that the adults are generally omnivorous, able to eat particulate organic detritus and their own eggs and juveniles, in addition to their primary diet of phytoplankton. In combination, these two innovations substantially modify the response characteristics of the zooplankton compartment and contribute significant stability to the system. This scheme is intended to represent the dominant populations of smaller zooplankton in Narragansett Bay, especially the seasonally abundant copepod species, *Acartia tonsa* and *A. clausii*.

The equations of adult zooplankton (Table 2) determine possible reproduction from a metabolic carbon budget. The daily weight-specific respiration rate of adults (XRESP) is an exponential function of temperature assuming the familiar $Q_{10}$ relationship. The maximum daily ration under no food limitation (XRTNMX) is similarly determined. Assuming a hyperbolic density dependence (from Ivlev, 1945), a food limitation term (XRTN) estimates the fraction of the maximum ration that may actually be ingested (RTN). This calculation is based on the total available food (PCMG), which includes total phytoplankton (PTOT), any non-living particulate organic matter assumed to be available (POM), zooplankton eggs (EGGTOT), and a portion of developing juveniles assumed to be subject to adult cannibalism (ZJCANN).

The initial estimate of RTN is derived from the organismic considerations and ignores the competitive effect of total population size. Further, instantaneous rates are most appropriate for the explicit method of integration (see below). Thus, the anticipated ration is derived by using a filtering rate that would capture the predicted amount if the food concentration remained unchanged (RLF). When evaluated explicitly in combination with all other processes acting on the abundance of food, a realized ration is calculated which is available to meet adult metabolic needs. If the assimilated fraction of the realized ration exceeds the calculated respiration (URA>0), eggs are produced and released into the water. If the unrespired assimilation is negative, adult biomass is reduced, and no reproduction occurs.

The hatching time of the eggs (H) is an empirical exponential function of temperature (McLaren *et al.*, 1969; Landry, 1975). During this delay, eggs are subject to adult cannibalism, predation by other carnivores, and tidal mixing. Upon completing the incubation period, newly hatched juveniles begin development. The duration of this growth period (D) is initially determined from temperature corrected for the anticipated variation in the future. As the simulation proceeds, however, actual development of each cohort is slowed by any unavailability of food that may be encountered. Throughout development juveniles are also subject to cannibalism and predation, although a cutoff may be designated which excludes adult cannibalism above a certain developmental stage.

The juvenile growth rate is again based on metabolic considerations (Table 2). Weight-specific respiration is estimated from

Table 2.  Basic equations of the zooplankton formulation in
the bay model.  See text for further explanation.

---

I.    Adult Metabolic Balance and Reproduction

$$XRESP = XRESP_0 \cdot e^{(Q10RSP \cdot Temp)} \qquad (day^{-1})$$

$$XRTNMX = RMX_0 \cdot e^{(Q10RMX \cdot Temp)} \qquad (day^{-1})$$
$$RTNMX = XRTNMX \cdot ZAMGC \qquad (mgC/day)$$
$$PCMG = PTOT + POM + EGGT\ OT + ZJCANN \qquad (mgC/L)$$

$$XRTN = 1.0 - e^{(-IVLEVK \cdot PCMG)}$$
$$RTN = XRTN \cdot RTNMX \qquad (estimated) \qquad (mgC/day)$$
$$RLF = RTN/PCMG \qquad (day^{-1})$$

(Simultaneously, evaluate rates and compute rations.)

$$RTN = RTNP + RTNZJ + RTNEGG + RTNPOM \qquad (realized) \qquad (mgC/day)$$
$$URA = XASSIM \cdot RTN - XRESP \cdot ZAMGC \qquad (mgC/day)$$
$$(URA > 0 \text{ goes to eggs})$$

II.   Egg Hatching and Juvenile Development Time

$$H = H_0 \cdot e^{(-0.11 \cdot Temp)} \qquad (days)$$

$$D = D_0 \cdot e^{(-0.10 \cdot Temp)} \qquad (days)$$

III.  Juvenile Metabolic Balance and Grwoth

$$XRESPJ = RJ:RA \cdot XRESP \qquad (day^{-1})$$

$$GROMAX = GMAX_0 \cdot e^{(-0.11 \cdot Temp)} \qquad (day^{-1})$$
$$RTNJMX = ZJTOT \cdot (GROMAX + XRESPJ)/XASSIM \qquad (mgC/day)$$
$$PCMG = PTOT + POM \qquad (mgC/L)$$

$$XRTNJ = 1.0 - e^{(-IVLEVK \cdot PCMG)}$$
$$RTNJ + RTNJMX \cdot XRTNJ \qquad (estimated) \qquad (mgC/day)$$
$$RLFJ = RTNJ/PCMG \qquad (day^{-1})$$

(Simultaneously, evaluate rates and compute rations.)

$$RTNJ = RTNJP + RTNJPM \qquad (realized) \qquad (mgC/day)$$
$$GRO = XASSIM \cdot RTNJ / ZJTOT - XRESPJ \qquad (day^{-1})$$

$$ZJ(i) = ZJ(i) \cdot GRO \cdot e^{(-RLF - FCARN)} \qquad (each\ age\ class)$$

---

the adult rate, using development-time-averaged allometric factor
(RJ:RA).  A maximum instantaneous growth rate (Heinle, 1966) is
fixed as a function of temperature (GROMAX) and then combined with
the respiration rate, assimilation efficiency (XASSIM), and biomass
to project the maximum ingestion for all juveniles (RTNJMX).  A
food limitation term (XRTNJ) is calculated similarly to that of the
adults, except the available food includes only phytoplankton and
particulate matter.  The estimated ration (RTNJ) is converted to a
filtering rate (RLFJ), as with adult grazing, and evaluated with
the other instantaneous rates.  A realized ration is determined
that is available to meet respiratory demands, and the final realized growth factor (GRO) is calculated, taking into account carni-

vorous predation rates.

## NUTRIENTS

Conservation of mass for nutrients is carefully maintained throughout all compartments and transfers. This is relatively straightforward for benthic fluxes, external inputs, tidal mixing, carnivore excretion, and phytoplankton uptake. Zooplankton excretion is somewhat complicated, however, since different carbon-to-nutrient ratios are assumed for the copepods and their various food sources. An expression was derived that maintains the appropriate C:N and C:P ratios in the zooplankton, regardless of the relative balance of phytoplankton (potentially of two types), detritus, or animal tissue in the diet. The result is a reasonable scheme whereby the assimilation and excretion rates and ratios are also variable (see Kremer and Nixon, 1978).

## NUMERICAL METHOD OF SOLUTION

Because the goal of our model was to simulate eight spatial regions over long time periods, numerical integration methods requiring short time steps were undesirable. Since the instantaneous rates appropriate for uptake, growth, and grazing in planktonic systems have exact analytical solutions, this approach permitted a time step of one day with no numerical integration error. Further, to minimize the error due to changing values of the interacting state variables over the time step, a two-step scheme was employed that defined the rates for each day, first, as the average of an initial estimate and, second, as based on the projected changes. Finally, an analytical expression was derived to separate the effect of each of several component rates that may be simultaneously taking place. Thus, if the net phytoplankton growth increment is calculated in terms of instantaneous rates of growth (K), adult zooplankton grazing (GZA), and juvenile grazing (GZJ), the implicit rates may be separated as follows:

$$P = P_0 \cdot e^{[(K - GZA - GZJ) \Delta t]}$$

$$\text{Total Growth} = \frac{K \cdot P_0 \cdot (e^{[(K - GZA - GZJ) \Delta t]} - 1)}{K - GZA - GZJ}$$

$$\text{Adult Ration} = \frac{GZA \cdot P_0 \cdot (e^{[(K - GZA - GZJ) \Delta t]} - 1)}{K - GZA - GZJ}$$

$$\text{Juvenile Ration} = \frac{GZJ \cdot P_0 \cdot (e^{[(K - GZA - GZJ) \Delta t]} - 1)}{K - GZA - GZJ}$$

## General Stability of the Bay Model

Early in the development of the bay model, it became apparent that the tight coupling of the components resulted in a stable model system. During the coefficient selection process and sensitivity analyses, plausible values simulated plankton patterns consistently within reasonable bounds for the natural bay, although some departure from appropriate seasonality occurred for certain combinations. Despite some periods of rapid oscillation, extreme boom-bust cycles, leading to extinction of phytoplankton and then zooplankton, were notably absent from the characteristic responses of the model. We consider this a desirable feature that is interesting since no explicit, arbitrary constraints forced this stability.

In some cases, this stability seems attributable to what are in a sense redundancies in the control mechanisms within the model. For example, three major nutrients are considered, even though nitrogen is in shortest supply most of the time. This makes the model less sensitive to the value of a single half-saturation constant, or even to the fundamental nature of the nutrient limitation formulation. Phytoplankton are potentially limited by any nutrient, by herbivorous grazing of the zooplankton and benthos, or by tidal flushing. Zooplankton are simultaneously subject to food limitation and predation yet may balance these against alternative food resources, including previous reproductive products that serve somewhat as storage for times of need.

"Stability" is a term which has received substantial attention in ecology, and some comment on its various interpretations is in order. Analytical investigations are often concerned with identifying points to which a system may converge under certain specified conditions, or with characterizing cyclic patterns of continuous, damped, or undamped oscillations of the state variables. Special cases of these behaviors are especially interesting--such as limit cycles, stationary points, and global stabilities--because they represent persistent conditions allowing species to coexist indefinitely.

In natural systems and in models designed to represent them more closely, such ideal states of stability are rarely, if ever, observed. Rather, the complex interactions of many environmental factors and multispecies communities reveal only clues that their long-term persistence may involve some or many of the features identified as potentially important in analytical studies.

In the present case, our numerical model moves seasonally through periods which represent different types of dynamic stability (Fig. 1). Because temperature, light, and other factors are constantly changing, the phase plot of the two plankton compartments is not directly comparable to its conventional application. For example, lines of constant phytoplankton and constant zooplankton, which are at the core of graphical analyses of the phase plane (Rosenzweig and MacArthur, 1963), are impossible to specify, due both to the complexity of the model's equations and to the changing forcing functions. Nevertheless, characteristic patterns emerge

which are similar to special analytical cases and which suggest
shifts in the dominance of the mechanisms' underlying behavior of
different seasons.

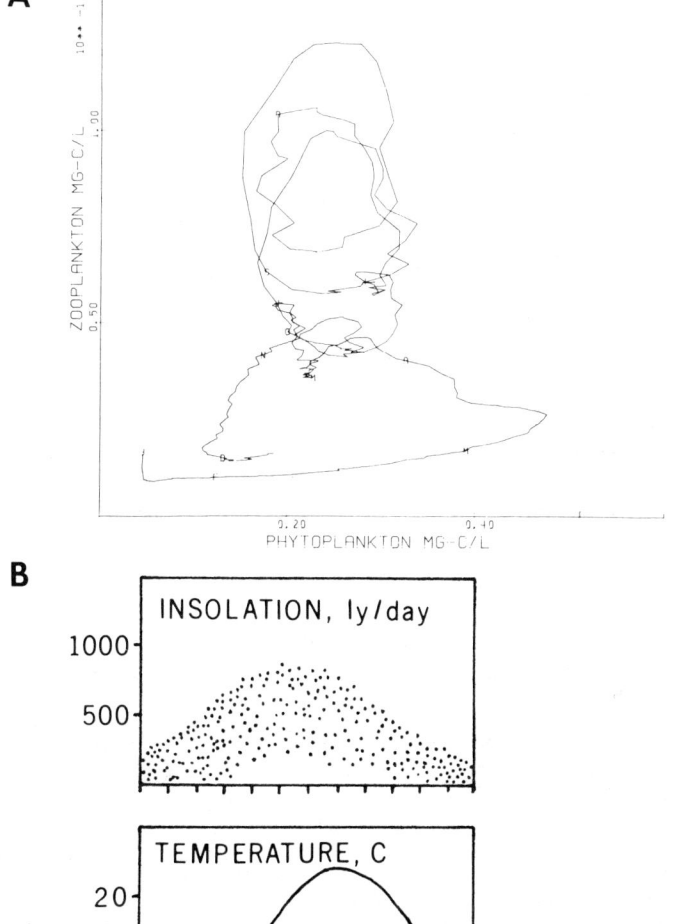

Fig. 1.  Phytoplankton-zooplankton interactions simulated by
the Narragansett Bay model.  The phase plane plot (A) reveals
strong stability in late Spring followed by a period of undamped
stable cycles in the summer.  The temporal patterns of light, temperature, and simulated nitrogen concentration (B) suggest that
resource limitation of the phytoplankton is closely related to the
shift in stability characteristics.

After the initial dramatic bloom of phytoplankton from a winter minimum, the zooplankton respond, bringing the system to a point resembling the classical unforced case of strong neighborhood stability (Fig. 1). From April through June, brief excursions occur, but the ensuing response repeatedly drives the system back to a single point. During July, a new state is reached, which produces broad, undamped oscillations of roughly constant duration. This behavior resembles stable limit cycles, although it is not completely comparable since the entire domain has not been mapped. Following this, in October, the system again moves to a new position, where it remains almost dormant in preparation for a repeat of the annual pattern.

Since the formulations of the model are fixed, we must look to changes in other variables for indications of the causes of the distinct patterns observed in the phase plot (Fig. 1B). Solar radiation is highly variable throughout the year due to the stochastic cloudiness, but there is no clear correlation with the two periods of interest, April through June and late July through September. Similarly, the temperature extremes do not correspond to these times. However, the concentration of nitrogen reveals an interesting point: at the termination of the initial phytoplankton bloom, nitrogen is severely depleted and remains very low from April until well into July; thereafter, it remains high and unlimiting to the algae during the late summer and fall. Thus, the shift from strong "local stability" to broad "limit cycles" is accompanied by a change in the availability of a limiting resource of the phytoplankton.

The parallel between this observation and a general conclusion drawn by Maynard Smith for continuously reproducing predator-prey systems is clear: "If the prey is resource-limited, and not limited by the predator, this will tend to damp out the oscillations." (Maynard Smith, 1974, p.30). Here prey and predators are algae and herbivores, and although the continuously breeding constraint is not closely met by the zooplankton, nitrogen limitation during the late spring is extreme, and the zooplankton biomass is just beginning to respond. The relative lack of importance of zooplankton grazing early in the simulation was confirmed by a run with no zooplankton, in which the initial bloom terminates and reaches a stable lower level entirely due to the limiting effect of nutrients (Kremer and Nixon, 1978). Further, later in the summer the situation is reversed. Phytoplankton are no longer resource-limited; warm water temperatures stimulate rapid zooplankton metabolism, and the herbivores assume a dominant role, thus initiating closely coupled cycles.

Another interaction plays an important role during the summer months in the bay model. Various carnivores exert pressure on the zooplankton, significantly depressing their levels, especially during the late summer in the upper regions of the bay. Again, a parallel can be seen in the general conclusions of the theoretical analysis. Predator-prey oscillations should be damped if the predator is limited by factors other than its prey (Maynard Smith,

1974, p. 30). This effect can be seen clearly in a series of simulations in which the biomass of carnivores was sequentially increased (Fig. 2). In the computer runs, arbitrary constant biomasses of
menhaden, assumed to be feeding on all available zooplankton, were
forced in the upper half of the bay throughout the months of July
through October. While this is unrealistic, the effect of external
predation on damping the algal-herbivore oscillations is dramatic.

An apparent paradox is suggested by the two examples presented
above. In the first, oscillations are enhanced when the prey
(phytoplankton) are significantly affected by predators (zooplankton). However, in the second case, oscillations are damped when
the prey (zooplankton) are suppressed by heavy predation (menhaden).
The explanation for this becomes apparent when the details of the
formulations are considered. The predator-prey relationship of
phytoplankton and zooplankton is modeled with complete feedback
between both compartments, but the coupling of the zooplankton and
their predators is incomplete, with no chance for menhaden biomass
to respond within the context of the model. Thus, the carnivorous
pressure on the zooplankton is more accurately viewed as a factor
limiting them as predators. This is a feature predicted by Maynard
Smith to dampen the oscillations. (We are presently examining the
stability characteristics of a detailed mechanistic model including
the third trophic level.)

It is easy to overemphasize the apparent agreement between
analytical results and those of a mechanistic model. For example,
the presence of protective cover for the prey should theoretically
dampen the amplitude of oscillations in continuously reproducing
systems (Maynard Smith, 1974, p.30). In our formulation, low

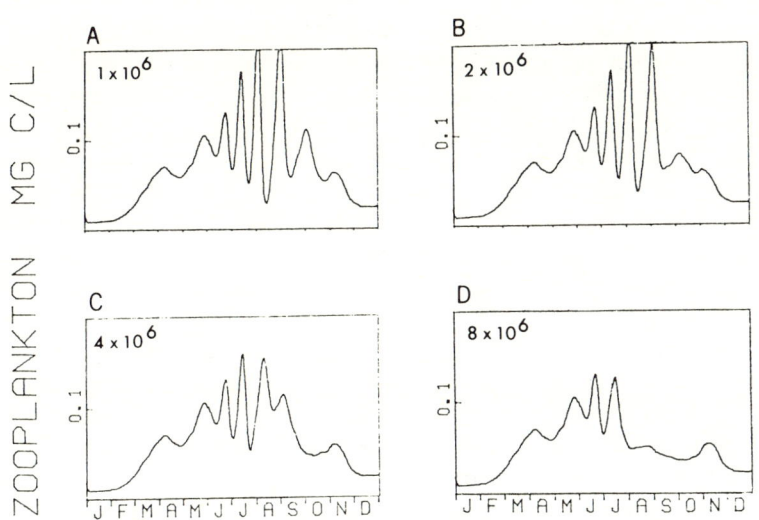

Fig. 2. The stabilizing effect of increased carnivorous pressure on the amplitude of plankton cycles. In four simulations, the
biomass of carnivores present from July through October was doubled:
A. $1 \times 10^6$ pounds of menhaden, B. $2 \times 10^6$ pounds, C. $4 \times 10^6$ pounds,
D. $8 \times 10^6$ pounds.

values of the Ivlev food-limitation exponent decrease the grazing
effectiveness of the herbivores at low phytoplankton concentrations,
indirectly simulating a refuge. And in our sensitivity runs, Ivlev
exponent values at the upper end of literature observations resulted in large amplitude, undamped oscillations (Fig. 3). Conversely,
low exponents dampened the cycles, in extreme cases effectively sup-
pressing the zooplankton population except in the warmest months.

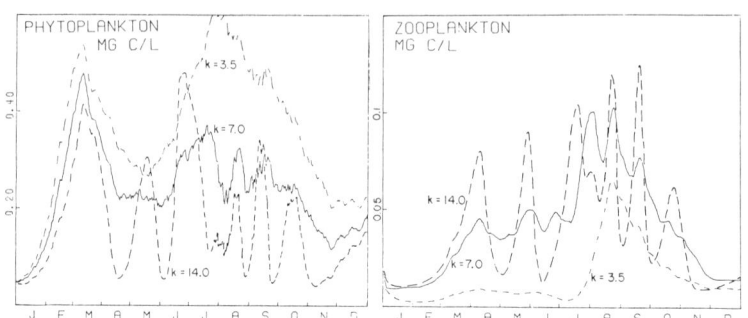

Fig. 3. Changes in the model resulting from doubling and halv-
ing the value of the Ivlev exponent for zooplankton grazing (from
Kremer and Nixon, 1978).

But the assumptions of the analytical model are not completely
met in our formulations. Zooplankton are not continuously reproducing
but are modeled with a temperature- and food-dependent development
time. Maynard Smith and Slatkin (1973) presented an analysis of the
role of cover in a system where the predator breeds in discrete seasons and requires a constant ingestion to maintain itself. This
approach, with its additional biological detail, might be expected
to parallel more closely our mechanistic model. Yet they found
that a prey refuge in this case need not increase stability in the
system and may even enhance any natural oscillations, making extinction of the predator more likely.

The resolution to this apparent conflict obviously must lie in
the difference in the other assumptions of the models and, more
precisely, in the additional detail of the mechanistic formulations.
A significant feature of both analytical models is that the prey
must be resource-limited, and we mentioned earlier that this is a
satisfactory approximation early in the simulations of the bay
model. However, a critical feedback exists in our model between
the predator and the prey's limiting resource. Regeneration of nutrients by the zooplankton, which is especially important at points
in the cycle when zooplankton are high and phytoplankton low, provides a stimulus to increased primary productivity as the predator
begins to decline. This coupling buffers the herbivore oscillations, avoiding the possibility of their extinction when food is
scarce. Additionally, the faster turnover of the phytoplankton
system is advantageous compared to the annual breeding seasons assumed by Maynard Smith and Slatkin. While the cyclic periods are
similar with respect to the generation times, starvation leading

to extinction is more likely in the case with discrete annual breeding than in the rapidly responding plankton system with overlapping
generations.

This topic brings up another fundamental feature of major importance in analytical models: the role of time lags.  In their simplest forms, coupled predator-prey equations of the Volterra type
assume instantaneous and continuous interactions.  It is for such
simplified and rarely realistic assumptions that many stability
analyses have been performed, such as that concerning resource limitation presented earlier.  Reviewing briefly, in systems where
the prey is resource-limited and the predator is prey-limited--
both reasonable assumptions for the bay model early in the yearly
simulation cycle--continuously reproducing systems should converge
to stable equilibria.  Further, increasing the time constant for
delayed feedback should introduce oscillations of increasing amplitude (Smith, 1972; Maynard Smith, 1974).  The bay model exhibits
stable oscillations, even with clearcut delays due to juvenile
hatching and development (Fig. 1), and might therefore be expected
to be even more stable when these lags are removed.  However, when
all age structure is removed from our model by allowing all adult
unrespired assimilation to become adult biomass directly, extremely
rapid cycles of large amplitude result even early in the run (Fig.4).

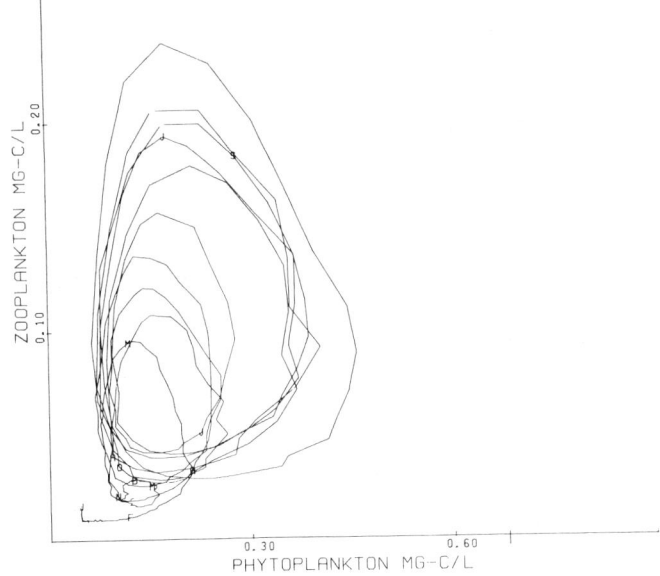

Fig. 4.  Changes in the model when eggs and juvenile age structure were removed from the model.

Again, the resolution is to be found in a subtle shift in the
balance of controlling influences within the mechanistic details.
In the absence of a time lag, total zooplankton biomass responds
rapidly to the early phytoplankton bloom, and a consequence of this
biomass is, again, increased regeneration of nutrients which poten-

tially limit the algae. Thus, shortening the time lag has resulted in the prey (phytoplankton) being no longer primarily resource-limited, a critical assumption for the stability of the original case. In fact, this system now approaches the case where the predator and prey are limited only by each other, as in Volterra's undamped equations which predict oscillations of constant amplitude (Volterra, 1926).

One more point deserves mention. The inclusion in our model of a more realistic life history for zooplankton is more than simply a time-delay term. Because of juvenile size differences, increased weight-specific respiration rate, and more restricted food supply, the age structure also represents differences in grazing effectiveness. In times of low phytoplankton abundance, juveniles will be hit the hardest. In addition, the assumption of cannibalism provides the adults with alternative food source. Within the system undergoing cyclic oscillations of phytoplankton and zooplankton, this combination of age structure and cannibalism significantly buffers the adult zooplankton population by reducing competition for food from newly hatched juveniles, while at the same time allowing the adults to reclaim this energy to meet their metabolic needs. The potential role of age-related hunting abilities in allowing adults to survive was recognized by Maynard Smith and Slatkin (1973) in their seasonally breeding model. Our model supports their interpretation, with the additional stabilizing role of cannibalism.

## Conclusions

The strategy of model building is predicated on the principle that a fundamental understanding of simplified abstractions is relevant to the behavior of a complex natural system. Maynard Smith (1974) emphasizes that our conclusions need not only apply to systems strictly conforming to the assumptions of the model. Rather, we may generalize concerning the kind of changes some given characteristic might contribute to the overall behavior of the natural system.

The comparison of models at two very different levels, designed and formulated with widely different goals and approaches, reveals examples of mutually consistent explanations of their behavior. The generalized character of coupled oscillations between predator and prey (herbivores and algae) with respect to the other factors limiting them emerges both in simplified analytical models and in our quite detailed, mechanistic simulations. Such agreement suggests that these are considerations fundamental enough to be meaningful, despite other subtle complex interactions.

On the other hand, some conclusions seem reversible, dependent upon other assumed interactions. The roles of prey refuge and of time lags in predator response depend not only upon the specific details assumed in these formulations, but also upon other interactions of the model. Thus, the addition of nutrient recycling, which is independent of prey cover or reproductive delays, may have consequences which dominate the stability characteristics of the model.

Based on this limited comparison of a mechanistic model and a few analytical examples, it is clear that even inferences between models of higher and lower degrees of abstraction may not always be drawn with confidence. Additional feedback controls and interactions not present in one case may well interfere with or mask the behavior predicted in the more restricted analysis. However, when consistent patterns emerge, despite variations in assumptions and mechanistic complexity, it is perhaps doubly encouraging.

## References

DiToro, D. M., D. J. O'Connor and R. V. Thomann. 1971. A dynamic model of phytoplankton populations in the Sacramento-San Joaquin delta. Adv. in Chem. Series. 106: 131-180.

Durbin, A. G. 1976. The role of fish migration in two coastal ecosystems: the Atlantic Menhaden (*Brevoortia tyrannus*) in Narragansett Bay, and the Alewife (*Alosa pseudoharengus*) in Rhode Island ponds. Ph.D. Thesis, Univ. Rhode Island, Kingston.

Durbin, A. G. and E. G. Durbin. 1975. Grazing rates of Atlantic Menhaden *Brevoortia tyrannus* as a function of particle size and concentration. Mar. Biol. 33(3): 265-277.

Eppley, R. W. 1972. Temperature and phytoplankton growth in the sea. Fish. Bull. 70(4): 1063-1085.

Goldman, J. C. and E. J. Carpenter. 1974. A kinetic approach to the effect of temperature on algal growth. Limnol. Oceanogr. 19(5): 756-766.

Heinle, D. R. 1966. Production of a calanoid copepod, *Acartia tonsa*, in the Pawtuxent River Estuary. Chesapeak. Sci. 7: 59-74.

Hess, K. and F. White. 1974. A numerical tidal model of Narragansett Bay. Univ. R. I. Sea Grant., Mar. Tech. Rep. 20, Kingston, R. I.

Ivlev, V. S. 1945. The biological productivity of waters. Uspekhi Sovrem. Biol. 19: 98-120.

Kiefer, D. A. and T. Enns. 1976. A steady-state model of light-, temperature-, and carbon-limited growth of phytoplankton. In: *Modeling Biochemical Processes in Aquatic Ecosystems* (R. P. Canale, ed.) pp. 319-336. Ann Arbor Science Publ.

Kremer, J. N. and S. W. Nixon. 1975. An ecological simulation model of Narragansett Bay -- the plankton community. In: *Estuarine Research* (J. Cronin, ed.). Vol. I, pp. 672-690. Academic Press, New York.

Kremer, J. N. and S. W. Nixon. 1978. *A Coastal Marine Ecosystem: Simulation and Analysis*. Ecological Studies, Vol. 24. Springer-Verlag, New York.

Kremer, P. 1975. The ecology of the ctenophore *Mnemiopsis leidyi* in Narragansett Bay. Ph.D. dissertation, Univ. R. I., Kingston, R. I.

Kremer, P. and S. W. Nixon. 1976. Distribution and abundance of the ctenophore *Mnemiopsis leidyi* in Narragansett Bay. Estuar. Coast. Mar. Sci. 4: 627-639.

Landry, M. R. 1975. Seasonal temperature effects and predicting development rates of marine copepod eggs. Limnol. Oceanog. 20(3): 305-496.

Lehman, J. T., D. B. Botkin and G. E. Likens. 1975. The assumptions and rationales of a computer model of phytoplankton population dynamics. Limnol. Oceanog. 20(3): 343-364.

Marchessault, G. D. 1974. The application of delayed recruitment models to two commercial fisheries. M. S. Thesis. Univ. of Rhode Island.

Matthiessen, G. C. 1973-74. Rome Point investigations, Quarterly Report to the Narragansett Electric Co. from Marine Research, Inc., East Warham, Mass.

Maynard Smith, J. 1974. *Models in Ecology*. Cambridge Univ. Press, London.

Maynard Smith, J. and M. Slatkin. 1973. The stability of predator-prey systems. Ecology 54(2): 384-391.

McLaren, I. A., C. J. Corkett, E. J. Zillioux. 1969. Temperature adaptation of copepod eggs from the Arctic to the tropics. Unpublished manuscript. Nat. Mar. Water Qual. Lib., EPA, Narragansett, R. I.

Nixon, S. W. and J. N. Kremer. 1977. Narragansett Bay -- The development of a composite simulation model for a New England estuary. In: *Ecosystem Modeling in Theory and Practice: An Introduction with Case Histories*. (C. Hall and J. Day, eds.), pp. 622-673. Wiley Interscience, N.Y.

Parsons T. R., R. J. LeBrasseur and J. D. Fulton. 1967. Some observations on the dependence of zooplankton grazing on the cell size and concentration of phytoplankton blooms. J. Ocean. Soc. Jap. 23: 10-17.

Riley, G. A. 1956. Oceanography of Long Island Sound. 1952-54. II. Physical Oceanography. Bull. Bingham Ocean. Coll. 15: 15-46.

Rosenzweig, M. L. and R. H. MacArthur. 1963. Graphical representations and stability conditions of predator-prey interactions. Am. Nat. 97: 209-223.

Smith, F. E. 1972. Spatial heterogeneity, stability and diversity in ecosystems. Trans. Conn. Acad. Arts and Sciences. 44: 309-335.

Steele, J. H.  1962.  Environmental control of photosynthesis in the sea.  Limnol. Oceanogr. 7: 137-150.

Volterra, V.  1926.  Variations and fluctuations of the number of individuals in animal species living together.  In:  *Animal Ecology* (R. N. Chapman, ed.) pp. 409-448. McGraw-Hill, N.Y.

Walsh, J. J.  1975.  A spatial simulation model of the Peruvian upwelling ecosystem.  Deep Sea Res. 22: 201-236.

# Qualitative Aspects of Estuarine Modeling

**Kenneth H. Mann**

Department of Biology
Dalhousie University
Halifax, Nova Scotia
Canada B3H 4J1

New information is provided on basic aspects of the nitrogen and carbon cycles in marsh-estuarine systems. Nitrogen fixation at rates up to 100 kg N per ha per annum are possible in *Spartina* marshes; the rates are apparently controlled by the concentration of dissolved N compounds in the sea water and by the degree of aeration of the soil. Sea urchins are shown to carry out N-fixation in their guts, at a rate which is inversely proportional to the N content of the food. The mysid shrimp, *Mysis stenolepis*, is shown to be capable of assimilating chopped hay or pure cellulose with efficiencies of 20-70%. It is concluded that these and other effects which are not incorporated in current marsh-estuarine system simulation models provide grounds for misgiving about the value of such models. The author emphasizes the need to focus attention on properties of the integrated system in addition to modeling the interactions of components.

## Introduction

An ecological simulation model can be regarded as a way of summarizing its author's knowledge about the complex set of interactions taking place in an ecosystem. If the author has an incomplete knowledge of important interactions, it is inevitable that the model itself and the predictions of the model will be incomplete and inaccurate. In this contribution I should like to draw attention to some recent developments in our knowledge of processes occurring in marsh-estuarine systems. They refer particularly to the roles performed by microorganisms in mediating transfers of material between trophic levels.

## NITROGEN-FIXATION IN SALT MARSHES

It is well established that nitrogen, more than any other element, tends to be the primary limiting factor to algal growth in coastal waters of temperate latitudes (Ryther and Dunstan, 1971). The annual cycle of nitrate and ammonia in surface coastal waters of the temperate zone has been documented many times (*e.g.*, Platt and Irwin, 1968). Values are high in winter when the water column is unstratified but fall to low levels in summer when there is thermal stratification and maximum utilization above the thermocline by the phytoplankton. Chapman and Craigie (1977) have recently shown that large algae, such as *Laminaria*, also adapt to this regime by growing rapidly in winter, when dissolved nitrogen is abundant, and building up internal stores that enable them to continue growth through part of the period of summer scarcity.

It has also been widely recognized that growth of *Spartina alterniflora* in salt marshes is limited by available nitrogen (Valiella and Teal, 1974; Broome *et al.*, 1975). Yet Sutcliffe (reported in Mann, 1975) noticed export of large quantities of total dissolved nitrogen in the summer from a Nova Scotia inlet with large areas of *Spartina* and *Zostera*. Subsequently, others have documented export of nitrogen from salt marshes (Dawson and Armstrong, 1975; Gardner, 1975). Obviously, if nitrogen supplies control primary production, an understanding of the nitrogen balance of salt marshes is an essential prerequisite to successful modeling.

In preliminary experiments, Patriquin (unpublished MS) placed belljars over *Spartina* and other salt marsh plants in the field and tested for nitrogen-fixation by the acetylene reduction technique of Hardy *et al.* (1968). Using short incubations, he found fixation of the order of $187–300\mu g\ N_2m^{-2}hr^{-1}$ at the edge of the marsh and $450–1875\mu g\ N_2m^{-2}hr^{-1}$ in the interior. With long incubations, the rates were higher. Nitrogen-fixation by blue-green algae on the mud surface is well-known (Jones, 1974; Van Raalte *et al.*, 1974; Whitney *et al.*, 1975), but when it was studied separately, it was found to account for only about 20% of the figures given above. Table 1 shows Patriquin's best estimates of potential nitrogen-fixation in the interior of a *Spartina* marsh in Nova Scotia during the growing season. It is seen that the total is close to the total accumulation of nitrogen in the *Spartina* plants during the same period. As will be shown below, it is not clear whether the system operates at its full potential, but if it does, there is about 100 kg N per hectare per annum which, in steady state, is available for export or storage in sediments.

The details of the nitrogen-fixing process are complex and not fully resolved (Patriquin, in press). At least two types of bacteria have been isolated which are suspected of being implicated in the nitrogen-fixation. One is a facultative anaerobe; the other is microaerophilic. Some bacteria have been observed on the surfaces

Table 1.  Comparison of potential nitrogen fixation during the
growing season in *Spartina alterniflora* in the interior of a Nova
Scotia salt marsh with the estimated accumulation of nitrogen in
the plant during the same period.  Data from Patriquin (unpub. MS).

| Potential N$_2$ fixation (gN m$^{-2}$, May-Dec.) | | N Accumulation (gN m$^{-2}$, May-Dec.) | |
|---|---|---|---|
| Mud surface mat | 2.23 | Total increase in leaves, May-Oct. | 5.55 |
| Roots and soil[1] | 9.25 | Accumulation in underground parts in growing season[2] | 7.75 |
| Total | 11.48 | Total | 13.3 |

[1]On 13 occasions between May 22 and Dec. 6, 1976, mud surface mats and
soil slices (roots and soil) were assayed for N$_2$ fixation by the
C$_2$H$_2$ reduction technique.  A 4:1 molar ratio of C$_2$H$_2$ reduced to N$_2$
fixed is assumed.

[2]The ratio of standing crop of leaves to standing crop of underground parts produced during the year was found to be 1.75:1.  This
ratio was applied to the study area.

of the roots, with concentrations in the depressions between the
cells, and others are located within the cells of the outermost
cortex.  Results of experiments suggest that the conditions favor-
ing N$_2$ fixation are provision of a carbon source by the plant roots
and of low oxygen concentration in the immediate environment of the
bacteria.  When roots are removed from the plants, then are washed
and tested for acetylene reducing activity in the laboratory, there
is a lag period of 12-24 hours during which no acetylene reduction
occurs (Fig. 1).  This is thought to be due to destruction of the
anaerobic microenvironments of the bacteria and their re-establish-
ment under experimental conditions by respiration of the plant
roots.  Nitrogen-fixation is low under totally anaerobic conditions,
but this is attributed to low production of suitable carbon sub-
strates by the roots in the absence of oxygen.  Maximum nitrogen-
fixation under laboratory conditions is obtained in an atmosphere
of 5%-9% oxygen, but fairly high nitrogen-fixation occurs in normal
air.  Under these conditions it is thought that low-oxygen microenvironments are being produced by the respiration of the roots.
When the experimental flasks are vigorously shaken, these microenvironments are destroyed, and nitrogen-fixation declines (Fig. 2).
Considerable nitrogen-fixation occurs in other parts of the *Spartina* plants (Table 2), but details have not been investigated.

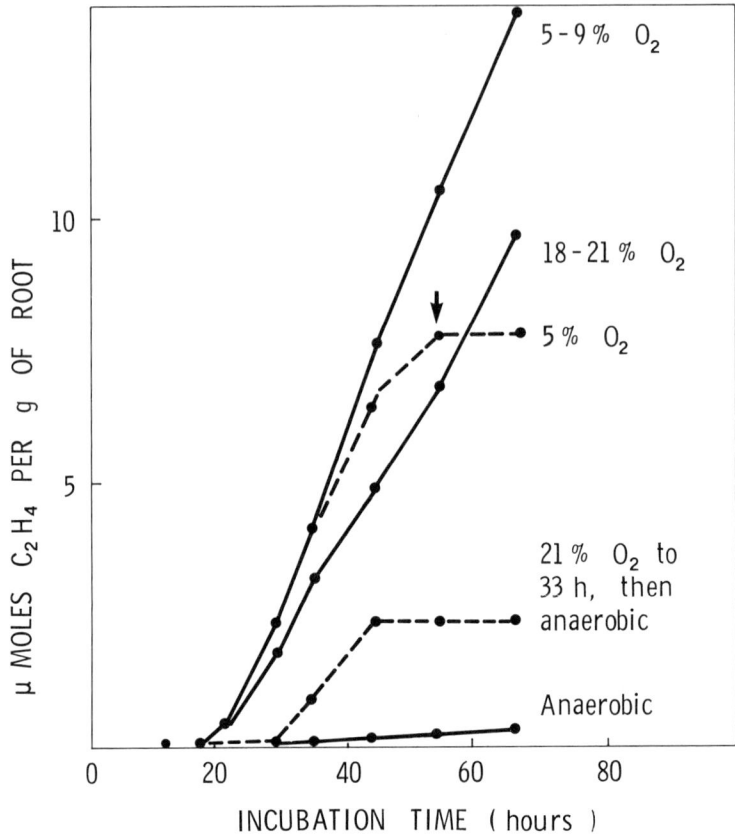

Fig. 1.  Time course of $C_2H_4$ production from $C_2H_2$ by washed, excised roots of *Spartina alterniflora* at various oxygen concentrations.  Data from Patriquin (unpubl. MS).

Table 2.  Nitrogen fixation ($C_2H_4$ production) in different parts of a clump of *Spartina alterniflora*, August 1974.  Rates are the maximum observed during time-course experiments in a 5% oxygen atmosphere.  (From Patriquin, in press).

| Plant parts | Rate, per unit weight (n moles $C_2H_4 \cdot g^{-1} \cdot hr^{-1}$) | Total for each part (n moles $C_2H_4 \cdot hr^{-1}$) | Percent contribution |
|---|---|---|---|
| Distal parts of leaves | 0.52 | 0.68 | 1.4 |
| Proximal 5 cm of leaves | 5.83 | 1.40 | 3.0 |
| Buds | 16.30 | 0.94 | 2.0 |
| Current year's underground stems | 3.42 | 1.83 | 4.0 |
| Previous year's underground stems | 24.20 | 16.00 | 34.0 |
| Roots on current year's stems | 21.40 | 8.15 | 17.3 |
| Roots on previous year's stems | 71.40 | 18.00 | 38.3 |

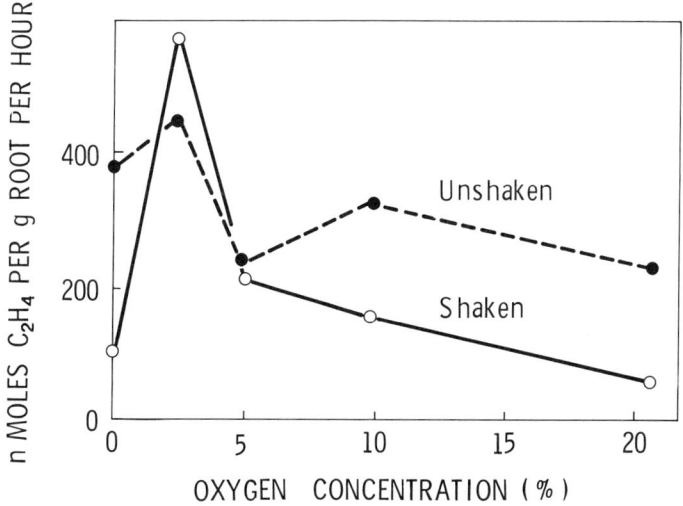

Fig. 2. Experiment to show the effect of shaking on $C_2H_4$ production by washed, excised roots of *Spartina alterniflora* at various oxygen concentrations. Roots were preincubated in seawater and 10% oxygen, water was drained off, and various oxygen concentrations were established. $C_2H_4$ production was determined for 2.2 hrs. Material was then transferred to a wrist-action shaker, and $C_2H_4$ production was measured for a period of 1.8 to 3.5 hrs. after shaking was initiated. Data from Patriquin (in press).

If there is a supply of dissolved nitrogen compounds from other sources, nitrogenase activity may be suppressed, and the system may even switch to denitrification (Patriquin, unpublished MS). Figure 3 shows that when the nitrogen-fixing activity was assayed on a number of angiosperm species from a high marsh, there was (with one exception) an inverse correlation between nitrogenase activity and ammonium concentration in the soil water. Suppression of nitrogen-fixation by dissolved nitrogen compounds would also explain why plants along the edge of a creek, having access to sea water on every tide, exhibit lower levels of nitrogen-fixation than those in the center of a marsh.

These results indicate that there is not any simple relationship between *Spartina* production and the nitrogen content of the tidal waters. There is a major nitrogen-fixing activity which is sensitive to both the oxygen regime of the marsh soils and the nitrogen content of the sea waters. A perturbation of the tidal regime, as by an impoundment, would probably affect nitrogen-fixation through its effect on soil aeration. All these factors must be taken into account in an effective model of salt marsh productivity.

## NITROGEN-FIXATION IN THE GUTS OF ANIMALS

This section deals with nitrogen-fixation in the guts of sea urchins. Sea urchin-kelp communities are common both on the open coast and in estuaries in many parts of the world. However, my point in reporting the work here is to raise the question: do

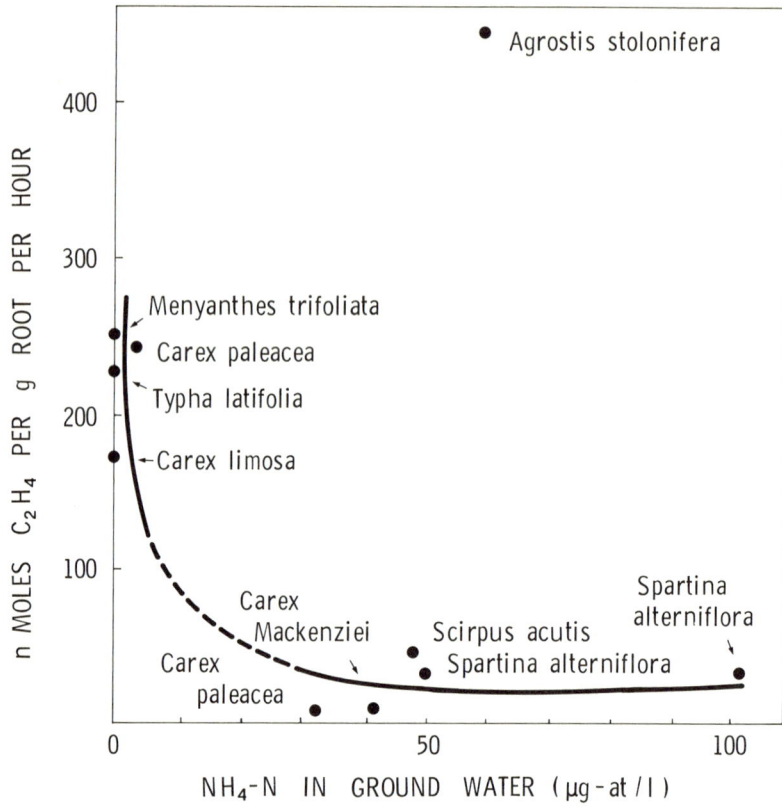

Fig. 3. $C_2H_4$ production of roots of various plants from high salt marsh plotted against $NH_4$-N in ground water. From Patriquin (unpub. MS).

analogous processes occur in the guts of salt marsh animals? If so, yet another component of a marsh-estuarine system model will be in need of revision.

In St. Margaret's Bay, Nova Scotia, an energy flow analysis (Miller *et al.*, 1971) indicated that the kelps, *Laminaria longicruris* and *L. digitata*, were major primary producers and the sea urchin, *Strongylocentrotus droebachiensis*, was the prime benthic herbivore. In a quasi-stable situation, the mean density of sea urchins in the kelp beds was 37 per $m^2$, but there were local aggregations of up to 500 per $m^2$ that were seen to eat their way through the kelp, consuming everything in their path. It was a puzzle to know how these urchins obtained sufficient nitrogen in their diets, for the C:N ratio of the kelp varied seasonally from 13.8:1 to 27.2:1 (Mann, 1971), and as a rough approximation, it is believed that most animals need in their diets a mean C:N ratio of 17:1 (Russell Hunter, 1970). We now know that sea urchins (with the help of bacteria) are capable of carrying out nitrogen-fixation in their guts (Guerinot *et al.*, 1977). Sea urchins incubated intact with acetylene showed evidence of strong nitrogenase activity (Fig. 4) although rates were very variable. When the rates were expressed as μ moles $C_2H_4$ produced per g dry weight of gut

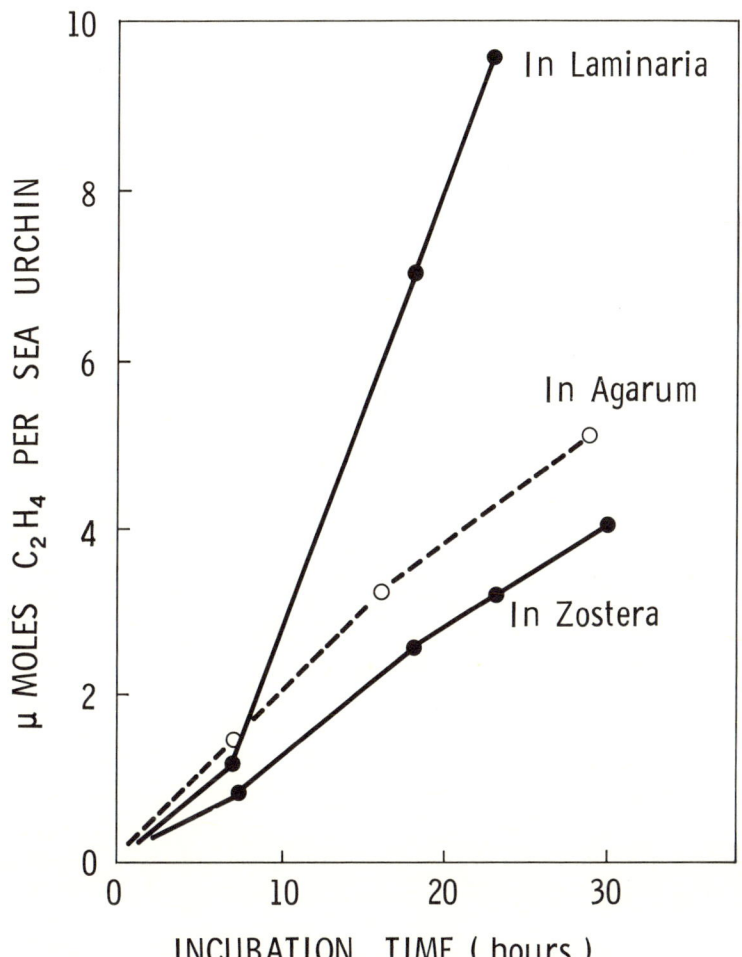

Fig. 4. $C_2H_4$ production by representative sea urchins taken from different areas in which they had been feeding December, 1975. From Guerinot *et al.* (1977).

(with contents), it was found that for urchins feeding on different types of food there was an inverse correlation between the nitrogen content of the diet and the rate of nitrogen-fixation (Table 3). Subsequent work (Guerinot, unpublished MS) has shown that for sea urchins feeding only on *Laminaria*, there is a change in rates of nitrogen-fixation with seasonal changes in the nitrogen content of the plants. Clearly, this is an adaptation which compensates for the low nitrogen content of the food.

When separate assays were made on the guts (with contents) and on the de-gutted bodies, only the guts displayed nitrogenase activity. There was no nitrogenase activity in the fresh plant material comprising the food. Bacteria have been isolated that are the presumed agents of nitrogen-fixation, but uptake by the urchins of the nitrogen fixed in this way has not yet been demonstrated. Fuji and Kawamura (1970) calculated a nitrogen budget for *Strongylocentrotus intermedius*. If the two species are comparable, nitrogen-

Table 3. Comparison of $C_2H_4$ production of sea urchins from different feeding areas. (From Guerinot *et al.*, 1977).

| Feeding area | % N in food | Mean sea urchin diam. (cm) | Nitrogenase activity (u moles $C_2H_4$ · g gut$^{-1}$· day$^{-1}$) |
|---|---|---|---|
| *Laminaria* spp. | 0.87 – 1.70 | 4.84 | 6.86 ± 5.23 |
| *Zostera marina* | 2.69 – 2.87 | 3.04 | 4.15 ± 10.40 |
| *Agarum cribrosum* | 2.80 – 3.00 | 2.88 | 1.21 ± 3.50 |

fixation in *S. droebachiensis* is equivalent to 8% to 15% of the animal's daily needs.

In general, it seems that judging by the C:N ratios, the food available to herbivores and detritivores in the near-shore marine environment contains abundant carbon whereas nitrogen is in short supply. Hence, for secondary as well as primary producers, it seems probable that nitrogen is a limiting factor. It would indeed be surprising if sea urchins were unique among marine herbivores and detritivores in having evolved (possibly by symbiosis) a capacity to fix molecular nitrogen.

## THE PROCESSING OF MACROPHYTE DETRITUS BY HIGHER TROPHIC LEVELS

The current view of the processing of marine macrophyte detritus (see review of Mann, 1976) recognizes that large algae tend to decompose quickly, while seagrasses, such as *Zostera*, decompose very slowly. *Spartina* is intermediate in this respect. It has frequently been observed that as the macrophyte particles are progressively reduced in size, their C:N ratio tends to decrease. This has been interpreted as an increase in the mass of microorganisms coating the surface relative to the mass of structural plant material. A change in C:N ratio can be brought about in several ways. One is that carbon compounds from the macrophyte material are converted to $CO_2$ or other soluble forms which are dispersed, while nitrogen is preferentially absorbed and retained by the coating of microoganisms. Another way is for the microorganisms to use the carbon of the plant material as an energy source but to take up nitrogen compounds from the water to build their own proteins. Gosselink and Kirby (1974) produced good evidence that the second process occurs. Dried ground stems and leaves of *Spartina alterniflora* were sorted into different sized fractions and incubated in flasks with sea water for 30 days at 30°C in the dark. Decomposition and build-up of microbial biomass were most rapid in the cultures with the smallest particles (67μ), and during the 30 days the total particulate nitrogen changed from 41 to 289 μg ml$^{-1}$. It was

inferred that uptake of nitrate and ammonium from solution had occurred at a rapid rate.  Hence, in yet another process in the marsh-
estuarine system, nitrogen appears to be a factor determining rates.

The inverse relationship between particle size of macrophyte
detritus and the rate of decomposition has been observed many times
(*e.g.* Newell, 1965: Odum and de la Cruz, 1967; Harrison and Mann,
1975).  It has been inferred that with decreasing particle size
there is an increasing ratio of surface area to volume, hence a
greater opportunity for microorganisms to colonize the surface and
attack the interior.  Newell(1965) first proposed that detritivores
obtained their nutriment primarily by digesting the coating of microorganisms and egesting the more refractory central core.  This
view has been confirmed several times and is widely accepted.  However, a recent study of a detritivore led to a very different conclusion.

Foulds (unpublished MS) studied the feeding of a mysid, *Mysis
stenolepis*, which congregates in subtidal beds of eelgrass, *Zostera
marina*, off the coast of Nova Scotia.  The mysids are capable of
both raptorial- and filter-feeding, and Foulds showed that in the
filter-feeding mode they effectively retain particles over the size
range 3-128μ, corresponding with the size range observed in suspension in the natural habitat.  Foulds then compared the assimilation efficiency of the mysids when fed sterile food and food
coated with microorganisms.  The efficiency was calculated by following the fate of radio-carbon in the food.  For this purpose he
used sterile and non-sterile $^{14}$C-labeled hay, and sterile and non-
sterile $^{14}$C-labeled pure cellulose.  Results are shown in Figure 5.
Surprisingly, the animals assimilated sterile food with an efficiency of 20-70%.  Even more surprising was the finding that non-
sterile food was assimilated with a lower efficiency than the
sterile food.

The almost inescapable conclusion is that mysids have their
own mechanism for digesting cellulose and do not rely on the di-
gestive activities of microorganisms living on the surface of the
particles.  There may be a cellulase secreted by the mysids themselves, or there may be a symbiotic microflora which performs this
function.  Whichever is the case, our model of the utilization of
macrophyte detritus by mysids must allow for a direct, highly efficient utilization of relatively refractory carbohydrates.  If
mysids, why not other common estuarine animals?  Once again, the
most fundamental processes in food chain dynamics are called into
question.

## Discussion

Since we have drawn attention to some little-understood aspects
of marsh-estuarine system functioning, it may be appropriate to
attempt to take stock of the present approach to modeling such systems.  An earlier discussion of the need for a holistic approach to
modeling (Mann, 1975) has been widely interpreted as a condemnation
of systems models.  However, Rigler (1976) described systems

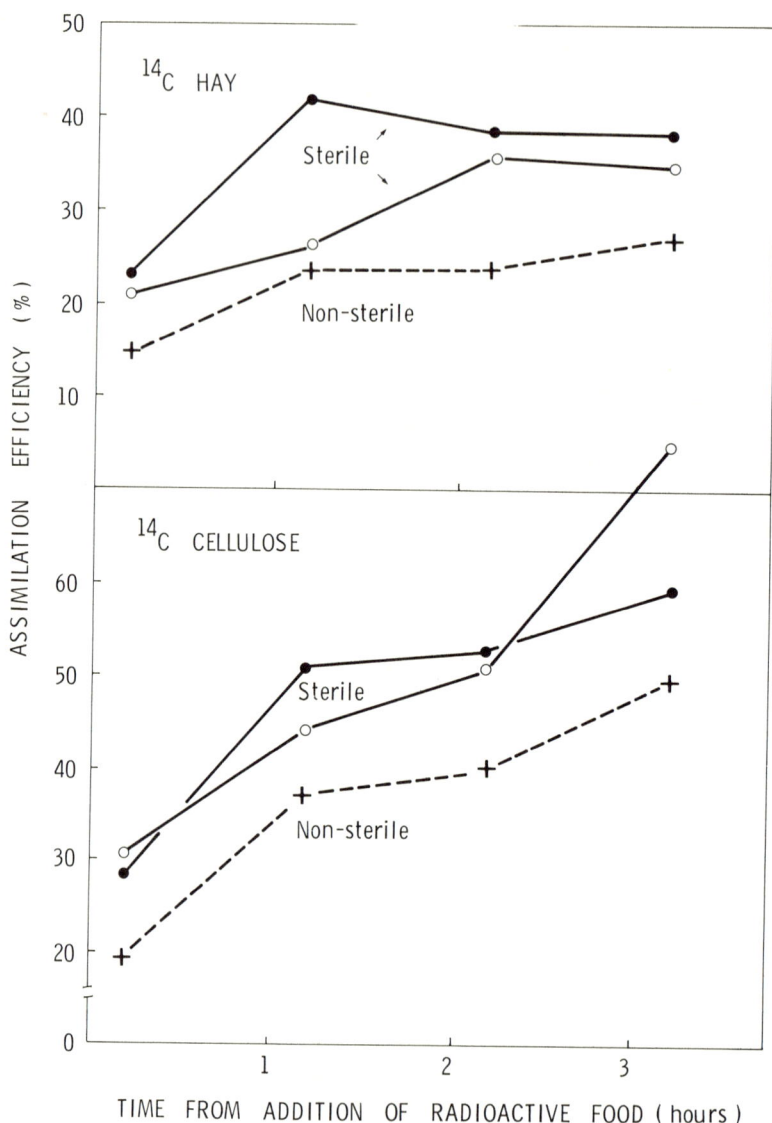

Fig. 5. Assimilation efficiency of *Mysis stenolepis* feeding on [14]C-labelled hay and [14]C-labelled cellulose, with and without microorganisms. From Foulds (unpub. MS).

analysis modeling as "the only approach we have thought of until now which has the potential to exploit studies of interactions between organisms." If we assume that he means complex interactions between several kinds of organisms and not merely predator-prey interactions, one is compelled to agree with him. Nevertheless, Rigler expressed himself as suspicious of the results on several counts. The first was that interactions between species are often difficult to quantify *in situ*. In other words, he was troubled by the old question of whether we can justifiably extrapolate laboratory results to field conditions or put them into a model which simulates the natural system. His second suspicion arose from the inevitable incompleteness of the models, with the necessity to choose between precision and generality. The questions raised earlier about the occurrence and significance of fundamental microbial processes, such as nitrogen-fixation and cellulose digestion, lead us to share Rigler's suspicions. If the magnitude of primary production in a salt marsh depends on levels of nitrogen-fixation, which are in turn dependent on the degree of aeration of the marsh soil, and if no one has studied this problem, how can we believe the output of a simulation model which leaves out the effect? Even if we immediately escalate the efforts to study the phenomenon, can we extrapolate to field conditions results of the type so far obtained in the laboratory?

Rigler's final suspicion was based on questions about whether we can draw boundaries around the systems under study without violating their properties as components of a larger system. This point was made in relation to three estuaries in Nova Scotia (Mann, 1975). It was shown that exchanges at the boundary with the open sea were qualitatively very different and extremely difficult to monitor. Even the question of whether estuaries are net exporters or net importers of organic matter is not yet subject to generalization. The basic idea that in an estuary bordered by extensive *Spartina* marshes the particulate organic matter in suspension is primarily derived from *Spartina* is open to question in light of the results obtained by Haines (1976), using $^{13}C/^{12}C$ ratios.

There are other aspects of systems simulation models as they presently exist that give grounds for misgivings about the value of their performance. When one reads the literature of mathematical ecology, there are several powerful influences on the behavior of systems which it seems virtually impossible to build into marsh-estuarine simulation models. The first of these is spatial heterogeneity of the environment. Smith (1972) drew attention to the stabilizing effects of spatial heterogeneity in ecosystem models, and many before him (*e.g.* Huffaker *et al.*, 1968; Paine, 1969) had made real-life observations supporting the idea. Spatial heterogeneity in an environment provides refuge for prey, increases the number of niches, and has very major effects on trophic interactions at all levels. Yet is is almost impossible to quantify, except in very simple situations.

A related, but distinct factor influencing trophic interactions is heterogeneity in the spatial distribution of organisms. A very striking example of this was observed on the coast of Nova Scotia (Mann, 1977). When kelp beds in St. Margaret's Bay were extensive and the sea urchins were fairly uniformly distributed, as mentioned earlier, with an average density of 37 per $m^2$, it was shown (Miller *et al.*, 1971) that the urchins consumed only about 3% of the annual productivity of the kelp. However, in a disturbed system those same urchins began to form dense local aggregations and consume all the kelp in their area, down to bare rock. They then moved on, leaving behind a few urchins which grazed any newly germinating sporophytes. In the course of about 8 years almost all the kelp in a bay of 140 $km^2$ was destroyed. The key factor was the build-up of local aggregations of the sea urchins. This led to loss of kelp productivity and also to loss of habitat for other species, such as crabs and lobsters, which declined in numbers dramatically. Events of this type have such drastic effect on ecosystem function that one doubts whether simulation models of the type now being built will ever contribute a predictive capability for management.

My difficulty in accepting the fact that whole ecosystem simulation models represent a major advance in our understanding of how ecosystems function stems from three major concerns. The first is that whenever I investigate an ecosystem, I discover new interactions that change in a basic way my idea of how that system functions. In other words, I judge that we need much more experiment and observation before we can construct a model that is even marginally acceptable. The second is that natural systems when performing in an integrated manner exhibit properties which we cannot even perceive by our reductionist, analytical approach. We have to spend a great deal more time observing the properties of whole systems. Simulation modeling, by its very nature, conditions us to take a reductionist approach. The third concern is that it seems very difficult, if not impossible, to apply the critical procedures of science to the results of simulation modeling of whole ecosystems. If we accept that a model is a complex hypothesis about how an ecosystem works, then as Rigler (1976) says, "we attempt to falsify the theory, or we search for an explanatory theory. We do not... verify our theory...verification is nothing more than an increase in our faith, and our faith is strengthened by our repeated failure to falsify the theory." However, my experience with ecosystem simulation models is that they can inevitably be falsified, and the authors readily admit this. The models are defended for the insights they give into possible interactions within the system. These are undoubtedly present, but how does one tell whether they are valuable insights or wrong insights?

I do not claim to have answers to the dilemma in which we find ourselves but wish to recommend strongly that we constantly try to focus our experiments and observations on properties of the total system. To this end, it seems to me that the current work with marine microcosms, *i.e.* captured water columns, tanks, etc., is more likely to reveal ecosystem properties. If ecosystems have an enormous number of interactions, including homeostatic mechanisms, of which we have no knowledge, one way to preserve as many as possible of them intact is to mimic the real system in a miniature living system. At the other end of the scale exists the possibility of observing and analyzing experiments with natural systems. These may vary in size all the way from coastal embayments to tide pools and form a continuous spectrum with the artificial microcosms. We urgently need for management purposes a set of testable generalizations about the behavior of ecosystems. While simulation models may serve to move our view step by step from the individual organism to the population and to population interactions, we need to set our sights even more firmly on identifying and quantifying the integrated properties of the total system.

## Acknowledgments

The author gratefully acknowledges the provision of unpublished material by Dr. David G. Patriquin and the following graduate students: W.C. Fong, J.B. Foulds, and M.L. Guerinot. Most of the work reported here was supported by a Negotiated Development Grant from the National Research Council of Canada, or by NRC operating grants to K.H. Mann and D.G. Patriquin. Dr. L.M. Dickie kindly criticized the paper. Mrs. Lyse Therriault assisted in the preparation of the manuscript.

## References

Broome, S. W., W. W. Woodhouse, Jr., and E. D. Seneca. 1975. The relationship of mineral nutrients to growth of *Spartina alterniflora* in North Carolina. II. The effects of N, P and Fe fertilizers. Proc. Soil Sci. Soc. Am. 39: 301-307.

Chapman, A. R. O. and J. S. Craigie. 1977. Seasonal growth in *Laminaria longicruris*: relations with dissolved inorganic nutrients and internal reserves of nitrogen. Mar. Biol. 40: 197-205.

Dawson, A. J. and N. E. Armstrong. 1975. The role of plants in nutrient exchange in the Lavaca Bay brackish marsh system. University of Texas Center for Research in Water Resources Tech. Rep. CRWR-129 EHE-75-06, pp. 1-115.

Fuji, A. and K. Kawamura. 1970. Studies on the biology of the sea urchin. VII. Bio-economics of the population of *Strongylocentrotus intermedius* on a rock shore of Southern Hokkaido. Jap. Soc. Sci. Fish. 36: 763-775.

Gardner, L. R. 1975. Runoff from an intertidal marsh during tidal exposure-recession curves and chemical characteristics. Limn. Oceanogr. 20: 81-89.

Gosselink, J. G. and C. J. Kirby. 1974. Decomposition of salt marsh grass, *Spartina alterniflora* Loisel. Limnol. Oceanogr. 19: 825-832.

Guerinot, M. L., W. Fong, and D. G. Patriquin. 1977. Nitrogen fixation (acetylene reduction) associated with sea urchins (*Strongylocentrotus droebachiensis*) feeding on seaweeds and eelgrass. Fish. Res. Bd. Canada 34: 416-420.

Haines, E. B. 1976. Stable carbon isotope ratios in the biota, soils and tidal water of a Georgia salt marsh. Estuar. Coast. Mar. Sci. 4: 609-616.

Hardy, R. W. F., E. J. Holsten, E. J. Jackson, and R. W. Burns. 1968. The acetylene-ethylene assay for $N_2$ fixation: laboratory and field evaluation. Plant Physiol. 43: 1185-1207.

Harrison, P. G. and K. H. Mann. 1975. Detritus formation from eelgrass (*Zostera marina* L.): the relative effects of fragmentation, leaching, and decay. Limnol. Oceanogr. 20: 924-934.

Huffaker, C. B., C. E. Kennett, B. Matsumoto, and E. G. White. 1968. Some parameters in the role of enemies in the natural control of insect abundance. In: *Insect Abundance* (T. R. E. Southwood, ed.). Blackwell Scientific Publications, Oxford.

Jones, K. 1974. Nitrogen fixation in a salt marsh. J. Ecol. 62: 553-565.

Mann, K. H. 1971. Ecological energetics of the seaweed zone in a marine bay on the Atlantic Coast of Canada. I. Zonation and biomass of seaweeds. Mar. Biol. 12: 1-10.

Mann, K. H. 1975. Relationship between morphometry and biological functioning in three coastal inlets of Nova Scotia. In: *Estuarine Research* (L. E. Cronin, ed.). Vol. 1, pp. 634-644. Academic Press, N.Y.

Mann, K. H. 1976. Decomposition of marine macrophytes. In: *The Role of Terrestrial and Aquatic Organisms in Decomposition Processes* (J. M. Anderson and A. Macfadyen, eds.), pp. 247-267. Blackwell Scientific Publications, Oxford.

Mann, K. H. 1977. Destruction of kelp-beds by sea-urchins: a cyclical phenomenon or irreversible degradation? Helgoland. wiss. Meeresunters. 30: 455-467.

Miller, R. J., K. H. Mann, and D. J. Scarratt. 1971. Production potential of a seaweed-lobster community in eastern Canada. J. Fish. Res. Bd. Canada 28: 1733-1738.

Newell, R. 1965. The role of detritus in the nutrition of two marine deposit feeders, the prosobranch *Hydrobia ulvae* and bivalve *Macoma balthica*. Proc. Zool. Soc. London 144: 25-45.

Odum, E. P. and A. de la Cruz. 1967. Particulate organic detritus in a Georgia salt marsh-estuarine ecosystem. In: *Estuaries*. (G. H. Lauff, ed.) Pub. no. 83 , pp. 383-388. A.A.A.S., Washington, D.C.

Paine, R. T. 1969. The *Pisaster-Tegula* interaction: prey patches, predator food preference, and intertidal community structure. Ecology 50: 950-961.

Patriquin, D. G. 1978. Nitrogen fixation (acetylene reduction) associated with cord grass, *Spartina alterniflora* Loisel. In: *Environmental Role of Nitrogen-fixing Blue Green Algae and Asymbiotic Bacteria* (U. Granhall, ed.), Bull. Ecol. Res. Comm., Stockholm, Vol. 26.

Platt, T. and B. Irwin. 1968. Primary productivity measurements in St. Margaret's Bay, 1967. Fish. Res. Bd. Can. Tech. Rep. no. 77, pp. 1-123.

Rigler, F. H. 1976. Book review of: *Systems Analysis and Simulation in Ecology*. ( B. C. Patten, ed.) Vol. 3 Limnol. Oceanogr. 21: 481-483.

Russell Hunter, W. D. 1970. *Aquatic Productivity: An Introduction to Some Basic Aspects of Biological Oceanography and Limnology.* MacMillan, N. Y.

Ryther, J. H. and W. N. Dunstan. 1971. Nitrogen, phosphorus and eutrophication in the coastal marine environment. Science 171: 1008-1013.

Smith, F. E. 1972. Spatial heterogeneity, stability, and diversity in ecosystems. Conn. Acad. Arts and Sciences Trans. 44: 307-335.

Valiela, I. and J. M. Teal. 1974. Nutrient limitation in salt marsh vegetation. In: *Ecology of Halophytes* (J. R. Reimold and W. H. Queen, eds.), pp. 547-563. Academic Press, N. Y.

Van Raalte, C., I. Valiela, E. J. Carpenter, and J. M. Teal. 1974. Inhibition of nitrogen fixation in salt marshes measured by acetylene reduction. Estuar. Coast. Mar. Sci. 2: 301-305.

Whitney, D. E., G. M. Woodwell, and R. W. Howarth. 1975. Nitrogen fixation in Flax Pond-Long Island salt marsh. Limnol. Oceanogr. 20: 640-643.

# Energy Quality Control of Ecosystem Design

**Howard T. Odum**

Department of Environmental Engineering Sciences
and Center for Wetlands
University of Florida
Gainesville, Florida 32601

If flows of energy generate characteristic designs for ecosystems, a relatively few kinds of energy analysis models may represent systems that succeed. Among the characteristics apparently found in all systems is a chain of increasing quality which converges energy flow in both space and time. High quality units can feedback more control action. Like the larger ecosystems, microcosms develop a spectrum of energy quality components depending on the quantity and quality of energy in the inflows. An energy quality ratio is defined as the ratio of Calories of one type of energy that generates another type under competitive conditions of maximum power transfer. Energy quality ratios quantitatively evaluate the cost and effect of the control units. Included are toxic substances and other amplifiers. Analyzing the spectrum of quality is suggested as a means of relating overall ecosystem characteristics to driving functions, both known and unknown.

## Introduction

The universe is composed of successful designs of nature, those that persist and survive. Whether large or small, these designs operate on flows of energy. If flows of energy are controlled by and at the same time are responsible for the self-designs of all systems, there may be common properties to designs of all systems. Models help us recognize principles relating energy and design. This essay suggests ways that energy diagrams help understand structure, organization, hierarchy, circulation, concentration, variation, and control.

The flows of energy build some systems larger than humanity

and some smaller than a human. In an intermediate size realm are
the systems of humanity and nature. Comparative modeling of various systems suggests common features for all systems (Odum, 1971).
For humanity to survive, the roles of humans must contribute to and
fit the common energy design requirements of all surviving systems.
Environmental management is application of control according to design for survival. Which are the allowable power patterns?

One technique for the study of ecosystem designs is the construction of microcosms, small, partially isolated realms which are
allowed to self-design. When microcosms are supplied energy flow,
materials, and genetic information, they rapidly build living ecosystems which have common design essence with larger ecosystems of
the biosphere. Energy analysis models help generalize about the
little systems and the big ones, increasing our confidence in generalization.

The rationale of the microcosm method given by Odum and Hoskin
(1953) still seems appropriate: to understand ecosystems, to in-
vent new ecosystems, and to engineer ecosystems to interface with
man. Models help the human understand the miniature ecosystems,
and these help us to understand the big wild ecosystems.

Understanding the principles of ecosystems is one of the reasons for starting microcosms that self-design a living and non-
living organization to match the energy sources (boundary conditions)
given. The study of the resulting designs helps test theories about
the way energy flow generates organization of ecosystems. The essence of ecosystems is conveniently analyzed in microcosm or macro-
cosm with diagrams of models using energy circuit language to show
simultaneously energy and kinetic relationships. One of the characteristics of ecosystems is the spectrum of energy quality and the
flow of energy that generates it. In this essay generalizations
about organization of energy quality are suggested to help understand ecosystems, their comparative anatomy and functions. When
microcosms are used to test for pollution stress, to test for ability to support man in space, or to miniaturize larger systems, the
energy quality theory may provide some guidance about scaling of
living models. Consider some definitions and theoretical proposals.

## Energy Quality

ENERGY COST RATIO (RATIO OF EMBODIED ENERGY)
The quality of energy may be defined by the energy cost of trans-
forming an energy flow of lower quality. For example, when one
animal eats another, it transforms the energy of the food to that of
the consumer. The ratio may be expressed as a ratio in units of
Calories per Calorie [Fig. 1(a)]. If the transformation were known
to be the most efficient possible at a competitive speed that yields
maximum power, it would represent a thermodynamically useful property of converting one kind of energy to another. Examining energy
cost ratios in systems which have been competing provides tables of
energy cost ratios as measures of quality. If all energies are

related to the solar energy required to generate them, an energy quality scale is generated, those with the largest ultimate energy cost in solar equivalents being the most costly and thus of highest quality in this sense. Whereas the illustrations cited from food chains are the familiar efficiency used by Lindeman and many others, the concept applies much more generally to energy transformations of non-living processes as well. For discussions of energy quality

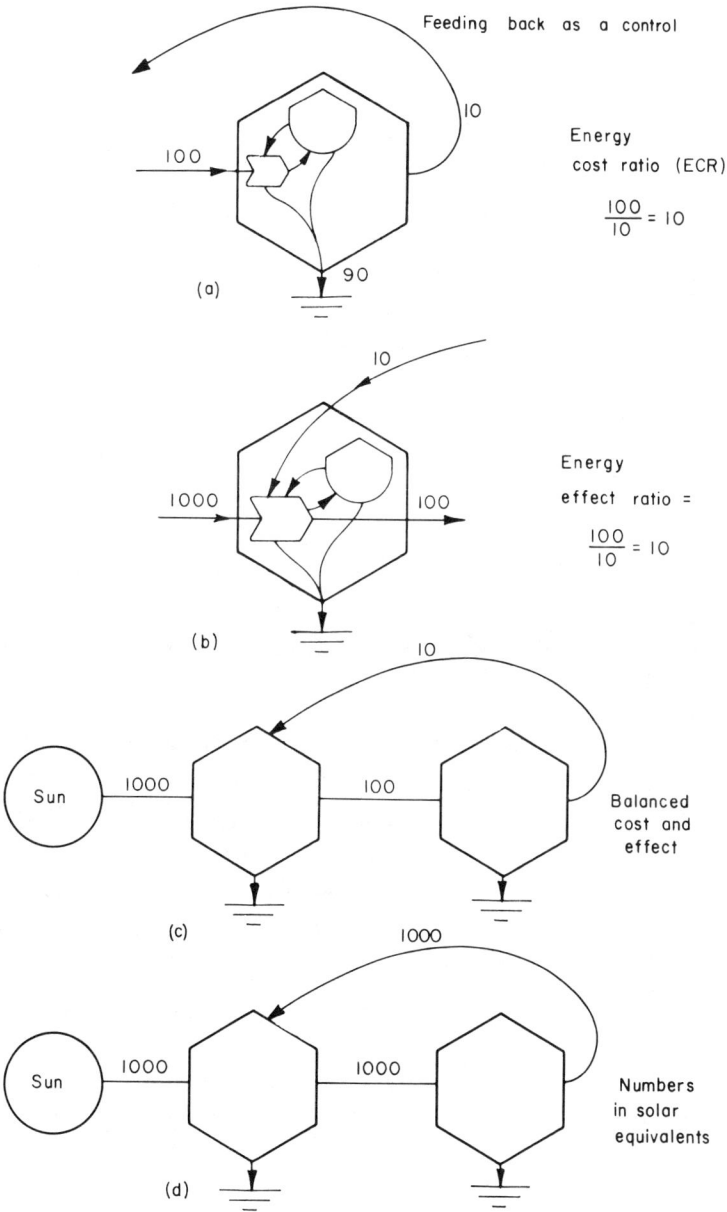

Fig. 1. Definitions of energy quality ratios: (a) cost, (b) effect, (c) cost effective balance. Flows are in Calories of heat equivalents: (d) energy flow expressed in Calories of cost in solar equivalents. Energy quality of the flows in Calories per Calorie of solar equivalents are the ratios of numbers in (d) to those in (c); 1, 10, and 100.

of energy flows of man's economy, see previous publications (Odum, 1975; Odum *et al.*, 1976; Odum and Odum, 1976). Embodied energy is another term for energy cost.

## ENERGY EFFECT RATIO

The quality of energy action may be measured in interactions that help generate new energy flow. For example, a Calorie of electricity may cause a much greater flow of Calories in a farm through amplifier actions. The ratio of Calories generated to Calories of control input provides an energy effect ratio [Fig. 1(b)]. Diagrams of observed data on energy chains can be evaluated to show these amplifier ratios. Units are also Calories per Calorie.

## BALANCE OF ENERGY COST AND EFFECT

The energy cost ratio and the energy effect ratio are two measures of the value of energy, one a measure of the resources used, the other a measure of the process stimulated.

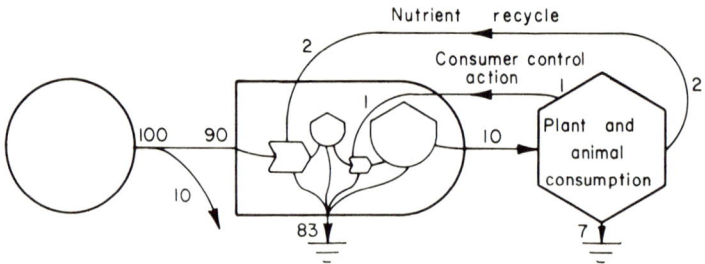

(a)

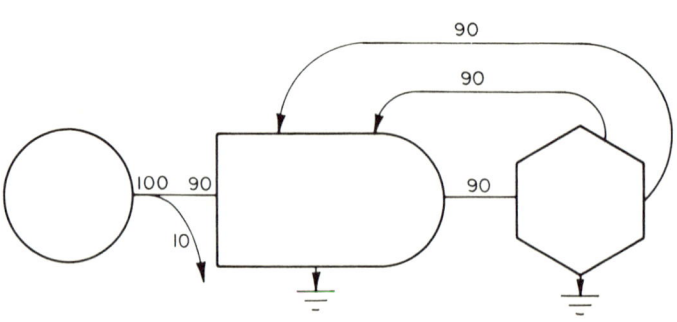

(b)

Fig. 2. Examples of no net production where there is cost-effect balance: (a) heat equivalents show declining values along the chain and feedback, (b) solar cost equivalents are unchanged as they pass if each stage is required for the next. The consumer unit generates two required feedbacks, each with the same cost and each with the same effect. Cost-and-effect values in solar equivalents are not additive.

It may be reasoned that good energy conservation leading to competitive survival of a system requires designs that use no more energy to upgrade quality than is ultimately delivered by that unit in its control effect. To use more energy is to drain the sources more than the process helps in feedback actions.

Where the energy effect generated is greater than the energy cost, there is a net production of value which constitutes a waste until it is fed back to achieve its potential in subsidizing secondary sources that would not alone yield net energy. After the subsidizing feedback is developed, the cost and effect are again balanced. The state of equality can be called "cost-effect balance" and is illustrated in Figure 1(c). How often cost and effect ratios are equal remains to be seen. Growing and transient systems depart from the cost-effect balances.

### DIAGRAMS IN SOLAR EQUIVALENTS

In Figure 1(d), the numbers for flows used are Calories of solar equivalents. Calories of solar equivalents are the Calories of solar energy flow upstream generating the flow downstream. Since costs were equal to effect in Figure 1(c), all flows have the same cost and effect, that of the Calories of sunlight flowing in. Although not used in this paper, cost equivalents and effect equivalents can be expressed in units of other types of energy such as coal equivalents, electrical equivalents, algal equivalents, etc.

### HIGHER QUALITY UNITS PREDOMINATE CONTROL PROCESSES

The higher the quality of energy, the greater the amplifier action it has if it can find a somewhat lower quality energy with which to interact. The very high quality actions of humanity in a fossil fuel culture controls and outcompetes the best of the high quality units of the lesser energized ecosystems. The highest quality units of every ecosystem may be regarded as the control agents of that system as they feed back their services. Whether we refer to soldiers, governments, managers, top carnivores, or storms, the actions are similar. They are often concentrated, pulsed, and capable of controlling and organizing the less concentrated elements into a closed loop pattern of support. That pulsing is a general tendency is postulated because more control energy can be delivered to high quality effect by saving energy and delivering it in a short time.

To establish a new system, it is the high quality energy control units that are required to develop the hierarchy and pattern. This occurs when new managers start new plants or new cultures colonize new countries or when new coral reefs are started with immigrants of complex members.

### EXAMPLES OF QUALITY EVALUATION IN MICROCOSMS

The theory of energy quality is readily observed in microcosms by the equivalent substitutions. In blue green algal mats without large consumers, the high quality energy generated is an electrical

voltage (Armstrong and Odum, 1963). With the water bug as a consumer, there is a change in algal membrane plant structure; the
electrical potential is lost and the high quality energy of the
system converges to the bug instead. There is a similar energy
cost for the bugs and the electricity. Both have similar solar
Calories per Calorie.

## NET PRODUCTION

To calculate net production taking into account the energy
costs, one multiplies all flows by their energy quality factors.
To obtain the net production of plants, the energy controlling
backflows to the plants must be subtracted. Examples are the
services of animals to the plants. Often these have not been included in net production calculations of ecosystems. In Figure 2
there is a balance of energy cost and effect when both are expressed
in solar equivalents and no net energy. In Figure 3, however,
there is a net energy with more solar equivalents produced by the
plants than used in the process.

## HETEROGENEOUS, DIVERGING PRODUCTION PATHWAYS

More than one type of production flow may be generated. For
example, an oyster puts out net yields to predators, pseudofeces,
feces, and mechanical work on water. The energy cost of each is
the same, that of the unified process. The diverging pathways can
have equal cost numbers placed on them. For example, in Figure
2(b), both outflows of the consumer units are given the solar
equivalent cost value of 90.

The diverging pathways labeled with the same cost differ in
energy quality, however, since the ratio of heat equivalents to cost
equivalents differs. For example, in Figure 2, taking ratio of
Calories in 2(a) to those in 2(b), one finds 1/90 as higher quality
than 2/90, since the cost per unit is higher.

## CONVERGING INTERACTION OF HETEROGENEOUS PATHWAYS

When flows feedback and interact, there is no implication that
the cost values are additive since the production functions are
interactive (often multipliers). If the feedbacks are having equal
effect as compared to their costs, all converging pathways will
have the same cost number--the cost number on the production outflow as in Figure 2(b). However, the interacting feedbacks are of
different quality since their cost ratios are different. In the
situation in Figure 2, the effect ratios are also different. (10/1
is higher quality effect than 10/2.)

## DIVERGING PATHWAYS OF THE SAME QUALITY

In Figure 3(a) the production pathway (20 heat calories per
time; 100 solar equivalent cost Calories per time) branches. Since
the two diverging flows are the same content and thus the same quality, the branches are additive and the cost is divided between them.
If they had been of different quality like those emerging from the

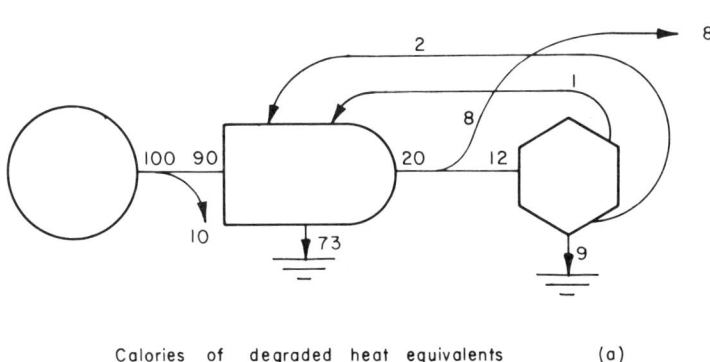

Calories of degraded heat equivalents          (a)

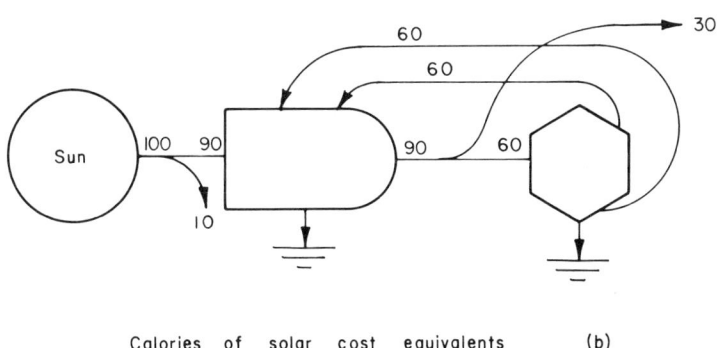

Calories of solar cost equivalents          (b)

Fig. 3. Net production when there is imbalance of energy cost
and energy effect: (a) Calories of heat equivalence, (b) solar
cost equivalents; net energy is 30.

consumer, it is postulated that their costs would be the same whether
one or both were considered. Each is a byproduct of the other. The
hypothesis is that processes selected for maximum power develop more
than one product as the most effective way to produce any one of the
products.

## Simple Chain With Single Source

POWER SPECTRUM
    Because transformation of energy to higher quality involves a
convergence with losses associated with the second energy law, the
quantity of energy drops as the quality of that moving up the scale
increases. The power spectrum describes the pattern as one of much
low quality energy and small amounts of high quality energy. If a
constant percent is dispersed for each step, there is an exponential
decline up the chain. (Mathematical integration of a percent is an
exponential function.)
    If there is only one source of energy of low quality such as
sunlight, all other entities are generated by stepwise upgrading.
A microcosm with sunlight as the only inflow is an example. The

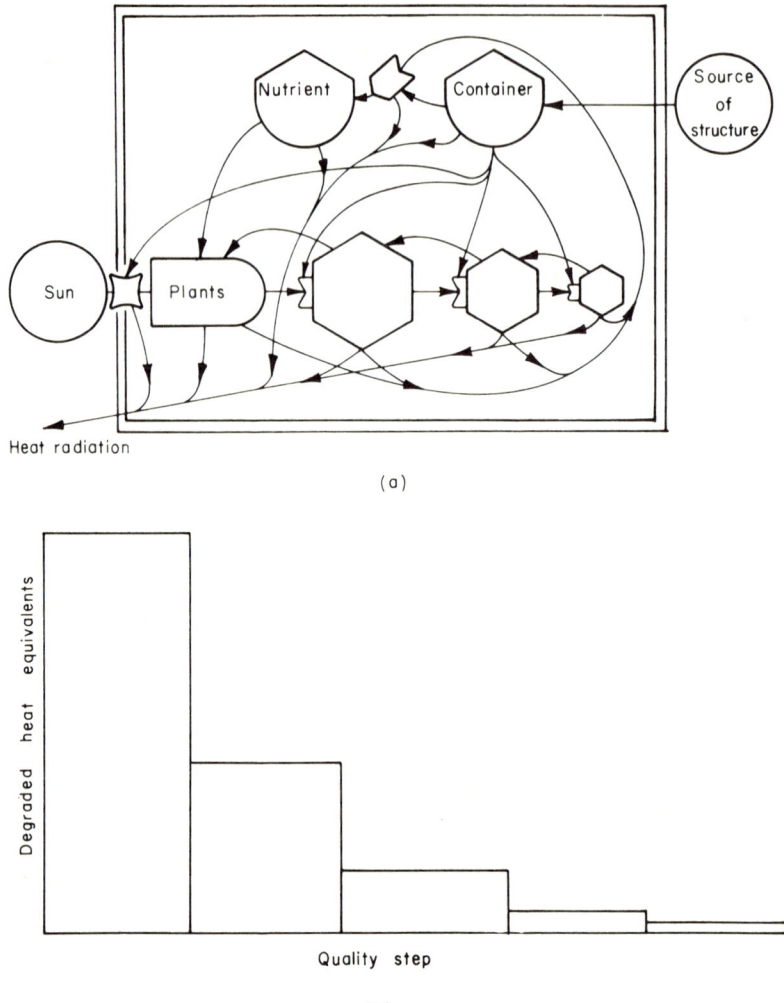

Heat radiation

(a)

(b)

Fig. 4. Energy pattern for a microcosm closed to matter, receiving only light, after initial seeding and provision of surfaces: (a) energy diagram, (b) energy quality spectrum.

model for such systems and its energy is given in Figure 4(a). Its discrete power spectrum is in Figure 4(b).

ENERGY QUALITY INCREASE BY SPATIAL CONCENTRATION

The upgrading of energy quality by energy transformation work often results in spatial concentration. Thus, solar energy is concentrated spatially by algae into organic packages; algal storages are spatially concentrated into zooplankton and benthos; such intermediate animals are spatially concentrated further into carnivores, etc. Similar principles apply to concentrations of food to towns in human affairs, to central governments, and finally to centers of information such as the nation s capital or the universities.

The feedback of the high quality units to amplify and control the lower levels is by divergence to greater space. Thus, each unit has a larger space from which it draws its energy and to which it

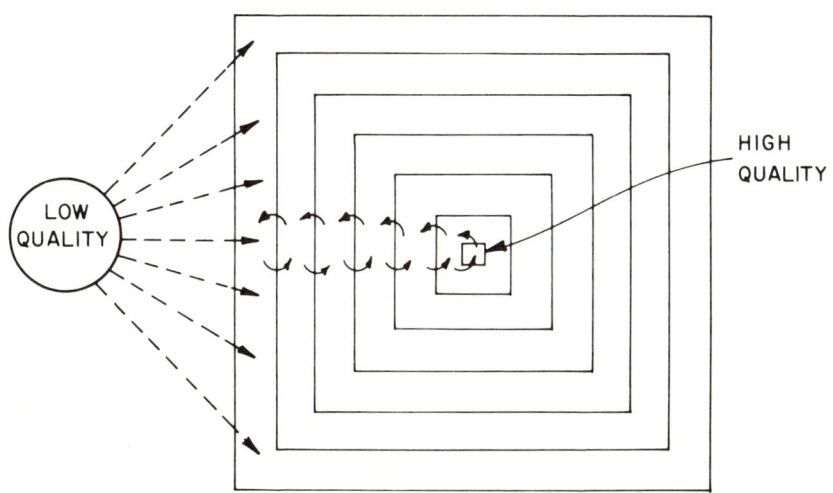

Fig. 5. Diagram showing convergence and divergence of energy flows in relation to energy chains.

directs its diverging services and amplifier actions and controls (Fig. 5).

## ENERGY QUALITY AND SIZE

Any high quality unit has two sizes. One is the dimension of its concentrated identity in which valuable qualities are located; the other is the space from which its energies are converged and the space back to which its control actions are dispersed. The core of a human tribe may be its chief, the core of the storm its highest wind vortex, and the core of the tree its trunk. Yet the size of the whole phenomenon, the human tribal zone, the storm, and the tree, is much larger.

The higher the quality of energy, the more concentrated is the core in space, but the larger is the space occupied by the unit served. Also, feedbacks cover more area (Figs. 5,6).

## CHAIN AS SEQUENCE OF DISCRETE UNITS

Each unit must store energy in order to control and feed it back as an amplifier. To return equal effect, observation shows there tend to be energy levels with discrete steps to the next ones up or down the scale.

## HIGH QUALITY UNITS AND MEANS OF
## COVERING THEIR SUPPORT AREA

Many kinds of convergence and divergence mechanisms are observed among the high quality units. Some, like the hawk, move around covering their area with their services and their drains of energy. Others, like the jellyfish, spread their protoplasm over a broader area by being diluted with low quality stuff. Others, like the spiders, build networks to cover space. Others, like the river, build a network of converging tributaries.

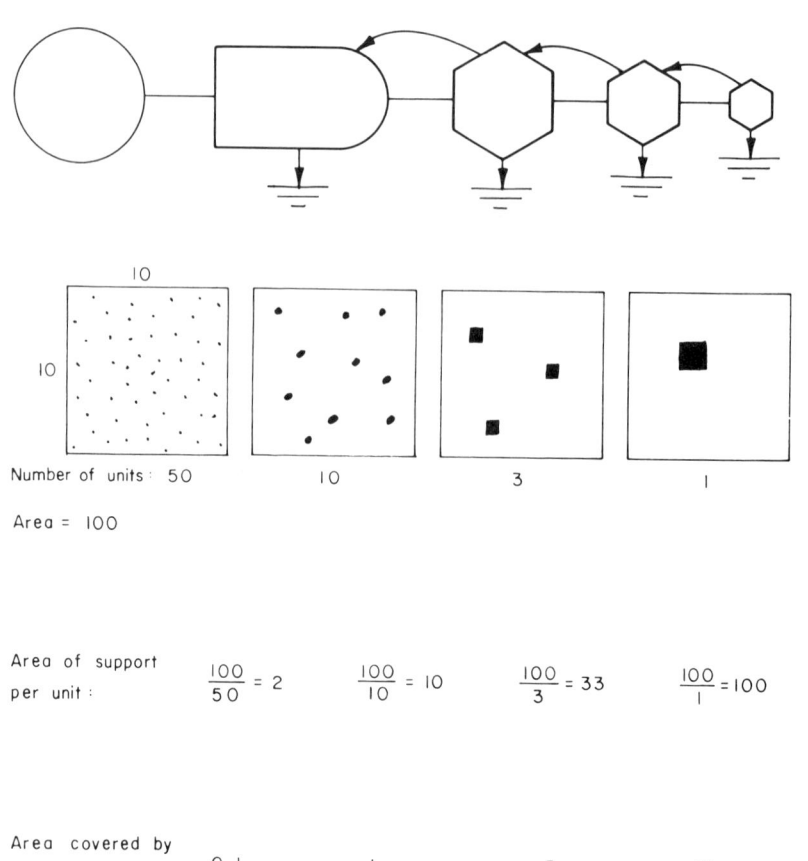

Fig. 6. Diagram showing two sizes to attribute to high quality units, one the support and influence area, the other the area of structure concentration.

## SIZE AND QUALITY RANGE OF ACTION

The range of interaction possible for a unit of a given quality is apparently limited to one or two steps more dilute and one or two steps more concentrated. Beyond this range the energy that can be amplified is not enough to justify the pay for the cost of the high quality unit. Humans cannot live directly on sunlight; bass cannot live on algae.

## QUALITY INCREASES WITH QUANTITY

For a flow of one kind of energy, the dependent chain can develop a longer sequence if the energy flow of the low quality base flow increases. More energy is stored and transformed at each step of increased quality, and thus at the top there is energy for development of higher quality units not found at a lower level. In other words, the power spectrum lengthens to the right. Thus, systems with greater energy resources can dominate other systems through the application of higher quality energy units, since they can amplify at a higher level of control.

## LARGER AREAS DOMINATE SMALLER ONES

Where energy is incoming on an area basis, a larger area can
converge more energy and thus develop a longer chain, developing
higher quality dominants that can control lesser systems. The
corollary is that smaller systems must join into larger ones in
order to survive.

Closed microcosms, being limited in dimension, develop lower
diversity and lower ability to control and compete as compared to
large systems or microcosms that are supplied control exchanges and
feedbacks from outside.

## CONCENTRATING THE TIME DELIVERY
## OF HIGH QUALITY ENERGY

High quality energy may be increased in its quality of effect
by concentrating its action in time. Instead of delivering its
energy continuously, it may be held and delivered in a short period
of time so that more action is possible. It becomes a higher
quality and capable of higher quality action. The systems with
property of saving and pulsing the high quality control actions
can dominate and prevail over other systems. If this principle is
correct, surviving systems will be found with pulsing delivery of
their high quality control actions back on themselves and in interaction with other systems.

The following are some examples of pulsing consumer actions
which delivered saved-up energies and caused their systems to prevail: fire, earthquakes, war, storms, migration, periodic harvest,
population oscillation, general human behavior, red tide, epizootics,
and epidemics.

## PULSING ENERGY IN SPATIAL SCANNING

The expression of pulsing in spatial pattern may occur as a
wave of action that spreads, using the stored energies as it goes
and stopping when the stores of energy are no longer remaining.
Then there must be energy accumulations before the wave can flow
again. The shifting of agriculture and the spread of conquests,
fire, disease, and energy-consuming fads are examples.

## THE ALTERNATIVE OF TIME- AND SPACE-CONCENTRATION

High quality energy can be generated through concentration of
quantities in space against the gradient of diffusion. One way is
to process potential energy of transportation against friction and
in accelerations and breaking so that some energy is concentrated
and thus increased in quality while more energy is degraded. Thus,
fish may concentrate into schools, zooplankton may be concentrated
by a circulating fluid, and fishermen may concentrate their catch.

High quality can also be developed by accumulating production
with time in one place without work on concentration, although
there is the inherent cost of depreciation of the storage during
the period of time. This makes production with time equivalent to
a concentration process in space. There is a space-time equivalency.

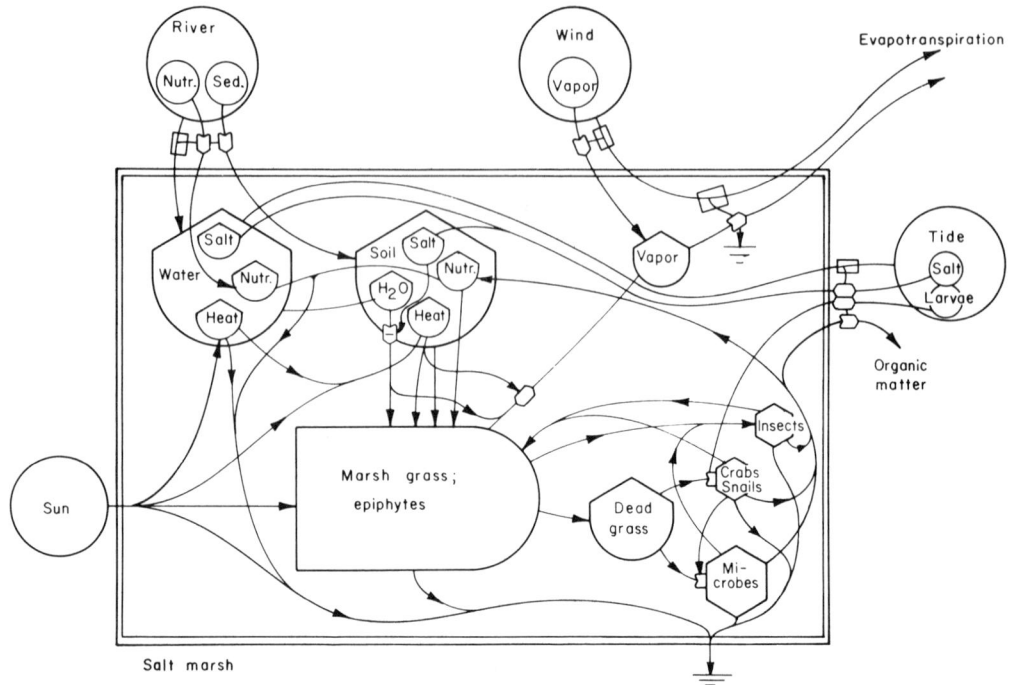

Fig. 7. Model of typical estuarine complexity with energy
inflow sources of varied quality. Order of quality from low to
high: sun, wind, river, tide, larvae.

## Multiple Energy Sources and Webs

When solar energy enters the earth, it is really a spectrum of
energy qualities which leads to a branching of energy interaction
that generates a web rather than a chain. In other words, the principles about energy quality and chains given in the previous section are oversimplified models of the real web that develops. In
any estuary, marsh, microcosm, or other real system, there is
usually more than one type of energy inflow, each with a different
quality. Given in Figure 7 is a model of a marsh as an example of
the web of interactions that develop when there are several energy
sources. High quality energies interact with low quality energies.
In Figure 8 are the kinds of spectra that can develop in ecosystems
and microcosms due to the inflow of different combinations of energies.

### ENERGY SIGNATURE
The set of energy inflows of several qualities constitutes an
energy signature of the biotope to which the biocoenosis adapts.
Putting all the flows in Calorie units and multiplying by their
energy quality units place the inflows in perspective as to their
costs and thus to the effects that must be obtained in order to
use the flows well. By this theory, the system should maximize
power by adjusting the amounts used to be equal when expressed in
equivalents of the same quality.

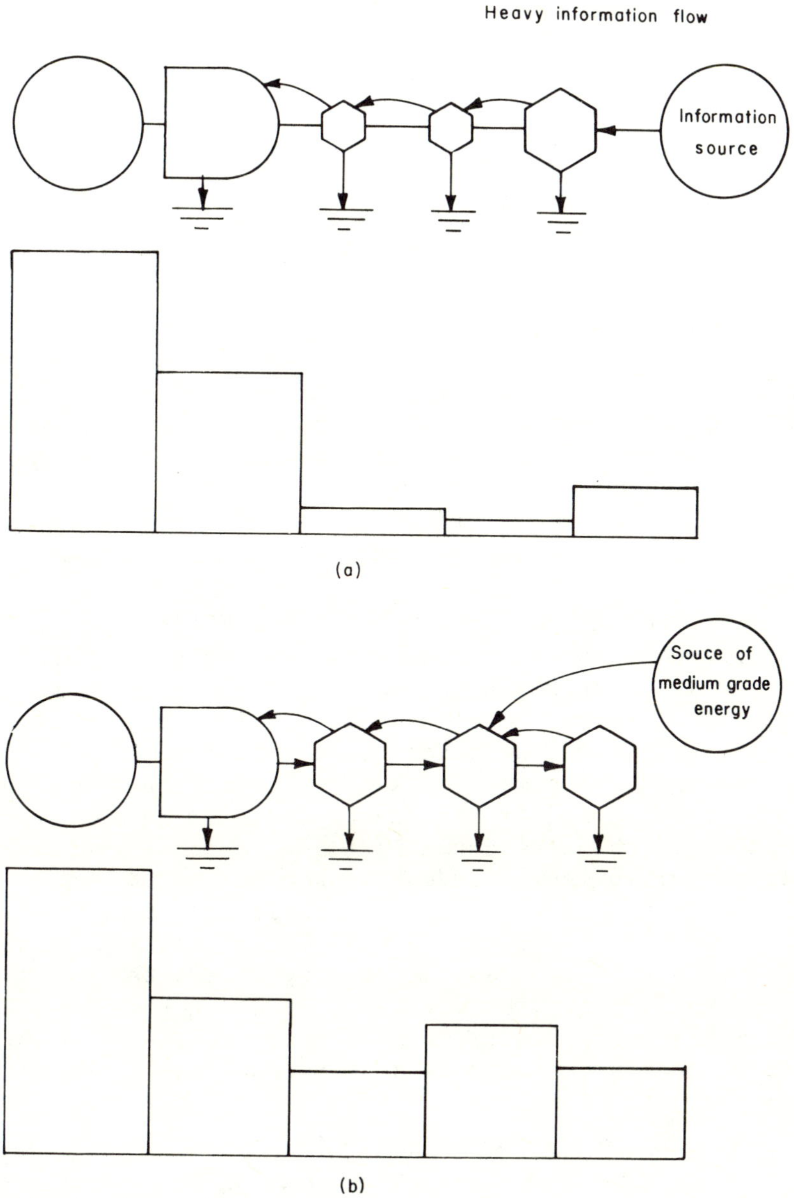

Heavy information flow

(a)

(b)

Fig. 8. Possible spectra for microcosms with combinations of
inflows of more than one quality: (a) large flow of information
such as human management, (b) large flow of migrating fish.


LAND AND SEA AND THEIR RELATIVE PRODUCTIVITIES

The relatively higher productivity of the land compared to
the sea as a whole begins to make sense as part of the understanding of the ratio of land to sea adjusted by the earth cycles. Land
is a very high-quality energy requiring convergence of much low-
quality energy of sunlight and heat-gradient energy to drive the
sedimentary and sea-floor spreading cycles. The land allows an
interaction of sunlight, wind, stable chemical surfaces, and rain,
further converging energies that were generated over the whole

earth on a relatively small area. The concentrating of the higher
quality energies generates a higher quantity production function
and ultimately higher quality units at the end of the chain, such
as human beings.

By its evolution of the land masses and mountains, the earth
controls the system of air and sea circulation that ultimately
supplies the high quantity of low-quality energy. The productivity
of the land is higher and its quality of energy higher, since this
allows the system as a whole to control and predominate.

## ESTUARINE ECOSYSTEMS

In the estuary model given in Figure 7, there are several
external energy flows, each with a different quality. As sketched
in the figure, the observed interactions in the estuary form a web
in which the highest quality energies pump the lesser quality energies. Because many of the energy inputs are moderately high-
quality, such as river and tidal flows, the total energy cost value
of the inputs to estuaries is very large and larger than most ecosystems.

## INFORMATION AS HIGH QUALITY ENERGY

Among the products of the energy chains are diversity and information which require large amounts of energy flow to maintain
and organize. One may model a whole web or aggregate the complexity as diversity in a downstream unit. If the pathways increase
as the square of the number of kinds of units that must be organized
with relationships and if the energy costs are proportional to
these pathways, then the drain is a quadratic one and provides a
non-detrimental logistic curve. Jensen (1975) discusses alternate
mechanisms for models which have logistic equations. This diversity model constitutes another (Fig. 9). Diversity is a means by
which the ecosystem can have a crowding effect that is positive
and also stabilizing. The diversity is a high-quality energy that
is far downstream, provides means for control, and is one of the
first to be lost when there is an interruption of the main energy
flow.

## TOXICITY AS HIGH QUALITY

There is a principle from pharmacology that active chemical
agents are very useful as controls at lower concentrations that
become very toxic at higher concentrations. In terms of energy
quality theory, a toxic substance may be regarded as very high
quality since it has high amplifier effect. What is detrimental to
one system has the means to be a strong positive effect as an amplifier to another system which is adapted. A system which is re-
sistant can profit by the toxic substance's channelizing effect in
eliminating competing pathways. Pollution may be regarded as the
release of high quality energy in excess without a design to
receive it.

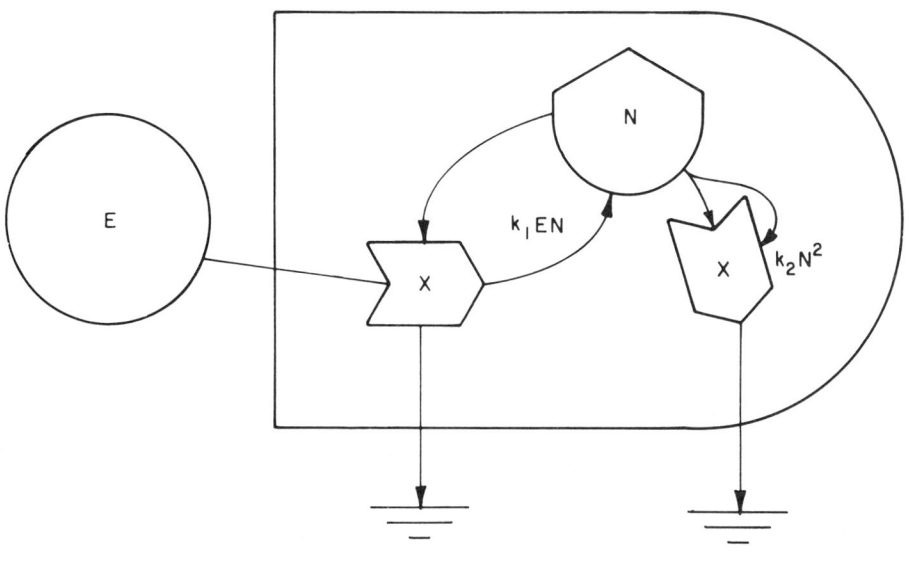

$$N = k_1 EN - k_2 N^2$$

Fig. 9. Diversity modeled as an information storage with quadratic cost and producing a special kind of logistic kinetics.

FOR THE FUTURE

The living microcosm models that are being developed in increasing numbers are the means for showing what designs go with what energy flows. The theories of energy quality, spatial concentration, spectral distribution, and control provide ways to identify ecosystem structures for each type of energy signature. This may help with old ideals of matching biocoenosis to biotope and explaining ecosystems observations with theories of causality. The realities of ecosystem design as they become understood through energy quality theory provide constraints on the creative minds in development of models. The ideal of a simple, characteristic basic pattern for all systems seems plausible.

## References

Armstrong, N. and H. T. Odum. 1963. A photoelectric ecosystem. Science 143: 256-258.

Jensen, A. L. 1975. Comparison of logistic equations for population growth. Biometrics 31: 853-862.

Odum, H. T. 1971. *Environment Power and Society*. John Wiley, N. Y. 331 pp.

Odum, H. T. 1975. Energy quality and carrying capacity of the earth. A response at prize awarding ceremony of Institute La Vie, Paris, June 18, 1975. Tropical Ecology 16; 1-8.

Odum, H. T. and C. M. Hoskin. 1953. Metabolism of a laboratory stream microcosm. Publ. Inst, Mar. Sci., Univ. Tex. 4(2): 115-132.

Odum, H. T. and E. C. Odum. 1976. *Energy Basis for Man and Nature*. McGraw-Hill, N. Y.

Odum, H. T., M. Brown, and R. Costanza. 1976. Developing a steady state for man and land: energy procedures for regional planning. In: *Science for Better Environment*, pp. 343-361. Proc. Int. Congr. Human Environment (HESC), Kyoto, Asahi Evening News, Tokyo, Japan.

# Necessary Conditions for Realism in Ecological Models

**Bernard C. Patten**

Department of Zoology and
Institute of Ecology
University of Georgia
Athens, Georgia 30602

A mathematical model serves to summarize certain useful aspects of experience in a formal, operational way. In ecology, mathematics are selected to represent specific phenomena without regard for the general knowledge base underlying those phenomena. It is suggested that models would better be selected to meet certain predetermined conditions for realism. The quest for such necessary conditions, and debates to accept or reject them, would contribute to development of a consistent body of theoretical ecology knowledge. Models that did not conform to agreed upon conditions would be considered unacceptable for ecological modeling. To illustrate how arguments might be developed, three necessary conditions are postulated and discussed: All models should be (i) explicitly forced, (ii) linear, and (iii) globally stable around a single equilibrium state when unforced, but locally stable or unstable when forced. Acceptance of these three conditions would sharply constrain the set of mathematics admissible for ecological modeling, and in addition would imply a general worldview of the way things are ecologically.

## Introduction

A mathematical model represents the endpoint of a long sequence of activities involved in the "mental experience" (Griffin, 1976) of observing, recording, and organizing human reality. A model is made in the context of a broader model, knowledge, and is intended to summarize certain useful aspects of experience in an operational form from which new inferences can be derived.

The purpose of this paper is to show that any mathematics selected to model specific elements of experience will ultimately be more useful if it is consistent with general knowledge underlying the particulars at hand than if it is not. If functions are unrealistic with respect to the broader frames of reference surrounding phenomena in nature, their derived properties may be inconsistent with general knowledge and thus of limited use.

In ecology, the selection of mathematical functions frequently is based on nothing more than correspondence to an empirical curve form observed in data. Thus, the logistic equation duplicates sigmoid growth of natural populations in limited environments. Michaelis-Menton formulations capture the hyperbolic shapes generated by saturation processes. And nonlinear limit cycles are frequently used to explain predator-prey oscillations. None of these models ties to an integrated central theory of ecological dynamics, yet each of the empirical curves is readily generated by a much simpler model--the forced, time invariant, linear differential system. If such a single model accounts for the data of a wider range of phenomena, then why should it not replace more specialized models that do the same but have no relationship to one another? The latter models developed from attention to isolated processes and, as it could be shown, were developed in violation of certain general criteria. Such criteria constitute necessary conditions for realism. Having met them, the modeler is free to go on and develop a sufficient model of the phenomena of interest for his purposes. But failing to satisfy them, the modeler may fail to link his model to the general knowledge base which is normal ecology, and worse, he contributes to further fragmentation of an already disconnected science.

The necessary conditions for realistic ecological modeling do not now exist, but I submit that they are developable as a matter of keen research in theoretical ecology. Discussion of the pros and cons of particular criteria could enliven the field, and resolution and acceptance would amount to establishing principles to serve as laws of modeling. To the extent that resultant models successfully represent reality, such principles would also become recognized as laws of ecology.

In the following sections three such laws will be proposed and examined as prospective universal constraints on realistic ecological models. All three will immediately be seen to be controversial, but the intent here is not to try to establish or reject them. My main purpose is didactic: to suggest that examination of necessary conditions would be a useful way to pursue theoretical ecology and to illustrate the nature of considerations that might enter dialogues on the pros and cons.

## Causality

PROPOSITION 1
*All ecological systems are causally open; therefore, dynamic ecological models should be explicitly forced at the system level modeled.*

Causal determinism, although its meaning and significance are unresolved in metaphysics (e.g., Sosa, 1975), is one of the fundamental assumptions of science. Systems are caused to change through interactions with other systems which are external to themselves. Aristotelian "efficient" causes originate outside the systems they move, and it is this exogenous character of causation that is significant for ecological modeling.

Patten, Bosserman, Finn and Cale (1976) have presented a general theory of causal flux in the systems ecology context. The systems model of causation distinguishes state variables, inputs, and outputs in the following way. Causation comes to a system from its environment via inputs. States pick up these influences at the system boundary and propagate them through the endogenous interactive structure. The system's response to inputs is output; the transfer of propagated influence from state to output variables also takes place at the system boundary. Output represents causality transmitted to the system level environment by the system; it is available as input to other systems.

All ecological systems conform to this input-state-output model. They are causally open at their boundaries, just as they are also thermodynamically open. Therefore, models of their dynamics should include input as a necessary property. A model that does not meet this property would be seen as violating a basic general condition and would serve long term purposes less well than a model that satisfied the constraint. Volterra population models,

$$\dot{x}_i = r_i - \alpha_i f(x_1,\ldots,x_n)\, x_i, \qquad i = 1,\ldots,n, \quad (1)$$

are an example of an important class of mathematical formulations that would be rejected under this criterion. There is no system level input to such systems, as would be represented in the differential equations context by nonhomogeneous terms:

$$\dot{x}_i = r_i - \alpha_i f(x_1,\ldots,x_n)\, x_i + z_i(t), \qquad i = 1,\ldots,n, \quad (2)$$

where $z_i(t)$ represents causality from outside the system acting on component i within the system.

When confronted with this criticism, proponents of Volterra models will often suggest that the input is implicit in the "innate capacity to increase," $r_i$, and that it is legitimate to internalize the causality in the process of abstracting the population from its natural surroundings. So clearly there is room for debate here, and refinement and resolution of the issue in the literature could significantly enhance the development of a consistent theoretical ecology. I think implicit modeling of input violates the causal principle for the following reason.

Consider the following form of the Volterra system:

$$\dot{x}_i = r_i x_i (K_i - \sum_{j=1}^{n} a_{ij} x_j)/K_i, \qquad i = 1,\ldots,n, \quad (3)$$

where $r_i$ is biotic potential, $K_i$ is carrying capacity, and $a_{ij}$ are interaction coefficients. The latter can be arrayed in an interaction matrix,

$$A = (a_{ij}), \qquad\qquad\qquad i,j = 1,\ldots,n, \quad (4)$$

each element $a_{ij}$ of which represents the causality propagated to population i from population j. In the terminology of Patten *et al.* (1976), A is a direct dependency matrix whose entries denote the direct effects of state variables $x_j$ on state variables $x_i$. From A it is possible to form a sequence of sequential dependency matrices (Patten *et al.*, 1976) denoting higher order interactions. Thus, $A^2$ specifies the indirect influence of j on i over paths of length 2, $A^3$ over paths of length 3, etc. By forming the infinite series,

$$A^* = \sum_{k=0}^{\infty} A^k, \qquad\qquad\qquad\qquad (5)$$

all dependency paths of all lengths k can be accounted for and the complete interactive structure of the system specified. However, this specification is only possible if the series (5) converges; then and only then does the matrix $A^*$ exist. The matrix is termed a transitive closure matrix from graph theory terminology (Ore, 1962) denoting completeness of a system's internal cause-propagating structure, i.e., all causal paths of all lengths accounted for.

If one assumes that such completeness is needed for the concept of a system to be meaningful, conditions for existence of $A^*$ illegitimatize the representation of externally derived causation by implicit means in mathematical models. The argument is detailed to greater degree and also illustrated in Patten *et al.* (1976, p. 521 ff.) and Proposition 2 (below). Partition the original interaction matrix A into m irreducible block diagonal submatrices $A_\ell$, $\ell = 1,\ldots,m$, so that $\Pi_{\ell=1}^{m}$ det $A_\ell$ = det A. The diagonalization process organizes the system into m subsystems, each of which must have causal contact with the system level environment in order for the transitive closure matrix $A^*$ to exist. In other words, each block diagonal subsystem defined by $A_\ell$ must receive explicit input from the environment of the system defined by A for the series (5) to converge, giving a completely specified endogenous propagating structure. Therefore, nonhomogeneous terms would have to be added to (3) to prevent the necessary condition of Proposition 1 from being violated. Implicit representation of input by $r_i$, a satisfactory approach taken at face value, is disallowed by a fundamental and more general understanding.

## Linearity

PROPOSITION 2
*Ecological systems are analyzable into components whose properties can be elaborated by experimentation which depends on the decomposition property of linear systems, and therefore both systems and their parts should be modeled by linear mathematics.*

The experimental isolation of systems from their natural surroundings is the common method of empirical science. It is not so

surprising, therefore, to have the practice carried to the extreme
of denying surroundings altogether in mathematical models, as
discussed above.  Ecologists believe in the experimental method and
in the transferability of experimental results from the laboratory
to nature.

Ecologists also believe that ecological systems are nonlinear,
and this is reflected in the preponderance of nonlinear models in
theoretical and systems ecology literature.  Only occasionally have
serious arguments in favor of a linear view of nature been advanced
(e.g., Patten, 1975).  Empirical evidence for nonlinearity of
processes is abundant in the form of thresholds, saturation points,
endogenous oscillations, limit cycles, etc.  Except for the many
examples of nonlinear behavior which could be cited, no concerted
attempts to provide linear interpretations have ever been made.
Therefore, nonlinear interpretations have never competed very directly
with linear ones.  Much of the seemingly nonlinear behavior of
natural phenomena is what I like to call pseudononlinearity--nonlinear
behavior elicited from linear systems because of the difficulty of
performing field or laboratory experiments to demonstrate superposition.

Figure 1 shows a linear system composed of six linear components,
three system level inputs, and two system level outputs.  Therefore,
if an input z is provided and a corresponding response y observed,
it is expected that if some linear combination $Lz$ of the same input
is applied, then the same linear combination $Ly$ of the original output will result as response.  This is a fallacy easily exposed with
the Figure 1 model, but few understand it.  A well-known systems
ecologist once asked me, in criticism of my claim that ecosystems
were linear, if I doubled the solar radiation input, would I expect
also to double system photosynthesis?  Of course not, but to say this
admits of a nonlinear input (solar radiation) and output

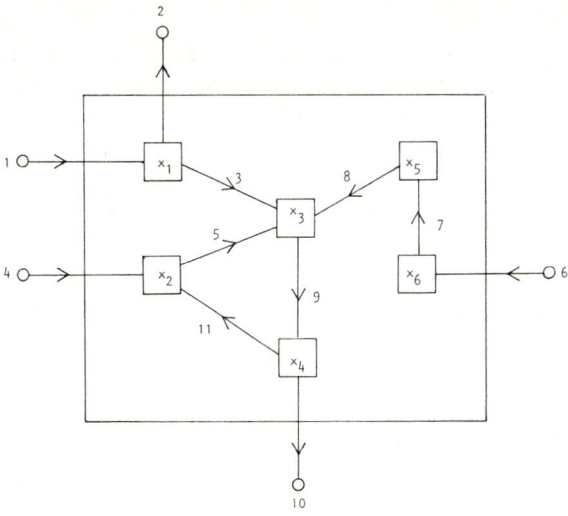

Fig. 1.  Hypothetical linear system, demonstrating pseudonon-
linearity.

(photosynthesis) relationship. Such a relation could also be pseudononlinear, however, because a true input-output experiment is not performable with an ecosystem.

In the system in Figure 1 if an experimenter excited input 1 and observed the system's response at output 2, superposition would be demonstrated. But if the response were observed at output 10, superposition would not be apparent and the system would be adjudged nonlinear. The reason is obvious: all the inputs which influence behavior observed at 10 were not manipulated in the experiment. All three system level inputs affect output at 10 and would have to be manipulated together in an experiment to demonstrate superposition. In the solar radiation-photosynthesis anecdote, if there were other inputs, such as nutrients, relevant to the output "photosynthesis," then all of these and not just sunlight would have to be doubled in order to double the output. There are numerous other input-output pairs $(z,y)$ in Figure 1 where experimental demonstration of super-position could not be made, e.g., $(4,5)$, $(4,10)$, $(6,9)$, $(1 + 6,10)$ and $(5,11)$. Extrapolation of the general point illustrated by this simple linear system to the vastly more complex interactive networks of natural systems makes one wonder about the possibility of demon-strating superposition anywhere, ever. The predominantly nonlinear world view of ecologists is then not surprising, given these circum-stances.

This nonlinear perspective on nature contradicts an equally strong belief in the validity of experimentation. To develop this incompatibility in systems terms, consider the global response func-tion $\rho$ (Mesarovic and Takahara, 1975) of a general time system:

$$y = \rho(x,z), \tag{6}$$

where $y$ is output, $x$ is state, and $z$ is input. This function defines the observed behavior $y$ of the system as a property of its state $x$ and input $z$. If this behavior can be separated into a component due to state and a component due to input, then the system is linear. This is the defining property of linear systems (Mesarovic and Takahara, 1975, p. 69), called the decomposition property:

$$\rho(x,z) = \rho_1(x) + \rho_2(z). \tag{7}$$

The function $\rho_1(x)$ is termed the zero input response; it represents the behavior of the system consequent to the stored energy associated with its state. The function $\rho_2(z)$ is the zero state response; it is the behavior due to external cause or input. In the general nonlinear system these behaviors are implicit in the total response (6), but it is impossible to isolate them as in the linear case (7). Consider the consequences of this in respect to experimentation with systems.

An experiment always consists of isolating, conceptually or actually, a system from its natural environment and then either (1) changing its state under controlled input conditions, (2) manipu-lating its input beginning with a known initial state, or (3) both.

The objective of the experimentation is to find out how the system behaves under a variety of inputs and states, i.e., to provide a behavioral description. This description constitutes knowledge about the system, a fairly concise characterization of its properties and propensities by which we know it and what to expect of it if we encounter it again. But consider the problems of assembling concise knowledge about nonlinear systems.

In the nonlinear case, state and input are forever confounded in system behavior that comes to the observer. Knowledge gained from particular observations cannot be isolated on system states to characterize the system as an entity, free of environmental influence. The system, per se, cannot be separated from its causes. To know a nonlinear system completely means to expose it endlessly to different inputs to generate different behaviors. There will never be an intrinsic behavior of the system, corresponding to $\rho_1(x)$ of the linear system, that can be distilled out and preserved as knowledge of the system's immutable character in different input environments.

The situation is readily illustrated by the Volterra model [Eq. (1)]. This model is nonlinear, and much literature in population ecology is devoted to developing its properties. But the model is also homogeneous so that the entity so described is subject to no input. The system behavior known is that corresponding to a response function [Eq. (6)] of the form $y = \rho(x,0)$. This behavior is state-related only, but it will not be conserved if nonhomogeneous input [Eq. (2)] is provided. The new behavior $y = \rho(x,z)$ will be both state- and input-related, but (1) these two components cannot be separated out, and (2) even if they could, the state-determined behavior would not be the same as that during the zero input case. For every possible input regime that could be provided, the system generates new behaviors, and no former knowledge of the system developed under different input conditions is of use in understanding the system in these new circumstances. By contrast, the state-related behavior $\rho_1(x)$ of linear systems is invariant over different environments, and it is only necessary to determine the specific input response $\rho_2(z)$ in order to know the system's total behavior [Eq. (7)] under different conditions of experimental input.

Ecologists believe in both the experimental method and in the nonlinearity of their systems. However, they have wider experience with and probably believe more deeply in the former property. Therefore, since ecological systems should admit of experimental manipulation, their representation should be confined to linear mathematics. This is the second postulated necessary condition for realism.

## Stability

PROPOSITION 3
*Ecological systems are intrinsically stable in the large but may be extrinsically caused to behave unstably in the small; therefore, their unforced models should exhibit global stability, and their forced models may be locally either stable or unstable.*

Two technical stability concepts are defined in physical science applicable to unforced (null input) systems. These are properties associated with equilibrium (resting) states $x^o$ of such systems. Let $x(t)$ be a vector of n system state variables, with $||x(t)||$ some norm such as $\Sigma_{i=1}^{n}|x_i(t)|$, where $|x_i(t)|$ is absolute value. Then, the equilibrium state $x^o$ is Liapunov stable if for every initial time $t_0$ and every real number $\varepsilon > 0$ there exists some real number $\delta(\varepsilon,t_0) > 0$, no matter how small, such that if $||x(t_0) - x^o|| \leq \delta$ then $||x(t)|| \leq \varepsilon$ for all $t \geq t_0$ (and $\varepsilon \geq \delta$ obviously). Thus, a system at equilibrium is Liapunov stable if, when it is displaced from equilibrium, its subsequent behavior is restricted to a bounded region of state space around $x^o$. Further, $x^o$ is asymptotically stable if it is Liapunov stable, and for any $t_0$ and deviation $||x(t_0) - x^o||$ sufficiently close to zero, $x(t) \rightarrow x^o$ as $t \rightarrow \infty$. The system returns in its unforced behavior to equilibrium following a displacement from that state.

By the causality principle, as discussed under Proposition 1, ecological systems are not without input. They are open and always forced. Therefore, nonequilibrium rather than equilibrium stability is the ecologically interesting case. Patten (1974) argued for the existence of one and only one resting state of ecological systems, the dead or zero state, $x^o = 0$. The resistance aspect of ecological stability was likened to Liapunov stability in the sense that the first law of thermodynamics (energy-mass conservation) resists unbounded growth away from $x^o = 0$ in a finite resource universe. The resilience concept of ecological stability corresponds to asymptotic stability in that the second law of thermodynamics guarantees return to $x^o = 0$ in absence of input. Thus, the core concepts of ecological stability connect consistently with those of physical system stability, and the mechanisms are seen as provided by universally held thermodynamic principles.

To force an ecological system off the zero state requires external causation in the form of energy and matter input. Given this input the ecological system moves into state space away from its zero equilibrium. Its nonequilibrium behavior can thus be viewed as a forced trajectory in state space, the nature of the trajectory being determined by the nature of the input and the topography of the state space. Assuming some nominal reference input and a time invariant (or time varying but structurally invariant) topography,[†] some reference trajectory $x(t) \neq x_0$ will result, deviations $x^*(t)$ from which can be induced by either direct state or input perturbations (Patten, 1974). The nonequilibrium trajectory $x(t)$ now serves as a reference in the same way that $x^o$ did for the equilibrium description of Liapunov and asymptotic stability.

With this model the situation in nature may now be explored: what can happen to a nominal ecological system? First, if its inputs are cut off, it will eventually die:

$$\lim_{t \rightarrow \infty} x^*(t) = x^o ; \tag{8}$$

---

†Meaning the topography maintains its basic characteristics with respect to the stability responses it will permit, as discussed below.

the system $x^*(t)$ so perturbed returns to equilibrium, $x^0 = 0$.
Second, if its state is perturbed directly at some t' to produce a
subsequent deviation from nominal behavior, $||x^*(t) - x(t)||$, $t \geq t'$,
but with the nominal input unchanged, two things may happen. The
system may return to its nominal trajectory following the disturbances:

$$\lim_{t \to \infty} ||x^*(t) - x(t)|| = 0, \tag{9}$$

or it may not:

$$\lim ||x^*(t) - x(t)|| > 0. \tag{10}$$

In the first case (9) the system is said to be structurally stable,
and in the second case (10) structurally unstable (Patten, 1974). A
third type of perturbation consists of varying the input from
nominal such that the state trajectory is caused to diverge, $x^*(t)$,
from nominal, $x(t)$. If the original input is then restored, the
same two things that could happen with a state perturbation are also
possible here. The system may return to its former behavior (9), or
it may continue (10) to exhibit some new behavior that was not
nominal before the input perturbation. In the first case it is
structurally stable and in the second case structurally unstable.

   What sort of state space topography would allow the types of
perturbation responses just described? The basic model exists everywhere in a three-dimensional space version, namely landforms that
allow water to flow downhill to the sea. As an example (and only
that, as other topographies are admissible), consider the equivalent
of a ridge and valley topography in multidimensional state space
increasing monotonically away from the single equilibrium point
$x^0 = 0$. Consider an ecological system in some forced nonequilibrium
nominal trajectory somewhere up one of the valleys. If its inputs are
zeroed, it will die, i.e., move down the valley through a sequence
of states to the zero state (8). If its state is perturbed, but
not too much, it will return from the hillside to its former trajectory in the valley (9). If its state is perturbed over a ridge,
however, it will move into another valley (10). If its input is
changed, the system may move up or down the valley, or over the
hillsides, or even over ridges to other valleys. When the original
input is restored, the system will settle in the valley where it is
at that time. If this is the original valley, then structural
stability (9) will be evidenced, but if not, the system will give a
structurally unstable (10) response.

   This "watershed" model of the state space topography of ecological systems is capable of accounting for just about any type of
stability or instability evidenced by natural systems, even with

appropriate modifications, the well-known "catastrophes" of René Thom (Thom, 1975). Because of their intrinsic zero state tendency without input, ecological systems in this model are seen as globally asymptotically stable everywhere in their state space. And, when extrinsically forced and perturbed, they may demonstrate either structurally stable or unstable responses as determined by local topography of the region of state space where their behavior is occurring. Thus, the characteristics stated in Proposition 3 are apparent, and a third necessary condition for realism of ecological models is that they be everywhere asymptotically stable in the unforced condition, and structurally stable or unstable locally, as appropriate to the system considered in the forced nonequilibrium condition. Mathematical formulations to represent ecological stability should be restricted to those that satisfy these properties.

## Conclusion

The basic idea pursued herein is to undertake the development of a consistent body of ecological theory by identifying, then debating, and then—if possible—resolving necessary conditions for realism in ecological models. Three such conditions have been stated as propositions. These are not falsifiable hypotheses in the usual hypothetico deductive sense, but are rather more like doctrines that could become generally held if the weight of evidence supported them or rejected certainly if the preponderance of information were against them.

An important consequence of this approach is the constraining of mathematics to conform to the general beliefs. This would have the effect of sharply reducing the kinds of mathematics and mathematical functions available to the ecological modeler, reflecting the limitation of worldview by the necessary conditions. Not all the mathematics ever invented should be expected to intersect human knowledge of nature. Mathematics is essentially self-generating in the logical domain and does not require for its existence or justification to be referenced to the ontological domain.

I have suggested in this paper (1) that all ecological models should be externally driven, (2) that ecologists should continue their faith that experimental results apply to nature by acknowledging that this belief makes a tacit assumption of system linearity, and finally (3) that ecological systems may be both stable or unstable but in certain definite ways. Each of the three propositions in which these properties were developed are bound to be controversial because they conflict with established norms. In each case the issues are complex, the arguments frequently subtle, and there always seems to be a central unresolved question that just eludes understanding: (1) Causality. Why is it necessary that a system have a completely specifiable internal cause-progating structure (transitive closure property)? This was the ultimate justification given for the need for explicit input. (2) Linearity. Why is the decomposition property of linear systems so fundamental to transferability of knowledge from one system to another? This was necessary because of inability to separate intrinsic (state) and extrinsic (environmental) contributions to nonlinear dynamics. (3) Stability. Why should the condition of being dead be equated to a single equilibrium state (zero state), and why should not other zero input conditions (spores, cysts, etc.) qualify as unforced equilibria? The first is a modeling choice and the latter an implication of the single equilibrium decision.

None of these central questions is resolved nor easily resolvable, and they are all basic to building a comprehensive know-ledge model of ecological phenomena. That is the point of the necessary conditions approach--to identify significant, difficult theoretical issues and frame them in a manner so that they can reasonably be researched and discussed for the future development of a consistent general theory of ecology. The mathematical models that attend such development will serve greater ultimate ends than those drawn expediently to meet limited, specialized objectives.

## Acknowledgment

University of Georgia, *Contributions in Systems Ecology*, No. 40.

## References

Griffin, D. R. 1976. *The Question of Animal Awareness*, Rockefeller University Press, New York.

Mesarovic, M. D. and Y. Takahara. 1975. *General Systems Theory: Mathematical Foundations*. Academic Press, New York.

Ore, O. 1962. *Theory of Graphs*. Colloq. Publ. 38. Amer. Math. Soc., Providence, Rhode Island.

Patten, B. C. 1974. The zero state and ecological stability. Proc. First Internat. Congress of Ecology, The Hague, September 8-14, 1974 (Suppl.).

_____. 1975. Ecosystem linearization: An evolutionary design problem. Amer. Nat. 109: 529-539.

_____. 1978. Systems approach to the concept of environment. Ohio J. Sci. (in press).

_____, R. W. Bosserman, J. T. Finn, and W. G. Cale. 1976. Propagation of cause in ecosystems. In: B. C. Patten (ed.) *Systems Analysis and Simulation in Ecology*, vol. IV, pp. 457-579. Academic Press, New York.

Sosa, E. (Ed.). 1975. *Causation and Conditionals*. Oxford University Press, London.

Thom, R. 1975. *Structural Stability and Morphogenesis*. W. A. Benjamin, Reading, Massachusetts.

# INDEX